Y0-BOE-422

WITHDRAWN

THE YEAR-BOOK OF FACTS

Printed in England by Antony Rowe Ltd., Chippenham

THE YEAR-BOOK OF FACTS IN SCIENCE AND ART

1839

Edited by
John Timbs

ROUTLEDGE/THOEMMES PRESS

This edition published by Routledge/Thoemmes Press, 1998

Routledge/Thoemmes Press
11 New Fetter Lane
London EC4P 4EE

The Year-Book Facts
45 Volumes : ISBN 0 415 14761 1

1st impression 1996
2nd impression 1998

Routledge / Thoemmes Press is a joint imprint
of Routledge Ltd and Thoemmes Ltd

British Library Cataloguing-in-Publication Data
A CIP record of this set is available from the British Library

Publisher's Note

The publisher has gone to great lengths to ensure the quality of this reprint but points out that some imperfections in the original book may be apparent.

This book is printed on acid-free paper, sewn, and cased in a durable buckram cloth.

THE
YEAR-BOOK OF FACTS

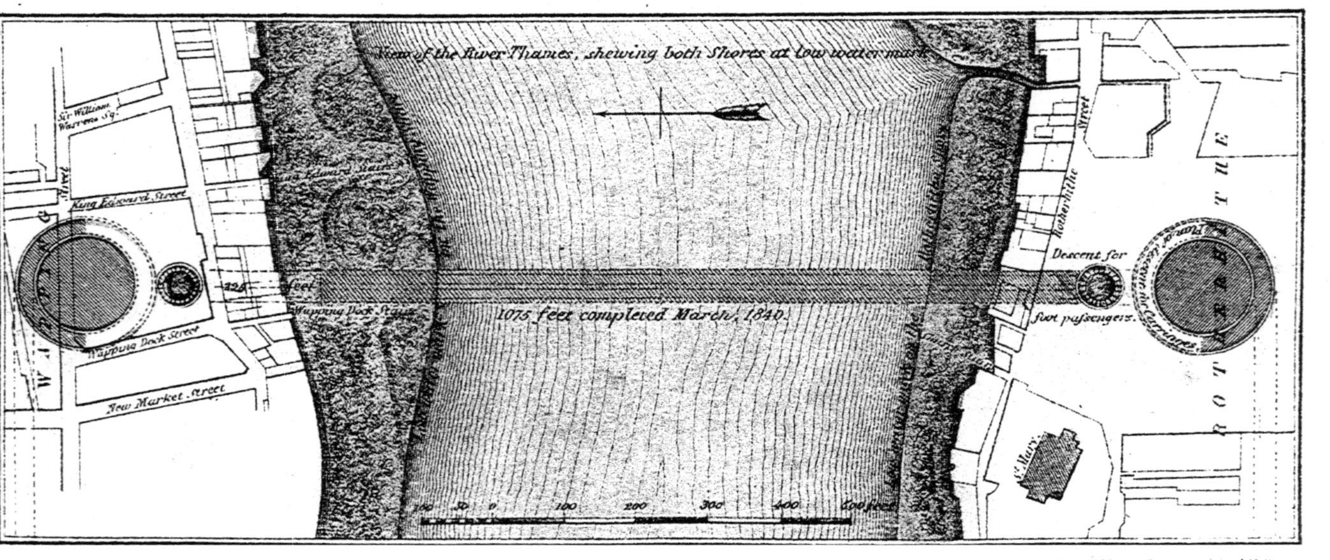

J.E. Jobbins Lith 3 Warwick Ct Holborn

PLAN OF THE THAMES TUNNEL,

Shewing its progress to March, 1840: distance to complete, 60 Feet.

For the Year book of Facts, 1840.

THE
YEAR-BOOK OF FACTS
IN
Science and Art:

EXHIBITING

THE MOST IMPORTANT DISCOVERIES AND IMPROVEMENTS OF THE PAST YEAR,

IN MECHANICS; NATURAL PHILOSOPHY; ELECTRICITY; CHEMISTRY; ZOOLOGY AND BOTANY; GEOLOGY AND GEOGRAPHY; METEOROLOGY AND ASTRONOMY.

Illustrated with Engravings.

BY THE EDITOR OF "THE ARCANA OF SCIENCE."

"Science pervades our manufactures, and Science is penetrating to our agriculture; the very amusements, as well as the conveniences of life, have taken a scientific colour."
PRESIDENT'S *Address to the Brit. Assoc.* 1840.

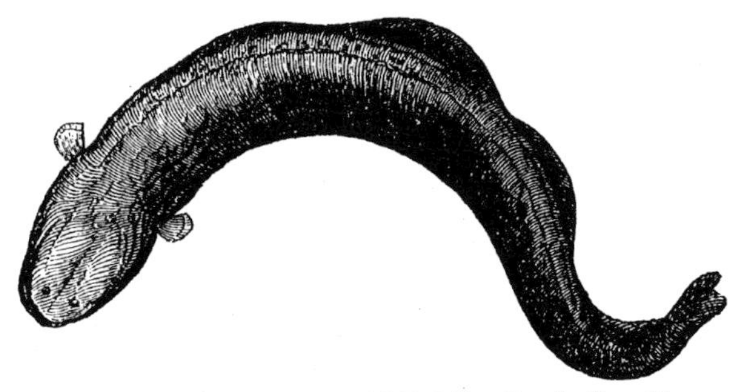

The Living Electric Eel at the Adelaide Gallery, Strand.—*See p.* 131.

LONDON:
CHARLES TILT, FLEET STREET.
MDCCCXL.

LONDON:
CLARKE, PRINTERS, SILVER STREET, FALCON SQUARE.

PREFACE.

THE extensive sale of the first "*Year-Book of Facts*" has naturally led to the production of a successor in the present volume; which, we hope, will be received as alike worthy of public approval. It is gratifying to state that the number of the *Year-Book* already sold is more than double that of the yearly sale of either of the recent volumes of the *Arcana of Science;* which circumstance must be regarded as the best proof of our extended design having been readily recognised by the public in general, as well as by the annual purchasers of the above work.

To popularize Science, by rendering its newest and most valuable facts accessible to all who read, has ever been our aim in providing healthful literature for the people. This design was, twelve years since, (upon the first appearance of the *Arcana of Science*,) generously commended by contemporaries, and as warmly encouraged by the reading public; and, it is not too much to add that few books published within the above period are better entitled, by the sterling character of their contents, to the respect of the reader. The whole series of volumes, including the *Year-Book*, 1839, is, indeed, a treasury of valuable facts; whereat the theorist and the practical man may meet, and into which they may, at all times, dip with pleasure and profit. This universal recommendation will, we trust, be found to pervade the *Year-Book*, 1840; in the compilation of which, neither labour, taste, nor judgment, have been spared, on our part, to secure so desirable an extension of popular favour.

A glance at the annexed "Contents" will remind the reader that "the progress of Science in its various applications to social improvement" is sought to be registered in the ensuing pages. Herein the mechanician and the student of natural philosophy; of electrical and chemical science; the naturalist and the lover of nature; the geologist and the meteorologist, the observer of the phenomena of the earth and the heavens—may alike be gratified with the novelty and variety of the several facts recorded; whilst their stock of useful knowledge in each section of Science will be refreshed and kept up by each accession of information.

Trusting, therefore, that "our labour of love" in the preparation of this volume has not induced us to over-estimate its attractiveness and useful merit, we commend it to the intelligent reader; in whose advantage lies our best "recompense of reward."

I. T.

Gray's Inn, March, 1840.

CONTENTS.

ENGRAVINGS.

THE

YEAR-BOOK OF FACTS.

Mechanical and Useful Arts.

THE THAMES TUNNEL.

(See Frontispiece.)

THE present state of the Tunnel will be best illustrated by the plan facing the title-page. From the annual meeting in March, 1839, to that in March, 1840, the rate of the progress of the works has steadily increased, with which there has also been a decreasing average cost per foot. In 1836, there were 117 feet completed; in 1837, only 28 feet; in 1838, 80 feet; in 1839, 194 feet; and from January 1 to March 1, 1840, have been completed 76 feet, being at the rate of 460 feet per annum; and the Tunnel is now completed to within 60 feet of the Wapping shore. Meanwhile, the public curiosity to inspect the Tunnel has increased with the progress of the works: in 1838, it was visited by 23,000 persons, and in 1839, by 34,000 persons; being an increase of 35 per cent. The works have now been in progress 15 years; the total sum expended, including the money advanced by Government, 363,000*l.*; and the Tunnel will be altogether completed for less than 500,000*l.* It is calculated that one archway will be opened throughout for foot passengers, towards the close of the present year.*

WATT'S INVENTIONS.

M. ARAGO, in his *Memoir of James Watt*, notes: "At a time when so many people are occupied with projects of rotatory steam-engines, it would be unpardonable were I not to state that Watt had not only thought of them, (of which we find proof in his patents,) but had actually constructed them. Mr. Watt subsequently abandoned them, not because they did not work, but because they appeared to him decidedly inferior in an economical point of view to machines of double power and rectilinear oscillations.

"There are, in fact, few inventions, great or small, among those so admirably combined in our present steam-engines, which are not the development of some of the original ideas of Watt. Examine his

* Abridged from the Report of the Annual Meeting, March 3, 1840.

labours, and, in addition to the principal points minutely enumerated in the text (of the memoir,) you will find that he proposed machines without condensation, in which, after having acted, the steam is dispersed in the air; and which were intended for localities where large quantities of water could not readily be procured. The operation of the principle of expansion in machines with several cylinders was also one of the projects of the Soho engineer. He suggested the idea of pistons, which should be perfectly steam-tight, although composed exclusively of metal. It was Watt who first had recourse to mercurial manometers for measuring the elasticity of steam in the boiler and the condenser; who conceived the idea of a simple and permanent gauge, by whose assistance might always be ascertained, with a glance of the eye, the level of the water in the boiler; and who, to prevent this level ever varying injuriously, connected the movements of the feeding-pump with those of a float; and who, when required, placed in an opening in the cover of the principal cylinder of the machine, the *indicator*, a small apparatus so constructed that it accurately exhibits the state of the steam, in relation to the position of the piston, &c. Did time permit, I could show that Watt was not less skilful and happy in his attempts to improve the boilers, to diminish the loss of heat, and to consume those torrents of black smoke which issue from common chimneys, however elevated they may be.

"The younger Mr. Boulton informs us, that in the year 1819, the establishment at Soho alone had manufactured of Watt's machines, a number, whose steady labour would have required not fewer than 100,000 horses; and that the saving result from the substitution of these machines for animal labour, amounted annually to more than 3,000,000*l.* sterling. Throughout England and Scotland, at the same date, the number of these machines exceeded 10,000. They effected the work of 500,000 horses, or of three or four millions of men, with an annual saving of from 12,000,000*l.* to 16,000,000*l.* sterling. These results must, by this time, be more than doubled."—Translated in *Jameson's Journal.*

PERFECT PARALLEL MOTION FOR THE STEAM-ENGINE.

On March 13, Mr. J. S. Russell submitted to the Society of Arts for Scotland, a very beautiful and scientific application of a mathematical proposition to the mechanism of the steam-engine. By this method, the piston-rod is carried up and down in a straight line, which is mathematically correct; while, by the method of Watt, the piston-rod is never at any one point carried in an absolutely straight line, and at top and bottom of the stroke it is very considerably deflected from the straight line, which gives rise to much friction and strain on the mechanism.—*Jameson's Journal.*

LEGISLATIVE ENACTMENTS FOR THE SAFETY OF STEAM-ENGINES.

A law has been promulgated in Belgium, dated June 24, 1839, which, among other enactments, determines the thickness of iron or copper which shall be used for various sizes of boilers, to be worked at certain pressures. The particular description of safety valve to be used is also described and figured. Every boiler must have two valves, one of which is to be kept locked, and inaccessible to any one except the head of the establishment to which the engine belongs. The weights upon the valve are to be examined and placed by a public officer entrusted with the matter, and are not to be increased without his intervention, on any pretence whatever. The boilers of stationary engines are to be supplied with a pressure gauge of mercury acting in the open air. Boilers of locomotive or marine engines are to have pressure gauges acting by compressed air, graduated so as to express the strength of the steam in the boiler. These gauges are to be so placed that the engineer or fireman attending the working of the boiler, shall be able to observe the pressure of the steam with facility.—*Mechanics' Magazine; abd.*

RUTHVEN'S PATENT STEAM BOILER.

MR. RUTHVEN, of Edinburgh, has patented a Steam-boiler, which practises slower combustion than usual, and thus insures less destruction of the boiler, and economy of fuel; the patentee being of opinion that "when the boiler is highly heated, the water is actually repelled from the surface of the plates by an atmosphere of caloric, from its not being able to absorb it with sufficient rapidity."

Mr. Ruthven has adopted a round boiler, about 2 feet diameter, and any required length; which form gives a strength that will sustain, with three-eighth plates, upwards of 800 lb. pressure per square inch; but the great desideratum, *surface*, was wanting; which has been gained by forming into a spiral form 500 to 1,000 feet, or more, (according to the size of engine,) of malleable iron pipe, of an inch diameter, more or less, as required: this spiral represents a still-worm of from 12 to 15 inches diameter; it is placed in the flue leading from the fire, and through which the heated air must pass in its way to the chimney; one end of the spiral is connected with the boiler under which the fire is placed, the other end being attached to the pump for supplying the water. Thus is exposed the whole surface of the spiral in the flue to the action of the heat passing through and around it; while the water is extracting the caloric in its passage to the boiler, the heated air passes into the chimney where the coldest of the water enters the spiral, and before the water enters the boiler, it is ready for converting instantly into steam. Besides the objects stated, this arrangement is stated to give strength, durability, and economy, with more security from steam power than has yet been obtained.

HEARNE AND DAVIES'S ROTATIVE DISC ENGINE.

MR. F. WHISHAW, C.E., having been requested to examine and report on the principle of construction of this engine, (also known as the Stoke Prior Engine,) has inspected six different ones, and reports them to have performed their duties satisfactorily. One of these engines, (Mr. Whishaw observes,) has been working for fifteen months, and has only required during this period, the expenditure of 3*s.* for repairs. In his examination, Mr. Whishaw has paid especial attention to two important points. First, he finds the moving parts of this engine to be so few in number, and their motion so uniform and regular, that the amount of friction must be very materially reduced; wherefore the wear of these moving parts, and their liability to derangement will be proportionally reduced. Secondly, with regard to the quantity of steam required to perform a certain amount of work, Mr. Whishaw reports that, from several trials made with an engine of this construction at the works of the British Alkali Company, near Bromsgrove, he found the work done by a 24 inch Disc Engine, working with the steam at 29 lb. pressure, to equal 20-horse power, with due allowance for friction; and the consumption of coal, (common Staffordshire), in general to 2 cwt. per hour, or rather more than 11 lb. per horse, per hour. This engine is worked by high pressure steam, which, after performing its duty, passes into the atmosphere: during the experiments, the average pressure was found by the mercurial steam gauge, to be equal to 29 lb. on the inch; but, in order to work the machine to the greatest advantage the pressure should be considerably increased. This engine equal to 20-horse power, with steam at 29 lb., would, with steam at 43½ lb., be equal to 30-horse work. It occupies a space equal to 4 feet square by 7 feet high, and its whole weight, including the frame, is 1 cwt. 43 qrs. 16 lbs.; but a requisite increase of the weight of the frame would probably increase the whole to 2½ tons; whilst the weight of a high pressure reciprocating engine of equal power would be not less than 20 tons. —*Midland Counties Herald; abridged.*

NEW ROTARY STEAM-ENGINE.

Mr. Rowley, of Manchester, has invented a Rotary Engine, combining simplicity of construction with rapidity of motion, greater power than a common engine of the same size, and a saving in fuel of upwards of 20 per cent. This engine consists of a circular steam-tight case, in which revolves a wheel of such a diameter as to leave a sufficient space round its circumference, for the action of the steam; this space being divided into two equal parts by a solid abutment, on the sides of which the induction and eduction of pipes are placed. This wheel is fitted with two or three pistons, which work within the wheel, so as to pass the abutment, afterwards projecting into the steam chamber, and forming a resisting surface against which the steam acts, and consequently propels the wheel, the axis of which passes through the case, and is connected with the shafting of the machinery to be put in motion. The action of this machine is extremely beautiful; no levers, beams, pistons, or crank plunging about, but all smooth, circular, rapid, and noiseless.—*Manchester Chronicle.*

CORNISH HIGH-PRESSURE EXPANSIVE CONDENSING STEAM-ENGINE.

On May 25, an engine manufactured in Cornwall by Messrs. Harvey and Co., of Hayle, from the specifications and plans of Mr. W. West, for the Carlisle Canal Company, commenced working, for the supply of the canal from the river Eden. The water has to be lifted 56 feet; the steam cylinder is 60 inches in diameter; that of the pump, 45; length of the stroke, 10 feet. In less than three minutes, the water was lifted to the pump-head, whence it was poured forth in a continuous and rapid stream, at the rate of 6,624 gallons per minute: consequently, working at twelve strokes, the quantity delivered in twelve hours into the canal is 4,769,280 imperial gallons of water, equal to 763,200 cubic feet, at an expense of fuel under *5s.*

Mr. West's engines, at the Fowey Consols Mine, in Cornwall, on a trial, lifted 125 million pounds weight one foot high with 94 lbs. of coal, and averaged upwards of 90 millions during 12 consecutive months. At the Portsmouth and Farlington water-works, the consumption has been reduced, (doing more work,) full 75 per cent. W. Wicksteed, the engineer of the East London water-works, computes their saving at 70 per cent., viz., 1,297*l.* 16*s.*, instead of 4,328*l.* 2*s.*—*Carlisle Journal.*

CONSUMPTION OF SMOKE.

MR. RODDA has patented a means of Consuming Smoke, by means of parting off a portion of the back of a furnace with fire-brick, so that when the coal has been coked in the forepart, it is thrust into the hinder division; and the smoke from the freshly-supplied coal being made to pass over the incandescent coked fuel, is consumed. The principal merit of this invention is its simplicity; consisting merely of a few fire-bricks, which may be placed in any furnace without expensive alteration.—*Mechanics' Magazine.*

CHANTER'S PATENT STEAM-BOILER AND SMOKE-CONSUMING FURNACE.

A STEAM-BOILER, of 20-horse power, built by Mr. Borrie, of the Tay foundry, has been so constructed as to consume its own smoke. The contrivance is as simple as it is efficient. There are no steam-jets, pipes for communicating heated air, blowing machines, or complicated apparatus, of any description, connected with it. No aid from a philosophical stoker is required, nor any attention is necessary more than is bestowed in the firing of common furnaces. A considerable saving of fuel is also effected, which, however, does not wholly arise from the consumption of the smoke, but is effected partly from the otherwise improved construction of the boiler. The additional expense of the boiler over those of the common construction is so small as scarcely to enter into the computation. The furnace is surrounded by water spaces (thus wholly within the boiler), and has two series of fire-bars. The first series, or those next the furnace door, is fixed at a considerable inclination. The second series is placed farther in, and below the first series, and has no inclination. The anterior extremities of the former project a few inches underneath the posterior extremities of the latter. Immediately above the posterior extremities of the first series of fire-bars, is a water partition guard, attached to the upper part of the furnace. Between the furnace door and this guard, a chamber is formed, which may be denominated the coal-chamber, that being the place into which the coals are put for feeding the furnace. Sufficient vertical space, between the first series of bars and the water partition guard, is left to admit of the coals falling down into the posterior or smoke-consuming chamber, after having been coked in the first or coal-chamber. The relative position of the lower part of the water partition guard and the surface of the coke on the bars of the smoke-consuming chamber, is so adjusted, that the smoke, before it can find egress into the flues, must pass over the surface of the coke. The construction, thus explained, will make the mode of operation easily understood. The furnace is fed in the ordinary manner; each successive quantity of coal is deposited in the anterior part of the coal-chamber, and, as the furnace in succession requires feeding, the fireman forces back the fuel in the chamber till he has sufficient room for the new feed,—the fuel in the posterior part of the coal-chamber being, by this process, projected into the smoke-consuming chamber. Thus, the latter chamber is entirely supplied with fuel in the highest state of ignition, emitting only a pure flame. Then the smoke arising from each successive feed in the chamber, having, from the relative position of the water partition guard, and the surface of the highly-ignited fuel in the smoke-consuming chamber, to pass under the former and immediately over the surface of the latter, is thus exposed to the intense action of the flame, and before reaching the flues is consumed. The boiler is in active operation on the premises of Mr. C. Smith, flax-spinner, Dudhope.—*Dundee Chron.*

BEST FORM FOR STEAM-VESSELS.

THE *United Service Journal* contains some interesting papers, by Lieut. Wall, R.N., "On the Construction, Proportion, and Power, best adapted to Sea-going Steam-vessels." In his last communication, he gives the following as the result of a careful examination of the experiments of Col. Beaufoy, Sir John Robison, and others:—

1. That the resistance of a body passing through a fluid is as the square of the velocity.

2. That an increased facility in passing through a fluid is imparted to a body by an increase of its length.

3. The resistances of small superficies are nearly as the surfaces; but the resistance is greater than a proportional increase of surface, and this deviation from theory increases with the extent of surface.

4. The resistance to the passage of bodies through the water is proportional to the angle at the head, other proportions being alike.

5. The resistances do not vary in the duplicate ratio of the angle of incidence.

6. For the higher rate of velocities, the flatter a vessel is built in the floor, the better, *when intended to be propelled by machinery.*

From which we infer: 1, that as double power will not propel the same body at double the speed, the larger steam-vessels are built, the better: 2, that the length of a vessel in proportion to her breadth may be increased beneficially, (for speed,) until the lateral resistance shall equal the increased momentum obtained by additional breadth, (which is at about eight times the breadth); that the extreme breadth ought to be exactly amidships; that the form of the bottom, *below the water-line*, ought to be as flat as possible for about three-fifths of the length, and from thence to draw in forward and aft for the entrance and run; and the bow to be as sharp as consistent with safety in a sea-way, but the run not to be too fine.

It is also recommended that the counters and harpings be kept full, the stern-post nearly upright, and the stern formed on the circular construction. In consequence of the shallowness of the vessel, the quarter-deck and forecastle should be covered in, by which much room will be obtained below; and the deck of the latter, being bevelled and carried well aft, will serve to turn any sea which may be shipped in steaming head-to-wind. The bulk-head of the quarter-deck cabin should be of a semicircular form, having ports cut at its junction with the vessel's side, for the escape of the wind.

Lieut. Wall also proposes to build large steamers with three instead of two engines; and sets forth advantages which would more than counterbalance the increased expense, weight, and friction of the third cylinder.—*Quoted in the Mechanics' Magazine.*

PROPORTION OF POWER TO TONNAGE OF STEAM-VESSELS.

MR. SCOTT RUSSELL has read to the British Association an important paper upon this subject of anxious inquiry with every proprietor and constructor of Steam-vessels. If, on the one hand, a large cargo is desired, which is generally the case, the power of the engine is made small, for the purpose of occupying small bulk, and consuming little coal; and although less velocity is then acquired than with greater power, still less fuel is consumed with a small engine than a large one, and thus it is supposed that greater economy is effected. In other cases, where velocity is absolutely required, a much larger proportion of power has been employed; and, of course, by a large power there is a much greater consumption of fuel than a small one in a given time; and not only so, but it is well known that this additional consumption is much greater than in the proportion of the velocity gained; so much so, that a consumption of four times as much fuel will not give more than about double the velocity. Thus, it has appeared that the use of very great powers, and great expenditure of fuel, has been made with only a very slight increase of velocity. All this, the usual reasoning on the subject, goes to prove the value of employing a small proportion of power to tonnage for economy. The advantage of low powers and of low velocities, in point of economy, would thus appear to be established: but this apparent advantage in theory has not been realized; on the contrary, it may be recollected, that at the Bristol meeting of the Association, Dr. Lardner remarked, that in a case which had come under his own knowledge, where the power had become gradually increased in the same vessel, and it was found that while the vessel was increased in speed, the consumption of fuel on the whole had diminished. This was a fact which Dr. Lardner attributed to Mr. Russell's Wave principle. Mr. Russell had himself attributed it at that time to the same cause; but he found afterwards, by examining the books of the expenditure of fuel in the steam-vessels of several companies, that they were aware that they had, on the whole, saved money by using high powers of steam, and high velocities, instead of low ones. This result he would not in the circumstances attribute to his Wave principle; and he had arrived at this very remarkable general result:—*That in a voyage by a steam-vessel in the open sea, exposed, of course, to adverse winds, there is a certain high velocity and high portion of power, which may be accomplished with less expenditure of fuel and of room, than at a lower speed with less power.* This principle he then proceeded to prove, and to illustrate by a case, in which the same vessel is taken with different powers of engine, and the result, as regards expenditure of fuel, determined first arithmetically, and then by a general formula, which will enable any one to determine any particular case.

Dr. Lardner remarked that the theory of Mr. Russell was quite new to him; but that he was satisfied of its accuracy. Mr. Russell added an important circumstance, in which great advantage might be derived from the application of high powers: thus, the owner of a vessel, which usually accomplishes her voyage in fifteen days, but owing to storms or adverse winds, occasionally takes twenty, must so regulate her stated periods of departure as to allow for these adverse circumstances; and four or five such vessels may be required to keep up a regular communication between places for which three or four would be sufficient, if, by increased power, the average voyage were reduced to ten, and the adverse to fourteen, days; and thus, a large capital would be saved, as well as more voyages accomplished. This principle is, however, only applicable to vessels making long voyages.—*Athenæum; abridged.*

AMERICAN STEAM NAVIGATION.

THE secretary of the navy at Washington has issued an interesting document of statistics on this subject, from which we may gather material for settling a question about the comparative Steam-boat force of England and America. In 1836, we had 600 steamers at home and abroad. Doubtless, the number has increased since then. The American number is rated at 800 *now*, of which 600 belong to the western waters,—where, in 1834, there were but 254. About 140 belong to the state of New York. In tonnage, America is more in advance; the total there being estimated at 155,000, to about 68,000 on our side, two years ago, at the same time. The American boats do not equal ours in size: the largest runs between New York and Natchez, and is of 860 tons; the next in size are on Lake Erie, and along the New England coast. The average is 200 tons.—*Athenæum.*

STEAM-BOAT NIGHT SIGNALS.

THE Society of Arts have awarded to Mr. Jennings a silver medal for his invention of Night Signals for Steamers, to indicate to the steersman in what direction to vary the helm. The plan consists of three differently coloured lamps, fixed on a horizontal rod attached to a swivel, which the man on the paddle-box varies in position relative to the steersman's eye, and thus conveys to him the requisite orders.

STEAM-BOAT ACCIDENTS.

THE following interesting and important tabular results are selected from the Report of Captain Pringle, and Mr. Josiah Parkes, C. E., appointed by Government to investigate the causes and means of prevention of Steam-boat Accidents.

Particulars of Explosions, and form of Boilers.

Date.	No. of Reference to the Schedule of Accidents.	No. of Accidents.	Name of Vessel.	Form of Boiler.	No. of Lives lost.	Personal and other Injuries.
1817	1	1	Norwich	cylindric....	9	many persons injured.
1824	4	2	Eagle	not known..	1	not known.
1824	5	3	Duke of Bridgewater	rectangular .	2	many persons injured.
1825	6	4	James Ewing....	ditto	1	not known.
1826	10	5	Graham	ditto	6	many persons injured.
1827	11	6	Fingal..........	ditto	2	not known.
1828	14	7	Corsair.....	ditto	1	ditto.
1828	17	8	Magdalene......	ditto	—	40 pigs scalded to death in the hold.
1829	20	9	Dumbarton Castle	ditto . ..	—	one person injured.
1830	24	10	Hercules	ditto	—	ditto.
1831	33	11	Royal William ..	ditto	—	no harm done, except to the boiler.
1834	45	12	Herald	cylindric ..	—	ditto ditto.
1835	48	13	Earl Grey	rectangular .	10	many persons injured.
1836	57	14	Freedom........	cylindric ..	1	the vessel was sunk.
1837	63	15	Union	rectangular .	24	many persons injured.
1838	73	16	Victoria	cylindric ..	5	others injured.
1838	74	17	Victoria	ditto ...	9	ditto.
1838	75	18	James Gallocher.	rectangular .	2	many persons injured
1838	76	19	William Stanley .	ditto . ..	—	ditto.
1838	77	20	Antelope........	ditto	—	18 cattle destroyed in the hold.
1838	79	21	Vivid	cylindric ..	2	boiler worn very thin in the ruptured part.
1839	91	22	Urgent	rectangular .	—	eight or nine persons severely scalded.
1839	92	23	Morning Star....	cylindric ..	2	boiler worn very thin.
				Total....	17	

It thus appears that 7 explosions have occurred in cylindric boilers.
" 15 ditto rectangular boilers.
" 1 ditto, the *Eagle*—form of boiler not ascertained.

Total.... 23

19 Explosions happened whilst the vessels were stopping, or on the instant of setting the engines in motion.
3 Ditto, whilst steaming.
1 Ditto, the *Antelope*, not ascertained.

23

ANTHRACITE IN STEAMERS.

Mr. Player has invented a Fire-place which is exactly suited to the peculiarities of Anthracite; enabling it to become slowly heated up to the burning point, and preventing its disturbance afterwards.

The arrangement is very simple, and easily applied to steam-boilers. As no smoke appears, at first sight it strikes the observer as if the coals to be burnt were thrown down its short chimney until it was completely choked; but on looking more closely, he perceives that this strange looking little chimney is the "feeding funnel" by which the Anthracite is propped up in a tall heap over the fire, and resting on it, where it remains with its lower stratum growing hotter and hotter till it kindles; then, as the burning mass on the grate beneath is gradually consumed, the layer just in contact with it sinks quietly into its place, where it in turn becomes fully ignited, and so on. The red-hot burnt air, (not flame,) is carried round about and through the steam-boiler by flues as usual; and then flies off, without a particle of smoke to mark its progress, through the chimney at the other end of the furnace. Experiments recently made have ascertained that one ton of Anthracite, burnt in a proper fire-place, will raise as much steam as one and a quarter ton of common coke. This fact, therefore, places the latter fuel at a serious discount, even at present rates, as it would require 30*s.* worth of coke to do the work of 27*s.* of Anthracite.

A small iron steam-boat has been built for experiments with this fuel, and has been appropriately named *the Anthracite*. The construction of the furnace is the main novelty. In the first experiment upon the Thames, although the boiler is small, it sufficed to generate an abundance of steam; in fact, the engine was working at 45 strokes per minute, which is said to be something above the proper speed. The peculiarities of the fuel were very striking. Its power of conducting heat is so trifling, that the upper surface of the mass in the feeding funnel right over the fire gave no indication of the heat beneath; and when the fire-door below was opened for an instant, black coals rested on the surface of the red ones. The fire-bars are laid sloping away from the funnel, so that as the fuel descends it spreads equally over the extent of the grate without the aid of a stoker. No slag or fire-cake results from the fuel, and a few cinders which fall through the narrow bars still contain so much carbon that they are thrown into the feeding funnel again. The little "stoking-room" is from the absence of all meddling with the fire only of a comfortable temperature—a great advantage on a tropical voyage. The "fire-doors" are never opened to feed the fire, as all that goes on upon deck through the funnel, (as millstones are fed through a hopper, but this does not hop;) and, consequently, the fire is never half extinguished (as in ordinary fire-places) by a fresh supply, with the necessary evil results in addition, of steam checked, power lost, and smoke emitted. The combustion of the Anthracite goes on smoothly and equally, like that of the oil in Parker's new lamp, which is heated almost to the boiling, or rather burning, point, before it is inflamed in the wick. Mr. George Rennie, who seemed to take a great interest in the experiment, caused the fuel to be weighed during the trip, and found that the quantity for producing the effect of one-horse power for an hour was, 6½ lbs.; but this, from the short duration of the experiment, can only be looked upon as an approximation.—*Mechanics' Mag.; abridged.*

THE FIRST STEAM-BOAT.

The honour of this discovery has been re-claimed for Scotland, in a paper lately read by Sir John Robison, to the Society of Arts for Scotland, containing "a Narrative of Experiments and Suggestions, by the late Mr. James Taylor, of Cumnock, for the application of Steam to Navigation." The circumstances are briefly as follow:

"In the year 1785, the late Mr. Miller, of Dalswinton, was engaged in speculations on the practicability of propelling vessels by paddle-wheels; and, in 1787, he constructed a double boat of sixty feet long, with paddle-wheels in the space between the two vessels: these were worked by capstans, turned by men. The speed of this vessel was tried against a row-boat belonging to the Custom-House, which was distanced in the race.

"Mr. Miller became desirous of substituting some other mechanical power for that of the men, and consulted Mr. Taylor, who proposed *that a steam-engine should be applied to give motion to the paddle-wheels.* After making some objections, Mr. Miller consented to be at the expense of an experiment, and authorized Mr. Taylor to employ a clever mechanician, of the name of Symington, to make a small engine with a four-inch cylinder. This was accordingly done, and, on the 14th October, 1788, this engine having been erected on a twin-boat, *the first steam-boat voyage ever made** was successfully performed on Dalswinton Lake.

"Encouraged by this success, Mr. Miller agreed to make another trial on a larger scale, when Mr. Taylor, and his engineer, Mr. Symington, proceeded to Carron Foundry, where an engine with a cylinder of eighteen inches diameter was prepared and fitted to a vessel, which was tried on the Forth and Clyde Canal, for the first time, in November, 1789. This vessel moved at the rate of seven miles per hour—a rate which has hardly been exceeded by the most improved modern steamers in canal waters.

"The expense of this experiment having exceeded the estimates Mr. Miller declined to proceed.

"In 1801, Lord Dundas, then Governor of the Forth and Clyde Canal Company, employed Mr. Symington to make an engine for an experimental vessel for that Company. The vessel was soon after completed, and made many experimental trips on the canal; but being found to create a wave destructive to the banks, was, on that account, laid aside.

"It was this latter vessel, (the third in succession from the first trial), which was visited and studied by Mr. Fulton, the American engineer, whose first steam-boat was launched in 1807, nineteen years after the successful trial at Dalswinton Lake. Mr. Henry Bell, of Helensburgh, who constructed the steam-boat *Comet*, on the river Clyde, in 1812, accompanied Mr. Fulton in his visits to Lord Dundas's vessel.

"The above facts are supported by vouchers from various persons, several of whom are alive; and are fully corroborated by accounts pub-

* A claim has lately been made to the credit of this application of steam-power, by M. Jouffry, in 1775 or 1776, on the Saone in France.—J. R.

lished in the *Dumfries Journal* and the *Scots' Magazine*, of 1788; and in the Edinburgh newspapers of February, 1790.

"It is not generally known that the engines fitted in Fulton's first steamer, on the Hudson, were furnished by Messrs. Boulton, Watt, and Co., of Soho, who were applied to by Mr. Fulton, in person, to undertake to construct them."

THE ARCHIMEDES STEAMER.

THE *Archimedes* was noticed in the *Year-Book of Facts*, 1839, p. 11. The following additional details are abridged from two papers contributed to the *Railway Magazine*, by Mr. George Rennie, the engineer, and Mr. F. P. Smith, the adapter of this new application of the screw to propelling vessels. The main dimensions are: length between the perpendiculars, 107 feet; length over all, 125 feet; extreme width, 22 feet; depth in the hold, 13 feet; burden in tons, 237; length of engine-room, including boiler, 19 feet. She is built of English oak, and sheathed with patent sheathing; the lines both fore and aft are beautifully run; she is furnished with three masts, and is schooner-rigged; Mr. Wimshurst was the builder. The engines are united in one frame, the upper part of which is supported by columns of wrought iron: they consist of two cylinders, 3 feet in diameter each, and one air-pump between them; the length of stroke is 3 feet; the power is estimated at 90 horses; the motion is given to the cranks by double-side-rods, the joint length of which is 8 feet, by which the advantages of long connecting rods are obtained. The hand-levers, the blow-off, and injection cocks, are so arranged as to be under the immediate command of the engineer; the wheel-work, consisting of two cogged wheels and two pinions, is placed in a strong iron frame independent of the vessel. The shaft which drives the propeller passes from the lowermost pinion under the cabin-floor, and through a water-tight stuffing-box in the inner stern-post, to the propeller or screw. The screw is made of plates of iron fastened to arms of wrought iron, keyed upon a wrought iron shaft; and, when the engine is at work, makes 5½ turns for every complete revolution of the crank-shaft. The boiler is made to suit the shape of the vessel, upon the usual principle of low pressure; with the exception of containing a double set of flues, so arranged as to obtain the greatest quantity of surface. The weight of the engines and boiler, chimney, coal-boxes, driving-machinery and propeller, is 64 tons 8 cwt.

The merit of this new application of the screw is due to Mr. Smith; for, although it is not a new invention, the failure of all previous experiments, and his perseverance in bringing the principle forward once more, entitle it to be considered as a novel application. Mr. Smith's first trial was made upon a small boat of 6 tons burden, with the screw fixed in the dead wood of the stern; when the screw was driven with sufficient rapidity to give a velocity to the boat of 7 or 8 miles per hour; she ran from Blackwall to Margate in 8½ hours; and towed the *British Queen* steamer into the West India South Dock. The screw which propelled this experimental boat measured only 2 feet in diameter and 2 feet in length; and her success encouraged the formation of a company to make the experiment upon a larger scale, with the vessel already described.

The details of her performances will be found in Mr. Rennie's paper in the *Railway Magazine*, No. 40; p. 312. The fact of her performing upwards of 10 miles an hour, for 25 strokes per minute of the engine, is considered as an earnest of the effective properties of the screw as a propeller: she increases her speed half a knot per hour for every extra stroke of the engine per minute. The trials were made both by the common and Massey's patent log, from 7 knots up to the highest speed of the vessel, and the result was perfectly uniform. The immersed section of the Archimedes is about 140 feet, and the area of her propeller about 32 feet; yet she moves forward within one eighth of the same speed as the screw, in consequence of the obliquity of its action; for this oblique form being continued the whole of the circle, it throws off the water at right angles from every portion of its surface in a column, which then assumes the form of an inverted cone, the base of which is far greater in area than the midship section; as the screw is turned quicker, the base of the column is proportionally enlarged, whereby a greater resistance is obtained; and the *Archimedes* is propelled at a rate of 10 miles per hour, at a loss not exceeding one-eighth, when in smooth water; and in rough, the loss is not materially increased, as the screw remains wholly immersed, and not affected by the rolling of the vessel, as is the case with the paddle-wheel. Mr. Smith likewise considers the speed of propelling by the screw to be unlimited, provided you do not spare the power to give it the requisite number of revolutions.

The efficiency of the above principle for naval warfare has been generally admitted. By this arrangement, the engines are completely protected from shot by the coal-boxes on either side; whilst the propellers, from being wholly immersed in the water, are out of the reach of shot. The engines, screw, and the whole of the machinery, were made at Messrs. G. and J. Rennie's manufactory.

We regret to add, that in an experiment with the *Archimedes*, on May 30, the boiler exploded just as the vessel was about to get under weigh; in which accident the second engineer was scalded to death. The cause was overloading the boiler; and the accident was entirely unconnected with the new propeller or arrangement of the engines. The boiler has been removed, and replaced by others of different construction.

From the various accounts of subsequent trials, it would appear that the average of the *Archimedes'* speed has been about 8 miles an hour, with between 22 and 23 strokes per minute of the engine; each stroke producing $5\frac{1}{3}$ revolutions of the screw, which is 5 feet in diameter. The *Archimedes* had her three masts and rigging standing.

THE CYCLOPS STEAM-FRIGATE, THE LARGEST STEAM MAN-OF-WAR IN THE WORLD.

On the 10th July was launched, at Pembroke Dockyard, the above-named magnificent Steam-ship. Her dimensions are as follows:—

Length	225 feet.
Beam between paddles	38
Depth of hold	21
Tonnage	1,300

She is, therefore, 200 tons larger than the *Gorgon*, launched from the same slip about two years ago. Her equipments as a man-of-war, are the same in all respects as a frigate, having a complete gun or main-deck, as well as upper or quarter-deck; on the main-deck she will carry 18 long 36-pounders, and on the upper-deck four 48-pounders, and two 96-pounders; tremendous guns, on swivels, carrying a ball 10 inches in diameter, and sweeping round the horizon 240 degrees.

Like the *Gorgon*, she is commanded by a post-captain, these being the only two steamers taking frigate rank; her crew consists of 210 men, 20 engineers and stokers, and a lieutenant's party of marines, who has the charge of the guns.

All the guns move upon slides and fixed pivots, therefore taking a much wider range than the ordinary carriage can give. She is schooner-rigged, but her fore-mast is of the same scantling and height as a 36-gun frigate. Her draught of water, with all on board, including six months provisions, completely armed, and 20 days fuel, is 15 feet; this quantity of fuel, 400 tons, is carried in the engine-room, but there is space in the fore and after-hold for carrying 10 days more coal, making in all sufficient for 30 days run. She has an orlop-deck below the gun-deck, of such magnificent dimensions that there is space to stow 800 troops and their officers with comfort; so that taking the *Cyclops* all in all, she may be considered the most powerful in her Britannic Majesty's service.

This vessel has been built in six months, under the immediate inspection of Mr. Wm. Edye, master builder, of Pembroke Yard, from drawings and plans prepared by his brother, Mr. John Edye, assistant to Sir Wm. Symonds, the Surveyor of the Navy; upon whose principles of building she is constructed; namely, the uniting and combining the sailing and steaming properties together in a most eminent degree, as is the case with the *Gorgon*, also built after his plan.

Her engines are of 320 horse-power, and are constructed upon the same principle as those already in the *Gorgon*.—*Mechanics' Magazine; abridged.*

THE PRESIDENT STEAM-SHIP.

The *President* steamer was laid down towards the close of 1838, and has been completed as regards her hull. Her builders are Messrs. Young and Curling, of Limehouse; in whose dock-yard was also constructed the *British Queen*. The proprietors of both vessels are the British and American Steam Navigation Company; the *President* being intended to run, with the *British Queen*, between London and New York.

The main-deck, wherein is the round-house, or saloon, is 8 feet 6 inches in height; length, 78 feet. The vessel is built of oak, with fir planking, and has three masts and three decks; her upper-deck being flush from the bows to the stern, without a poop. The stern is ornamented in sufficient taste; the window-line being surmounted with the arms of England and America, supported by the lion and eagle, appropriately painted. The whole is crowned with a neat dentelled cornice, and a boldly-carved wreath gracefully hanging at the ends.

The figure-head of the *President* is not yet completed: it will be a bust of Washington, after Canova. The makers of the engines are Messrs. Fawcett, Preston, and Co., of Liverpool. The commander is Captain Kean.

The *British Queen* and *President* are believed to be the longest ships in the world. Their comparative dimensions are:

	British Queen.	President.
	Ft. In.	Ft. In.
Length extreme from Figure-head to Taffrail	275 0	268 0
Ditto on Upper Deck	245 0	243 0
Ditto on Main Deck	.. ..	224 0
Ditto of Keel	223 0	220 0
Breadth within Paddle-boxes	40 0	41 0
Ditto over bends	40 4	41 4
Ditto over all	64 0	68 0
Depth	27 0	.. ..
Ditto from Spar Deck	.. ..	32 9
Ditto from Main Deck	.. ..	23 6
Tonnage	2016 Tons	2366 Tons
Power of Engines	500 Horses	600 Horses
Diameter of Cylinders	77½ In.	80 In.
Length of Stroke	7 Ft.	7 Ft. 6 in.
Diameter of Paddle-wheels	31 Ft.	30 Ft.

We have omitted to notice the extraordinary size of the paddle boxes, each of which is decorated with a star of five points, measuring fifteen feet from point to point. This magnificent vessel was towed out of the builders' dock on December 9.

THE STEAMER NICHOLAI.

A SUPERB steamer, thus named, has been built in Gordon's dockyard, Deptford, for plying as a packet between Lubeck and St. Petersburgh. She was built by Mr. Taylor, in four months from laying the keel, after the design of Mr. Carr, surveyor to the East India Company. The steam machinery, by Messrs. Seaward and Cassil, is precisely similar to that of H. M. Steam-frigate, the *Gorgon*: the engines are of 240 horse power, and the engine-room is only 45 feet long, whereas, on the ordinary plan it would have exceeded 62 feet. The *Nicholai* is of 800 tons burden, and will carry 150 passengers, a number equal to that of the *Great Western*, of 1,400 tons.

EXPORTATION OF IRON STEAM-BOATS.

THE materials, from the manufactory of Messrs. Laird and Woodside, of Liverpool, have been shipped, for the construction of three iron steam-boats, in large pieces of plate riveted together, each forming a section of the respective boats; so that the whole may be with facility put together on arrival at the port of their destination, Monte Video, South America. The plates are from a quarter of an inch to three-eighths in thickness.

LENGTHENING A STEAMER.

A NOVEL operation has been effected in Chatham dockyard; that of lengthening the *Gleaner* Steam-vessel, which had been taken into dock for that purpose. She was sawn in two a little more than one-third of her length from her stern, and ways were laid from the fore part of her to tread on; the purchase falls were rove, and brought to two capstans, and the order being given by the master shipwright, the men hove away, and in five minutes the fore section was separated from the after part a distance of eighteen feet. The space between will now be filled up by new timber. There is no record of any ship or vessel having been lengthened in this dockyard before the *Gleaner*.—*Kentish paper*.

ANTHRACITE COAL.

AN Association has been formed in South Wales for extending the applicability of Anthracite Coal. America took the lead in its use in steam-boats, and also for domestic purposes; the quantity of carbon which it contains giving out a greater heat than that description of coal where combustion is more rapid. Satisfactory experiments have been made with Anthracite, both with marine and locomotive engines in this country, and it is to be hoped that its superior economy will not be lost sight of.

NEW SYSTEM OF INLAND TRANSPORT.

AN experiment has been made on the Forth and Clyde Canal, in Scotland, which seems likely to affect seriously the relative value of property in canals and railways. On some canals in Scotland, light iron vessels, capable of carrying from 60 to 100 passengers, are towed along by a couple of horses, at a rate of 10 miles an hour; and this is effected by what is called riding on the wave. This new system of wave navigation, the theory of which has been fully explained in the reports of the meetings of the British Association, has hitherto been limited in its use by the speed of horses. The experiment to which we now allude shows that the locomotive engine is capable of performing feats equally astonishing in water as in land-carriage. *A locomotive engine, running along the banks of the canal, drew a boat, loaded with sixty or seventy passengers, at a rate of more than nineteen miles an hour!* and this speed was not exceeded only because the engine was an old-fashioned coal-engine, whose maximum speed, without any load, does not exceed twenty miles an hour; so that there is every reason to infer that, with an engine of the usual construction employed on railways, thirty, forty, or fifty miles an hour will become as practicable on a canal as on a railway. The experiments to which we refer, were performed under the direction of Mr. Macneill.—*Athenæum; abd.*

ERICSSON'S PROPELLERS.

ON Jan. 29, the R. F. *Stockton* towed the American packet-ship *Toronto* from Blackwall to the lower point of Woolwich, a distance of $3\frac{1}{4}$ miles, in 40 minutes, against the flood-tide, then running from 2 to $2\frac{1}{2}$ miles; thus towing her through the water at the rate of upwards of six miles an hour. The *Toronto* is 650 tons burden, she measures 32 feet beam, and drew, at the time of the trial, 16 ft. 9 in. water; thus presenting a sectional area of more than 460 square feet. Now, the fact of this body having been moved at a rate of upwards of six miles an hour, by a propeller, or piece of mechanism, measuring only 6 ft. 4 in. in diameter, and occupying less than 3 feet in length, is one which, scientifically considered, as well as in a practical and commercial point of view, is of immense importance.—*Times.*

DRAINAGE OF LANDS BY STEAM-POWER.

In the fens of Lincolnshire, Cambridge, and Bedfordshire, a steam-engine of 10-horse power has been found sufficient to drain a district of 1,000 acres of land, and the water can always be kept down to any given distance below the plants. If rain fall in excess, the water is thrown off by the engine; if the weather be dry, the sluices can be opened, and the water let in from the river. The expense of drainage is about 2*s.* 6*d.* an acre; the first cost of the work generally amounts to 1*l.* an acre for machinery and buildings.—*Durham Advertiser; abd.*

PROGRESS OF RAILWAYS.

During the past year, the following lines of Railway have either been partially opened, or opened throughout their whole extent:—

The *London and Southampton* Railway (to which the name of *South Western* has been given,) was opened on June 10, eight miles onward, to Basingstoke, and between Southampton and Winchester; leaving thus only the 18 miles between Winchester and Basingstoke for completion.

The *Great Western* line was farther opened on July 1, to Twyford, five miles from Reading; it having been previously opened as far only as Maidenhead.

The *Eastern Counties* Railway is intended to have its termini at Webb's Square, between Bishopsgate-street Without and Shoreditch church, London, and at Great Yarmouth—a distance of 126 miles: the engineer is Mr. Braithwaite. Of this line, $10\frac{1}{2}$ miles, between the Mile-end road and Romford, were opened on June 18. The distance is a series of viaducts and bridges; there being 50 bridges, one of which crosses the Lea, with a span of 70 feet.

The *Manchester and Leeds* Railway was opened from Manchester and Littleborough, a distance of 16 miles, on July 3.

Of the *York and North Midland* Railway, 14 miles, (between York and the junction of the Railway with the Leeds and Selby lines,) were opened on May 30.

The *Midland Counties* Railway was opened between Derby and Nottingham, nearly 16 miles, on June 4.

Of the *Glasgow and Ayr* Railway, the eleven miles between Ayr and Groine were opened on August 5.

The *Birmingham and Derby* Junction Railway, $38\frac{1}{2}$ miles, was opened throughout on August 12. This work has been completed within three years, at a cost not exceeding the estimate. This line is most important, as opening a direct communication from the north to Birmingham, Gloucester, Bristol, and Exeter; the Bristol and Gloucestershire Act, completing the continuous line, having passed in the last session of Parliament.

The *London and Croydon* Railway, 10½ miles, 16 yards, crossed by 18 bridges, was opened throughout on June 1.

The *Newcastle and North Shields* Railway, 6½ miles, was opened on June 18. In this short line, however, is a tunnel 70 yards long, and 24 bridges; two of the latter were described in the *Year-Book of Facts*, 1840, p. 19.

The *Brandling Junction* Railway is intended to connect the Tyne and the Wear, and by its junction at Gateshead with the Newcastle and Carlisle lines, to complete a railway communication between the German Ocean and the Irish Sea. This line was opened on Sept. 5.

The *Aylesbury Branch* of the Birmingham Railway, a straight line, 7½ miles long, has also been opened: the junction is made at Cheddington, 35 miles from London.

RAILWAY POST-OFFICE CARRIAGE.

THE letter-bags are conveyed along the London and Birmingham Railway in a Carriage constructed for the purpose, the interior being fitted with nests of pigeon-holes, drawers, desks, and pegs; and is attended by one or more clerks and a guard, the former to sort the letters, during the journey, and the latter to tie up and exchange the mail-bags. For taking and delivering the bags, during the passage of the train, to obviate stoppages for this purpose, attached to the near side of the office is an iron frame, with a piece of net, which is expanded to receive a bag from the arm of a standard at the side of the road. At the same moment that a bag is delivered into the net, another is let down from the office by the machine, and thus an exchange of bags is instantly effected. This ingenious contrivance is the invention of Mr. J. Ramsay. The post-office carriage is 15 ft. 3 in. in clear length; 7 ft. 7 in. in width; and 6 ft. 10 in. in height: its cost, 600*l.*—*Literary World.*

RAILWAYS AND TURNPIKE TRUSTS.

THE Select Parliamentary Committee appointed for the purpose of ascertaining how far the formation of Railways may affect the interests of Turnpike Trusts, have agreed to their Report; in which they are of opinion that the proportion between the scale of duty imposed on these modes of conveyance by land should be brought nearer to an equality. As far as an approximation can be made, by comparing the scale of duty on the average number of passengers conveyed by railroads, or by public carriages and posting on common roads, it appears the duty is as follows:—

For every passenger by the railway, one-eighth of a penny per mile.
For every passenger by a stage-coach, one-fourth of a penny per mile.
For every person travelling by post, three-fourths of a penny per mile.
While the conveyance of passengers by water is free from duty,

It appears that the use of Steam-carriages on turnpike roads might be encouraged by lessening the tolls payable on such carriages, which, (as appears from the evidence of one of the witnesses examined,) in some trusts are equal to a prohibition, although such carriages are considered to cause less injury to the roads than those of less weight drawn by horses.

RAILWAYS IN GERMANY.

THE line from Leipzic to Dresden is now completed, and has been opened the whole distance. The distance between Mayence and Wiebaden is proceeding rapidly; more than two-thirds of the distance is finished. The same may be said of the line between Frankfort and Hattersheim; but, from Hattersheim to Cashel it is proceeding slowly. The line from Frankfort is now opened as far as Hochst; but the continuation from thence to Cassel will not be completed before the next spring. The railway from Berlin to Potsdam has also been completed: the journey now occupies three-quarters of an hour, whereas, by the old road system, it required nearly a day to pass from Berlin to the royal palace and gardens at Potsdam.—*Foreign Quarterly Review.*

RAILWAY IN ITALY.

ONE of the most stupendous works of modern times is a projected Railway from Venice to Milan, connecting the seven richest and most populous cities of Italy with each other—Venice, Padua, Vicenza, Verona, Mantua, Brescia, and Milan; the most gigantic portion will be the bridge over the lagoons, connecting Venice with the main land. The length of the railroad will be 166 Italian, (about the same in English) miles, passing through a population of 3,500,000, the seven cities having alone a population of 500,000: *viz.*, Venice, 120,000, Padua, 44,000, Vicenza, 50,000, Verona, 46,000, Mantua, 34,000, Brescia, 42,000, and Milan, 180,000 inhabitants, to which may be added 20,000 foreigners in Venice and Milan. It is calculated the transport, when completed, will average 1,800 persons, 1,500 tons of goods, and 1,000 tons of coals daily.—*Foreign Quarterly Review.*

RESISTANCE OF AIR TO RAILWAY TRAINS.

DR. LARDNER has presented to the British Association an elaborate Report of his experiments in that department of Railway Constants which relates to resistance. This report, which took nearly four hours in the delivery to the Association will be found ably condensed in the *Athenæum*, (No. 619.) Dr. Lardner states that the results of this extensive course of experiments corroborate and fully establish a doctrine which he ventured to advance before a committee of the House of Lords in the year 1835, but which was then and subsequently pronounced to be paradoxical and absurd. That doctrine was, that a railway laid down with gradients, from sixteen to twenty feet a mile, would be for all practical purposes, nearly, if not altogether, as good as a railway laid down, from terminus to terminus, upon a dead level. The grounds on which he advanced this doctrine were, that a compensating effect would be produced in descending and ascending the gradients, and that a variation of speed in the train would be the whole amount of inconvenience which would ensue; that the time of performing the journey, and the expenditure of power required for it, the expense of maintaining the line of way, and supplying locomotive power would be the same in both cases; that, therefore, he thought, no considerable capital ought to be expended in obtaining gradients lower than those just mentioned. Dr. Lardner concluded by stating in detail a number of conclusions which he considered to be warranted by the experiments; care having been taken to lay nothing before the section, except what had been fully borne out by the experiments themselves. He regards the following conclusions as established by such experiments:—

1. That the resistance to a railway train, other things being the same, depends on the speed.

2. That at the same speed the resistance will be in the ratio of the load, if the carriages remain unaltered.

3. That if the number of carriages be increased, the resistance is increased, but not in so great a ratio as the load.

4. That, therefore, the resistance does not, as has been hitherto supposed, bear an invariable ratio to the load, and *ought not to be expressed at so much per ton.*

5. That the amount of the resistance of ordinary loads carried on railways at the ordinary speeds, more especially of passenger trains, is very much greater than engineers have hitherto supposed.

6. That a considerable, but not exactly ascertained proportion of this resistance is due to the air.

7. That the shape of the front or hind part of the train has no observable effect on the resistance.

8. That the spaces between the carriages of the train have no observable effect on the resistance.

9. That the train, with the same width of front, suffers increased resistance with the increased bulk or volume of the coaches.

10. That mathematical formulæ, deduced from the supposition that the resistance of railway trains consists of two parts, one proportional to the load, but independent of the speed, and the other proportional to the square of the speed, have been applied to a limited number of experiments, and have given results in very near accordance; but that the experiment must be farther multiplied and varied before safe, exact, and general conclusions can be drawn.

11. That the amount of resistance being so much greater than has been hitherto supposed, and the resistance produced by curves of a mile radius being inappreciable, railways laid down with gradients of from sixteen to twenty feet a mile have practically but little disadvantage compared with a dead level: and that curves may be safely made with radii less than a mile; but that farther experiments must be made to determine a safe minor limit for the radii of such curves, this principle being understood to be limited in its application to railways intended chiefly for rapid traffic.

DEVIATION OF RAILWAY LINES.

Dr. Lardner has described to the British Association an instrument which he has constructed for ascertaining the deviations of lines of rails. He employs a truck with wheels, without flanges, and perfectly cylindrical, and on which a platform is placed. An iron tube crosses this, terminating at each end at right angles, into which is introduced mercury; so that it is, in fact, a mercurial level. On placing this along the lines of railway, the mercury in the columns is found to go up and down; and into each is introduced a float and a piston, to the top of the rod of which a pencil is attached, which, on the occurrence of any incorrectness of the line, describes a curve on a sheet of paper, the ordinate of which gives the variation of the rail. Dr. Lardner has tried it on several parts of a line where the variation is perceptible. The instrument was checked so as to show that this curve was doubtlessly the real representation of the line; and being simple and easily applied, it would, no doubt, be found useful to contractors on new lines of rails.—*Literary Gazette.*

IMPROVED MACHINE FOR CUTTING RAILWAY BARS.

Mr. J. Glynn has communicated to the Institution of Civil Engineers the following means of Cutting the ends of Railway Bars square, so that they may truly abut against each other. In general, the ends, rough and ragged as they come from the rolls, are separately re-heated and cut off by the circular saw; but the accuracy in this case depends on the workman presenting the bar at right angles to the plane of the saw. As this cannot be insured, the difficulty is thus obviated: The axis of the saws and the bed of the machine, which is exactly like that of a slide lathe, are placed at right angles with the line of the rolls in which the rails are made; the saws are fixed in headstocks and slide upon the bed, so as to adjust them for cutting the rails to the exact length; they are three feet in diameter and one-eighth of an inch in thickness, with teeth of the usual size in circular saws, and make 1,000 revolutions per minute; the teeth are in contact with the hot iron too short a period to receive any damage; but to prevent all risk, the lower edge of the saw dips in a cup of water. The saw plate is secured between two discs of cast-iron faced with copper, and exposed only at the part necessary for cutting through the rail. The rail on leaving the rolls is hastily straightened with wooden mallets on a cast iron plate, on which it lies right for sawing, and sufficiently hot; thus a considerable saving of time, labour, and heat being effected. The rail is brought into contact at the same time with the two saws, and both ends are cut off by one operation. If the saws be sharp, and the iron hot, the 78 lbs. rails are cut through in twelve seconds. The rail on leaving the saws is placed in a groove planed in a thick cast-iron plate; thus all warping is prevented.

PNEUMATIC RAILWAY.

On May 14, was made a second series of experiments with models, of Clegg's Atmospheric Principle of propelling carriages by means of exhausting a tube laid down the line of road, of the air contained in it, and creating a vacuum. The tube being exhausted by means of an air-pump, the models, the leading one having a piston which forced open the valve of the tube, proceeded at a rate of extreme velocity along the line, a distance of thirty or forty yards, the ascent being one foot in thirty; the models were heavily laden, each carrying a couple of persons, and upwards of 15 cwt. of ballast being disposed over the whole. The machinery performed to perfection and gave general satisfaction. The advantages that this system proposes, both for the public and the railroad proprietors, are very obvious — cheapness, security, speed, and no danger of explosion.—*Examiner.*

ADAMS'S BOW-SPRING RAILWAY CARRIAGES.

MR. ADAMS's application of Bow-springs to Railway Carriages will be attended with the following very prominent advantages; —

1. A great diminution of friction.

2. Diminution also of weight — because the elasticity of the springs, and the equable motion they produce, will admit of considerable reduction in the weight of almost any part of railway vehicles, and also in the fitting up of the locomotive engines.

3. Security of position on the rails. It has hitherto been deemed necessary to keep the axles in their positions by means of side guides; which, however, prevent them from accommodating themselves to any of the unavoidable inequalities of the railway.

4. Adaptation to all changes of circumstances. It has been found by exact measurements, that the axles of many railway carriages are not placed accurately parallel, and cannot run true on the same line; the consequences of which are, increased friction, — increased wear and tear of the rails, the wheels, and carriages,—great additional weight in every part of the carriages, to enable them to withstand the violent oscillations and concussions which even a small deviation, (at high velocities,) from true parallelism in the axles must occasion, while the power given by the Bow-springs to each wheel to accommodate itself to every ordinary inequality and impediment, is a remedy for all, or nearly all, of the evils to which reference has been made.—*Railway Times.*

GALVANIC TELEGRAPH AT THE GREAT WESTERN RAILWAY.

THE space occupied by the case containing the machinery, (which simply stands upon a table, and can be removed at pleasure to any part of the room,) is little more than that required for a hat-box. The Telegraph is worked by merely pressing small brass keys, (similar to those on a keyed bugle,) which, acting (by means of galvanic power) upon various hands placed upon a dial-plate at the other end of the telegraphic line, point not only to each letter of the alphabet, (as each key may be struck or pressed,) but the numericals are indicated by the same means, as well as the various points, from a comma to a colon, with notes of admiration and interjection. There is likewise a cross (X) upon the dial, which indicates that when this key is struck, a mistake has been made in some part of the sentence telegraphed, and that an "erasure" is intended. There are wires communicating with each end, as far as completed, passing through a hollow iron tube, not more than an inch and a half in diameter; which is fixed about six inches above the ground, running parallel with the railway, and about two or three feet distant from it. The machinery and the mode of working it are so simple that a child who could read would (after an hour or two's instruction,) be enabled efficiently to transmit and receive information.

This telegraph is the invention of Mr. Cook and Prof. Wheatstone, of King's College. Between Drayton, Hanwell, and Paddington, it has been in operation for a year, and not the least obstruction to its working, by any of the wires failing, has yet taken place. Should this accident occur, especially when the whole line is open to Bristol, (from the wires being enclosed in a tube about an inch in diameter,) it might be expected to be difficult of repair, or to ascertain the precise point of injury throughout the 117 miles; but this apparent difficulty has been met by Mr. Cook's invention of a piece of mechanism, in a mahogany case, not more than eight inches square, by which means the precise point of injury would be indicated in an almost incredibly short space of time.

PNEUMATIC TELEGRAPH.

THE following is the description of the Pneumatic Telegraph, invented by Mr. S. Crossley; a model of which may be seen at the Polytechnic Institution.

"Atmospheric air is the conducting agent employed in the operation of the Pneumatic Telegraph.

"The air is isolated by a tube extending from one station to another; one extremity of the tube is connected with the gas-holder or other collapsing vessel, as a reservoir, to compensate for any diminution or increase of volume arising from compression or from changes in the temperature of the air in the tube, and for supplying any casual loss by leakage; the other extremity of the tube terminates with a pressure index.

"It will be evident to every one acquainted with the physical properties of atmospheric air, that if any certain degree of compression be produced and maintained in the reservoir, at one station, the same degree of compression will speedily extend to the opposite station, where it will become visible to an observer by the index.

"Thus, with ten weights, producing ten different degrees of compression, distinguished from each other numerically, and having the index, at the opposite station, marked by corresponding figures, any telegraphic numbers may be transmitted, referring in the usual way to a code of signals, which may be adapted to various purposes and to any language. The only manipulation is that of placing a weight of the required figure upon the collapsing vessel at one station, and the same figure will be represented by the index at the opposite station.

"In establishments where the telegraphic communications do not require the constant attendance of a person to observe them, and where periodical attendance is sufficient, the signals may be correctly registered on paper, by connecting with the air tube an instrument called a pressure register, invented by the projector of the Pneumatic Telegraph, which has been successfully employed in large gas-light establishments upwards of fourteen years, for registering the variations of the pressure of gas in street mains. The same instrument produces also an increased range of the index scale, by which means the chance of errors from minute divisions is obviated.

"The introduction of railways has not only created an additional use for telegraphic communications, but the important difficulty which previously existed in the expense of providing a proper line and safe foundation, is at once removed by the site of the railway itself, possessing as it does, by its police, the most ample security against injury, either to the tubes or electric wires.

"The time occupied in transmitting intelligence by the Pneumatic Telegraph will depend on the capacity of the air-tube, the degree of compression given, and the distance between the stations: but should greater despatch be required than is afforded by one air-tube, and the cost be of minor importance, several tubes may be employed, each fitted in the manner above described, so that all the figures contained in one telegraphic number may be communicated at once; and, with four tubes, 9999 different signal numbers may be communicated, referring to so many words or sentences, and these numbers may be multiplied four-fold by letters A, B, C, &c., as indices to distinguish each series.

"There has been upwards of twenty years' experience in the transmission of gas for illumination through conduit pipes of various dimensions. In several instances, the gas has been supplied at the distances of five to eight miles by low degrees of pressure. As one proof of great rapidity of motion, it has been observed, that when any sudden interruption in the supply has occurred at the works, the extinction of all the lights, over large districts, has been nearly simultaneous. Another instance of the great susceptibility of motion which frequently happens, is the flickering motion of the lights at great distances when water has accumulated in the pipes.

"The only experience in the transmission of atmospheric air through conduit tubes, which applies more particularly to this subject, may be referred to at three railway establishments; *viz.*, Edinburgh, Liverpool, and Euston-square, London. In these establishments, air-tubes, from 1¼ to 2 miles in length, have been employed for the purpose of giving notice when a train of carriages is ready to be drawn up the inclined plane by the stationary engine at the summit, so that it may without delay be put in motion. This notice is communicated by blowing a current of air through the tube at the foot of the inclined plane, and sounding an organ-pipe, a whistle, or an alarm-bell, at the stationary engine. It will be satisfactory to know, that this operation has been regularly performed from two to four years without one single failure or disappointment.

"It may be farther noticed, that a trial was made with a tube of one inch in diameter, very nearly two miles in length, returning upon itself, so that both ends of the tube were brought to one place: the compression applied to one end was equal to a column of seven inches of water; and the effect on the index at the other end appeared in fifteen seconds of time.

IMPROVEMENTS IN LOCOMOTIVES.

The following new invention, applicable to Locomotive Engines, is considered by a number of scientific men well calculated to reduce the expense, and increase the safety, of internal intercourse:—The advantages of it are: 1st, the condensing the steam after it escapes from the cylinders, and the water produced thereby returned to the boiler to be wrought over again and again; by which means the boiler is rendered more durable, being kept perfectly free of incrustation or deposit of any kind; and no stoppage is required to take in water; of course freeing the engine of the burden of carrying a supply along with it. 2nd. The air that supports the combustion of the fuel is considerably heated previous to entering the ash-pit; by which the smoke is completely consumed, although fresh coal be used in the furnace. Consequently, a great saving in the consumption of fuel is effected. A successful experiment has been made with the apparatus. The inventors are William and Andrew Symington, sons of the late William Symington, the introducer of practical steam navigation.—*Edinburgh Chronicle.*

SIR JAMES ANDERSON'S STEAM CARRIAGE.

A steam Passenger Carriage, upon Sir James Anderson's plan, has been built for "the Steam-carriage and Wagon Company," by Mr. Dawson, of Dublin. The front body, which is entered at the side in the usual way, contains more than ample space for six passengers, each having an arm-chair, and as convenient, if not better and more comfortable accommodation, than the first-class railway carriages. The back body, which is entered at the rear, is intended for ten passengers, although affording sufficient room for twelve. It is so ample in its dimensions that one may walk perfectly erect, from end to end, without incommoding the passengers at either side; it is admirably ventilated and lighted, and is to be furnished with a peculiarly constructed table. The outside passengers sit round the roof, fourteen in number; the carriage altogether containing 30 passengers. The front boot contains a cistern for water, and a space for coke or fuel for a stage of from ten to twenty miles; and there is room at different parts of the carriage for stowage of about $1\frac{1}{2}$ tons of luggage, if necessary.—*Abridged from Saunders's News Letter.*

NEW PENTOGRAPH.

Prof. Wallace has exhibited to the Institution of Civil Engineers a Pentograph of a novel construction, by which drawings may be copied or reduced and etched with great facility. Mr. Macneill bore testimony to the advantages of this construction over every other which he had seen; and stated that he had been enabled to finish a plan in $3\frac{1}{2}$ hours, which could not have been done by an ordinary Pentograph in less than 12 hours.—*Athenæum.*

LARGE SHEET OF PAPER.

There has been lately sent from the manufactory at Colinton, a single sheet of paper, weighing 533 lbs., and measuring upwards of a mile and a half in length; the breadth being only 50 inches. Were a ream of paper composed of similar sheets made, it would weigh 266,500 lbs., or upwards of 123 tons.—*Scotsman.*

NEW SOLDERING APPARATUS.

An apparatus for Soldering, on a principle entirely new, has lately been introduced. It consists of a chamber containing hydrogen gas, which, issuing therefrom, passes through a long elastic tube, and terminates in a curved pipe; but, before it escapes, a small portion of air is mixed with the gas, through another elastic tube, connected with a small pair of bellows, worked by the hand or foot. To solder anything, it is sufficient to direct the flame on the object, when the melting will solder the fracture. This mode of soldering is applicable to all situations and places; to any angles, either projecting or receding, or even conveniently overhead. The inventor of this apparatus is Baron Desbassins, of Richmond: it will, doubtless, be found eminently useful in connecting tubing for locomotive boilers, and in various ways connected with mechanical engineering.—*Surveyor, Engineer, and Architect, No. 1.; abridged.*

BEET ROOT SUGAR.

The manufacture of Sugar from Beet Root is flourishing in France. The works are stated actually to number 600, employing 175,000 agricultural and manufacturing labourers. The yearly produce of Beet Sugar is already 50,000 kilogrammes; the culture is established in 37 departments; in 47 it may take root; in 3 only it has refused.—(*Oct.* 30.)

Bread has been made in France from Beet Root, mixed with a small portion of potato-flour.

EAST INDIA SUGAR.

On Nov. 16, Mr. Solly read to the Asiatic Society a short report on some Sugar manufactured at Dindoree, from the juice of the Mauritius or Otaheite sugar-cane.—Mr. Solly stated that since the year 1792, when, from the limited supply and high price of West Indian sugar, the attention of the East India Company was drawn to the importation of East India sugar, numerous attempts had been made to improve the cultivation and manufacture of East India sugar, so as to bring it into competition with West India sugar. Notwithstanding this, its quality remained very inferior, and, till latterly, it was believed that India could not supply good sugar; lately, however, sugar has been brought over from India, of a very superior quality, similar in kind to some of the better sorts of West Indian. Mr. Solly exhibited some sugar manufactured under the direction of Dr. Gibson, and read a report on it from Mr. Travers, the eminent wholesale grocer, who described it as being of superior quality, good grain, desirable complexion, and as likely to find a ready market in this country. Mr. Solly stated that he had analyzed a portion of the sugar, and found it to contain a very fair per-centage of crystallizable sugar, but that its colour was difficult of separation; this, though objectionable for its uses by the refiner, does not interfere with its ordinary uses.—*Athenæum.*

IRON SAILING SHIP.

THE first sailing vessel ever built of iron has been constructed in Liverpool, and named the *Ironsides*. She first sailed for Pernambuco, which she reached in forty-seven days. The compass was correct throughout the whole passage; so that no fear need be entertained as to its general correctness on board of iron-built ocean-going vessels. Her hull is literally water-tight; tonnage, 264 tons; draught of water, after, 8 feet 7 inches—forward, 8 feet 3 inches.

THE LARGEST IRON SHIP IN THE WORLD

Is now building by Messrs. Ronalds, Fortdee, Aberdeen, for a Liverpool company. Her length of keel is 130 feet; breadth of frame, 30 feet; depth of hold, 20 feet; length over all, 137 feet; tons register, 537.

NEW LINE-OF-BATTLE SHIP.

"THE Nile," of ninety-two guns, has been launched at Devonport. She was designed by Sir R. Seppings, and was laid down in 1827: she is sister-ship to "the Rodney," ninety-two, launched at Pembroke, in 1833, which vessel has proved to possess admirable qualities as a man-of-war. "The London," ninety-two, building at Chatham, is also a sister-ship of "the Nile." The dimensions of the latter are: length, from figure-head to taffrail, 240 feet 6 inches; length of the gun deck, 205 feet 6 inches; height of figure-head above the under part of the keel, 51 feet 2 inches; ditto taffrail ditto, 58 feet; extreme breadth of main wales, 54 feet $3\frac{1}{2}$ inches; moulded breadth, 52 feet $11\frac{1}{2}$ inches; depth in hold, 23 feet 2 inches; burden in tons (new measurement), $2,622\frac{5}{8}$; ditto by the old measurement, $2,545\frac{3}{4}$.

EFFECTS OF LIGHTNING ON SHIPS.

ON April 16, was read to the Electrical Society a paper "On the use of black paint in diverting the effects of lightning on ships," by Capt. John Arrowsmith. In the *Philosophical Transactions*, vol. xlvii., will be found a relation of lightning passing over the parts of the masts painted with lamp-black and oil, without the least injury, while it shivered the uncoated parts, tearing out splinters so as to render the masts comparatively useless. The experience of thirty voyages, during which, on reference to his journals, the vessel he commanded had been at ninety-eight different periods within the vortex of the electric fluid, and escaped without having been once struck, has confirmed Capt. Arrowsmith in the belief of the efficacy of the precautions adopted by him at the commencement of those voyages, on reading the singular facts related of piebald cattle and horses struck by lightning. The affinity of the fluid to those parts of the streaks in the animals which were white, being very remarkable, led him to adopt the use of black paint on the mast heads, yards, caps, and trucks; and to take in and furl the upper and light sails, whenever forked lightning approached the vessel he commanded. Capt. Arrowsmith adds, from this experience: "any part of a body composed of wood, sufficiently coated with black, or lamp black and oil, possesses a property of resisting the destructive effects of the fluid;" and that, had he possessed this knowledge during his sea-life, he would have used the *black* in the fullest extent, covering with it the whole of the masts and hull of the vessels.—*Literary Gazette.*

PRESERVATION FROM SHIPWRECK.

CAPT. Dansey, R. N., has invented a contrivance of the kite and messenger for this humane purpose. The kite is to be flown from the ship; and, as soon as it is over the land with which it is desired to communicate, the messenger is sent up, and is carried along by the wind with sufficient force to disconnect one of the fastenings of the kite. The kite immediately turns on its edge and falls, carrying with it the line, to which, afterwards, hawsers may be sent, and communication kept up with the shore. This contrivance was practically exhibited at the United Service Institution, on April 22, by Capt. Saumarez, R.N.—*Literary Gazette.*

NEW LIFE-BUOY.

CAPT. Henvey, R. N., has invented a life-buoy of the simplest but most efficacious description, and one which has already proved useful in practice. This life-buoy consists of a light wooden frame, and in shape of a horse-shoe, but sufficiently wide to admit of a man's body, and rendered buoyant by plates and disks of cork attached to it. The hinder part is open; but the front is fashioned into the shape of a beak, for more readily cutting the water.

The "Lincolnshire Association for the Preservation of Lives in Shipwreck," having satisfied themselves of the utility of Capt. Henvey's invention by trying one, lost no time in ordering three more; and they have it in contemplation to combine the use of these life-buoys with Capt. Manby's apparatus in such a way that the lives of mariners and of passengers on board stranded ships, they hope, may be saved, when, by all the means heretofore used, nothing could be done to relieve them. A word or two will make this plain. By means of Capt. Manby's apparatus a cord, attached to a shell, forced from a howitzer, is thrown over the wreck, and then the crew, by getting hold of the cord, are generally enabled to pull a boat to them from the shore. But cases sometimes occur in which the ship is so placed, or the sea is so high, that the boat is either swamped or dashed to pieces; the night closes in upon the sufferers, and all perish. Now, it seems quite possible that in many cases where no boat could live, a series of Capt. Henvey's life-buoys might be drawn off and made use of in succession till every person was removed from the wreck. Cases, too, not unfrequently occur in which the life-boat from the shore is enabled to get within a few yards of the ship, but owing to the agitation of the water in the intervening space, both crew and passengers are lost. Capt. Henvey resides in Guernsey, and a post-paid letter so addressed would, we are well assured, meet with every attention.—*Times.*

RE-SHIPPING A RUDDER AT SEA.

MR. M. J. Roberts has communicated to the Edinburgh Society

of Arts a new method of re-shipping a rudder at sea, and that with ease, in the strongest weather. The apparatus for this purpose is simple and cheap. Before the ship leaves her dock, a hole must be bored in the heel of the stern-post, of sufficient size to allow of two small ropes being rove through it: these ropes should be of copper wire (or, the committee suggest, of American hides), laid up as a common hemp rope. Let two ropes be rove through the hole in the heel of the stern-post, and both parts of each rope be brought on board through the rudder-case: being of a small size, they will offer but little obstruction to the ship's progress; or, to prevent them altogether from doing so, a groove may be made in the sides of the stern-post for the ropes to lie in. Suppose the ship's rudder to be carried away: if it be only unshipped, recover it by means of the rudder chains, and bring it upon deck. But, if the rudder be totally lost, make use of a spare one, which can easily be carried in separate pieces; or, in default of this, rig up a rudder of spars and ropes. Bore a hole through the heel, so as to correspond with the hole in the stern-post, and through the rudder-hole work a grommet of rope, or wire, to be very strong, as it must traverse firmly in the hole. Now bring over the quarter an end of each rope rove through the stern-post, and make both fast to the grommet in the heel of the rudder. Then drop a guy rope through the rudder-case, bring the end in over the quarter, and make it fast to the head of the rudder. All is now prepared. Heave the rudder overboard, haul up the guy made fast to the rudder-head (this will lead it to the rudder-case), and rouse in the slack of the rudder heel-ropes. Bring the rudder-head up the rudder-case, and, when high enough, haul taut and belay your heel-ropes. The rudder-head guy-rope, (which should be a strong one), may be now made fast to a spar going across the deck, or a framework may be made above the rudder-head, to which the head-rope may be fixed: this is to support the weight of the rudder, and thus take off the strain from the heel-ropes. But, in a two-decked ship, as in a ship of war, the head-rope may be fixed to a carline between the beams of the deck, immediately above the rudder-head. Every thing is now in its place; and the rudder is quite as secure, if not more so, than when the ship left her port.

The great objection to all plans before suggested for re-shipping a rudder is the difficulty of keeping the heel of the rudder down, from its known buoyancy, the ship's forward motion, and the action of the waves. These causes combine to throw up the heel of the rudder, render it useless in steering the ship, and tear away the cumbrous guys, &c., usually led under the bottom of the ship to keep the heel of the rudder down. But, by Mr. Roberts's method, the rudder will be maintained in a good position, and in perfect security.—*Abridged from Jameson's Journal.*

CONSTRUCTION OF LIGHTHOUSES ON SANDS.

It having been represented to the Corporation of the Trinity House that Mitchell's patent screw moorings might be advantageously employed in constructing lighthouses on sands, an experiment to ascertain its practicability has been made under the superintendence of their engineer, Mr. J. Walker. The spot selected is on the verge of the Maplin sand, at the mouth of the Thames, about twenty miles below the Nore, forming the northern side of the Swin, or King's Channel, where a floating light is now maintained. This spot is a shifting sand, and is dry at low-water spring-tides. The plan is to erect a fixed lighthouse of timber framing, with a lantern, and residence for the attendants. For this purpose, in August, 1838, operations were commenced to form the base of an octagon, forty feet diameter, with Mitchell's mooring screws, one of which was fixed to each angle, and another in the centre ; each of these is four feet six inches diameter, attached to a shaft of wrought iron about twenty-five feet long, and five inches diameter, and, consequently, presenting an immense horizontal resisting surface. For the purpose, a stage for fixing the screws, a raft of timber thirty feet square, was floating over the spot, with a capstan in the centre, which was made to fit on the top of the iron shaft, and firmly keyed to it; a power of about thirty men was employed for driving the screws; their united labours were continued until the whole force of the thirty men could scarcely turn the capstan; the shafts being left standing about five feet above the surface of the sands. The fixing of the nine screws, including the setting out the foundation and adjusting the raft, which had to be replaced every tide, did not occupy more than nine or ten days.

Upon this foundation the superstructure of timber is to be placed; consisting of a principal post, strongly braced and secured, with angleposts made to converge until they form a diameter of about sixteen feet at the top, giving the superstructure the appearance of the frustrum of an octangular pyramid : the feet of the angular posts and braces are well secured and keyed down to the tops of the iron shafts, and the whole is connected at top and bottom with strong horizontal ties of wood and iron. The entire height of the superstructure will be thirty feet above the top of the iron shafts ; up to a point about twelve feet above high water-mark, spring-tides, the work will be open; the part above will be enclosed as a residence for the attendants; in the centre and above this will be erected a room or lantern of about ten feet diameter, from which the lights are to be exhibited.

The interval that has elapsed since the screws were fixed has fully proved the security of them, which, although driven into sand, seem as if fixed into clay.

The importance of this experiment certainly calls for a trial. The insecurity of floating-lights has been too manifestly productive of disastrous consequences not to call for a remedy, and it will be fortunate if by this means it be obtained. Within one month the Nore light was blown from her mooring; and the breaking away of the north-west light of the Mersey was supposed to have led to lamentable shipwrecks at Liverpool.

The progress of this work will be watched with interest. It is of much more importance than chain piers, as its success will enable us to obtain a foundation in positions where they cannot at present be used. It must be remembered that the screw can be employed where the pile is of no avail; that it possesses a much stronger hold, and has greater durability.—*Civil Engineer and Architects' Journal, abridged.*

THE LONDON DOCKS.

A GREAT improvement has been made in these docks, by the erection of a magnificent jetty, supported on massive piles, extending from the south-west quay, eight hundred feet across the large basin, affording a quay-frontage on both sides, for the loading of outward bound ships, of 1,600 feet. The jetty is sixty-two feet in width; and three lofty sheds, each two hundred and eight feet long, by forty-eight feet wide, have also been erected. The erection of the jetty is said to have cost the London Dock Company 60,000*l.* One million sterling has been expended during the last twelve years in enlarging and improving, including the excavation of the eastern basin and entrance.—*Civil Engin. and Arch. Journal.*

THE BUTE DOCKS.

THE splendid docks and ship canal at Cardiff, undertaken and completed at the cost of the Marquis of Bute, were opened on October 9th. The following details have been obtained from Messrs. Cubitt and Turnbull, the engineers:

"The river Taff, which falls into the sea at the port of Cardiff, forms a principal outlet for the mining districts, with which Glamorganshire abounds. The produce of these mines has hitherto found its way to market through the Glamorganshire canal; but its sea lock, constructed about forty years ago, has long been found inadequate to the demands for increased accommodation, consequent upon the extraordinary increase of trade since the canal was opened; some idea of which may be formed from the fact that, according to the Canal Company's Report, 123,134 tons of iron and 226,671 tons of coal passed down in 1837, making a total of 349,905 tons, or about 1,100 tons per day.

"The principal advantages of the undertaking are as follows:—A straight open channel N. N. E. and S. S. W. about three-quarters of a mile in length from Cardiff Roads to the new sea-gates, which are 45 feet wide, with a depth of 17 feet water at neap, and 32 feet at spring tide. On passing the sea-gate vessels enter a capacious basin, having an area of about an acre and a half, called the outer basin, calculated to accommodate vessels of great burden and steamers: the main entrance lock is situated at the north end of this outer basin, 152 feet long, and 36 feet wide, sufficient for ships 600 tons.

"Beyond the lock is the inner basin, which constitutes the grand feature of this work. It extends in a continuous line from the lock to near the town of Cardiff, 1,450 yards long, and 200 feet wide, an area of nearly 20 acres of water, capable of accommodating in perfect safety from 300 to 400 ships of all classes. Quays are built on each side for more than two-thirds of its length, finished with strong granite coping, comprising nearly 6,000 feet, or more than a mile of wharfs, with ample space for warehouses, exclusive of the wharfs at the outer basin. To keep the channel free of deposit, a feeder from the river Taff supplies a reservoir fifteen acres in extent, adjoining the basin. This reservoir can be discharged at low water by means of powerful sluices with cast-iron pipes five feet in diameter; and by ten sluices at the sea-gates, so as to deliver at the rate of 100,000 tons of water per hour."

"The feeder was commenced in 1834, the first stone of the docks laid on the 16th of March, 1837, and the last coping-stone laid on the 25th of May, 1839."

Some idea of the vastness of this undertaking may be formed from the fact of its having already cost about 300,000*l.*; and an additional expenditure of considerable amount will be incurred in the erection of warehouses, &c., along the quays.—*Abridged from the Times.*

THE DIVING BELL AT THE POLYTECHNIC INSTITUTION, REGENT STREET,

Is constructed of cast-iron, and weighs three tons; is about one-third open at the bottom, and has a seat around for the divers: it is lit by twelve openings, of thick plate glass, secured by brass frames screwed to the bell; six of these lights being triangular, and in the crown, and six square, in the side. The bell is "suspended by a massive chain to a large swing crane, with a powerful crab, the windlass of which is grooved spirally; the chain passes over four times into a well beneath, and to it are suspended the compensation weights," which, by acting upon the spiral shaft, accurately counterpoise the bell at all depths. It is supplied, by two powerful pumps of eight-inch cylinder, with air, conveyed by a leather hose to any depth.

The bell is constructed with all the improvements which modern science has suggested: the engineers being Messrs. Cottam and Hallan. It is slightly conical; five feet in height, and four feet eight inches in diameter at the mouth: its thickness is one and a half inch at the top, and two and a half inches at the bottom: the seat (which extends nearly round the inside) and the flooring, or support for the feet, are of wrought iron grating; both being covered with wood, and the seat carpeted, to suit amateur divers, of whom there is a *fair* proportion. Within the bell is affixed a knocker (such as we commonly see on street-doors), under which is painted:—

More Air—Knock Once.
Less Air—Knock Twice.
Pull up—Knock Three Times.

There is likewise affixed a written caution—"Visitors are requested to keep their seats, and their feet on the board." Instead of the strong lenses, or "bull's-eye lights," common in old bells, the windows are filled with plate glass seven-sixteenths of an inch thick. The leather hose is lined with caoutchouc cloth, and fitted inside with spiral wire. A peculiar provision is made for adding weights to the bell, and securing them with flanges to the outer rim; and six massive vertical straps meet on the crown in a double ring, by which the bell is suspended from the crane.

The bell is put into action several times daily: it will contain four or five persons seated; each pays one shilling for a descent; * and so universal is the public curiosity, that ladies and children are frequently occupiers of the seats.—*Literary World.*

PORTABLE PONTOONS.

In April, the following experiment was made by Colonel Macintosh, on the Serpentine river, Hyde Park: two pontoons, composed of inflated India rubber cloth cases, (prepared by Mr. Hancock), supporting a platform, were placed upon the water, upwards of forty of the Foot Guards standing upon the same. The weight of this number of men may be calculated at about a ton more than that of a heavy piece of ordnance; but, as the pontoons were not immersed quite half their depth, it is obvious that they could have supported a much greater weight. The raft thus formed was then towed out some distance in the Serpentine, and, having returned to the shore and landed the men, was taken to pieces, and the pontoons and other materials so disposed of as to show their convenience for transport by animals, without employing wheel carriages.—*Mechanics' Magazine; abridged.*

POWERFUL FIRE ENGINE.

Mr. Merryweather has constructed for the Liverpool Fire Police a large engine, which is equal in power to the combined force of two London engines, each having two working barrels seven inches in diameter, with an eight-inch stroke. In an experiment in Hyde Park, with nose pipes, one 1-8 inch and one 1-16 inch in diameter, two jets of water have been thrown to a nearly perpendicular height of eighty feet; and the branch, on being held at an angle of forty-five degrees, threw a stream of water one hundred and thirty feet distant. The cistern of the Liverpool engine, from end to end, measures eleven feet; the breadth, including the pockets, is three feet eight inches. The handles are nineteen feet six inches long; the working bar is two feet seven inches radius. The barrels are nine inches diameter, with an eight-inch stroke. The total weight is 35 cwt.—*Mr. Baddeley; Mechanics' Magazine; abridged.*

IMPROVEMENT ON BARKER'S MILL.

Mr. Stirrat, of Nethercraig, near Paisley, has patented this improvement, which consists, besides an ingenious water-joint and the application of a contrivance like the steam-engine governor, in a beautiful contrivance for preventing the friction arising from the centrifugal action of the water on the revolving arms of the machine. To remedy this, the patentee has the arms of the machine made with an eccentric curve, calculated according to the height of the fall, so that, when the machine is in operation, the water rushes out, at its

* "We understand the diving bell is a complete lion, and turns in nearly 1,000*l.* per annum."—*Railway Magazine*, October, 1838.

full speed, in a straight line from the centre to the extremity of the arena, where its power is wholly exhausted by action on the sides opposite the orifice by which it runs off. The advantages of this machine are stated to be very great. First, while by the common water-wheel, in some circumstances, only a small portion of the water-power can be used, and, under the most favourable circumstances, not more than sixty-five per cent., it is calculated that by this new machine not less than ninety-five per cent. of the motive power of water is available. Secondly, the most trifling rivulet, provided it have a good fall, can be taken advantage of by the new machine. And, thirdly, the expense of the improved Barker's Mill is not more than one-fifth of the expense of a water-wheel, to work in the same stream. —*Aberdeen Herald.*

ARTIFICIAL GRANITE.

M. D'HARCOURT's patent artificial granite blocks of Scotch asphalte have been laid down on the Southampton Railway. The sleeper was put in while the block was formed. It was usual to bore holes, and to fix the chains by bolts; but the above shorter method has been equally successful.

PAVING WITH BLOCKS OF WOOD.

MR. J. I. Hawkins has read to the British Association a paper "On paving roads and streets with blocks of wood placed end upwards." Although seven patents have been taken out in England for wooden pavements in little more than a year, Mr. Hawkins apprehends that no specimen in this country fairly represents this valuable means of road-making; and he considers the portions laid down in Oxford Street and the Old Bailey as very imperfect specimens. Mr. Hawkins then states the points on which the goodness and durability of such roads mainly depend:

1. That the wood be chosen from the heart of sound trees, not a particle of sap being laid down, lest early rottenness ensue. The resinous firs offer excellent materials; but any durable woods may be employed.

2. That the blocks, which are to be laid contiguously, be cut to an exact gauge, so as to fit closely and evenly together, and thus form a level floor.

3. That the depth of the blocks be at least one and a half the breadth; but twice the depth is preferable: each block, when rectangular, is supported by four others; when hexagonal, by six others; and this support will be the more effective in proportion to the accurate and tight fitting of the whole mass of blocks. Mr. Hawkins considers the hexagonal prism to be most advantageous, as affording the greatest quantity of wood from a tree, where the diameter of the prism is as large as can be cut out of the whole diameter of the sound part of the tree.

4. That the blocks be laid upon a firm bed of gravel, shingle, hard rubbish, or other hard material, well rammed down, and made even previously to laying down the block.

5. That a thin layer of only half an inch of fine gravel be spread evenly over the rammed and levelled surface of the bed at the time of laying the blocks, to favour their adjustment.

6. That the blocks be laid so as to present an even upper surface before any ramming of them is commenced, so that the ultimate making them even shall not depend so much on the rammer as on the evenness of the bed; the use of the rammer being to settle the whole firmly down, no block having to be rammed more than the rest. If one block be allowed to remain higher or lower than the contiguous ones, a jolting of the carriages will take place; and this jolting will press the lower blocks more than the higher ones: thus the evil, when commenced in the smallest degree, is found to be gradually growing worse; for the carriage wheel, when descending from the higher block upon the lower one, acts as a rammer; but, when ascending from the lower to the higher, it acts only as so much dead weight.

The blocks should be cut from dry wood, and used soon after being cut, lest their figure vary by warping. They may be roughly cut abroad while the tree is green, and re-cut after they have become dry.

The blocks may, perhaps, be cut by steam-power in England, at as little expense as by hand in foreign countries, where, from lowness of wages, it might be supposed England could not compete with them.—*Athenæum; abridged.*

STABILITY OF SUSPENSION STRUCTURES.

Mr. J. S. Russell, on January 16th, read before the Edinburgh Society of Arts, a paper on the means of preventing injury from the vibration of suspension bridges and other structures. Mr. Russell has, however, applied the principle mainly to suspension bridges, not only on account of their national and commercial importance, but also because, in these very light and slender structures, we are so near the limit of possible strength, that the mere addition of weight to the parts, instead of strengthening, is certain to weaken them, by increasing the load to be borne, at the momentum of the mass, and so increasing equally the danger of oscillation, instead of diminishing it; and, therefore, it is peculiarly necessary in this case to obtain the greatest possible strength to resist injury, with the smallest amount of weight. Mr. Russell's papers, illustrated with several diagrams, will be found in *Jameson's Journal*, No. 52, concluding as follows:—

"It is unnecessary to enter in this paper upon the practical details, which will immediately occur to the civil engineer, and suggest themselves readily to the intelligent mechanical engineer. The stays should be similar in construction to the main chains of the bridge, but much lighter; and each link should be kept in the straight line by a suspending rod from the chain, continued from the platform, so as both to keep the stay in the best position for resisting oscillation, and to distribute its weight more equally.

"The same principles which apply to a single arch apply to a series of arches, either in a wooden structure, or extensive stone arches, or a chain-pier, like that at Brighton. The arches should bear to each other the proportion of some of the numbers of the series of dividers already given, so as to prevent the propagation of oscillations from one to another.

"Stays in platforms, viaducts, and all wooden structures, when intended to prevent oscillations, should be placed at distances not perfectly equal, but in the proportion of the series of numbers already given; and all structures intended to prevent oscillation should be founded on the same principle."

BEST CONSTRUCTION OF BUILDINGS FOR SPECTATORS AND AUDITORS.

In almost every large room designed for an audience or spectators, or for both, because most people like to see as well as to hear a speaker, it may be noticed that certain seats are *the best*. These seats are neither too forward nor too far back; that is, they are not so far forward as, by being immediately under the speaker, to require to look up at a painful angle of elevation, and to permit his voice to pass over our heads; or, on the other hand, so far distant as to throw us behind a mass of people by whom vision would be intercepted, and over whose heads we should require to strain either to see or hear clearly. A perfectly good seat is one in which, without uneasy elevation of the head or eye, without straining or stretching, we can calmly and quietly take any easy position, or variety of positions, which we may be disposed to assume; and yet may in all of them see and hear the speaker with equal clearness and repose, so as to give him patient and undisturbed attention. The person who occupies such a seat feels as if the speaker were speaking principally to and for him; he finds that no one else stands in his way, and that he hears as well, and sees as well, as if there were no one else in the room but himself and the speaker. A room so constructed that every man in it should feel in this manner, that he had got one of the best places, and that no one else was in his way,—such a room would be perfect. Such a room, or rather approximations to such a room, we have sometimes, but very rarely, met with. On taking a particular seat, whether near the front or near the back of the audience, we have felt the comfortable assurance of having one of the best seats in the room.

The above observations are introductory to a paper communicated to the Edinburgh Society of Arts, by Mr. J. S. Russell; the object of which is to discover in what manner the interior of a building for public speaking should be formed, so that, throughout the whole range which the voice of a man is capable of filling, each individual should see and hear without interruption from any of the rest of the audience, with equal comfort, in an easy posture, and as clearly as if no other individual or spectator were present. The means proposed by Mr. Russell are explained, and illustrated with diagrams, in *Jameson's Journal*, No. 53.

STRUCTURE OF WOOD.

In a lecture, lately delivered to the Society of Arts, by Mr. Ducket, he described nearly every species of wood known among us, from the iron wood of Demerara, so hard as to defy the tool of the workman, down to the softest, which crumbles between the finger and thumb; observing that hardness, or softness, or colour, is so vague that botanists necessarily have recourse to microscopic tests for ascertaining the true characteristics of specimens. The tissues, cellular, vascular, and liqueous, were then described. Some of these fibres are so marvellously minute—pine-apple fibre, for instance, is only 1-7000th of an inch in diameter. The appearance of the medullary ray, by which change is effected, was next explained; then the simple, though exquisitely delicate, manufacture of rice paper by the Chinese—a process exactly like that adopted by our own cork-cutters. Next followed observations on the compression of wood, so wonderful in its effects, that boats employed in the whale fishery, sometimes dragged by a wounded fish into the deep, come up perfectly useless, and remain so, by the concussion and pressure of the waters, for a considerable time. Specimens of wood were submitted to the action of the oxy-hydrogen microscope: amongst them were sandal wood, satin wood, rosewood, oak, mahogany, the bramble, and the nettle; the magnified sections of the two latter surpassing all the others in symmetrical beauty.—*Literary Gazette.*

STRENGTH OF ROPE.

THE conservative power of mud, or clay, has been curiously exemplified in the case of vegetable fibre, by a portion of a cable lately brought up from the wreck of the Royal George. A length of junk, as it is called, consisting of a piece of cable twenty-three inches in circumference, and about eight feet long, was recovered; and, being unlaid, and the yarns taken out, they were found to be so good that a short piece of one of them bore a weight of 193lbs., and broke only with a weight of 195lbs. The following very curious experiment was then made:—

A piece of cablet, or rope, measuring two inches and a half in circumference, was made out of the yarns from one of the strands taken from the old junk of the Royal George, and marked No. 1; another cablet, also of two inches and a half, and twenty thread yarn, spun and tarred in 1838, was made, No. 2; and a third, of the same dimensions and number of yarns, spun and tarred according to the old method, in 1830, No. 3. These three cablets were then exposed to the usual test of strength by weights in the dockyard, when, to the great astonishment of every one, the following were the results:—

	cwt.	q.	lb.
No. 1. Made from the old junk saved from the Royal George, after being fifty-seven years under the mud, bore a weight of	21	3	7
No. 2. Spun and tarred in 1838	23	1	7
No. 3. Spun and tarred in 1830	20	1	7

Times.

PAPER MADE FROM WOOD.

M. DESGRAND, of Size-lane, London, has patented the following process for making paper from wood, stated to be the same with that recently patented by the MM. Montgolfiers, in France.

The fir and poplar are the only woods to be employed, barked, and cut into logs, four or five feet long, which are split into lengths, a few inches square, and carefully assorted in their various shades or colours, from the lightest of which white paper may be made. The lengths are chopped into pieces from two to four inches long, and from one to two inches thick; and as splintery as possible, so that the fibres may be the more easily separated. All knots of the wood, decayed parts, and crooked fibres, must next be picked out and rejected. A quantity of strips of one shade are deposited in a water-tight pit, with an outlet, and covered with lime water, in which the chips remain for a greater or less time, according to the temperature; in the south of France, from three to six weeks sufficing to dissolve the glutinous parts of the wood, when the chips will have sunk to the bottom of the pit. The lime water is then let off, and the fibres are washed in pure water, and then opened, divided, and flattened with stampers, or fulling mallets, so as to prepare them for the usual paper-engine, or cylinder, to reduce them to pulp, which is then, either by itself, or with other materials, made into paper or pasteboard, by a paper-machine, or frames. For white paper, the fibres are bleached before they are pulped.

MACHINERY AND THE WORKING CLASSES.

M. Arago, in his brilliant discourse "On Machinery considered in relation to the Prosperity of the Working Classes," observes:— "Thousands of workmen every day execute at the surface, and in the bowels, of the earth, prodigious labours, which it would be necessary totally to abandon, if certain machines were relinquished. One or two examples will suffice to make this truth sufficiently apparent. The daily removal of the water which rises in the galleries of the Cornish mines requires a power of 50,000 horses, or of 300,000 men. I ask if the wages of 300,000 workmen would not absorb all the profits which the mining operations might produce?" Again, "a single copper-mine in Cornwall, one of those known as the *Consolidated Mines*, requires a steam-engine of the power of more than 300 horses constantly at work, and thus every twenty-four hours realizes the labour of 1,000 horses. Concerning this, the assertion cannot be doubted, that no means could possibly be found beneficially and simultaneously to apply the strength of more than 300 horses, or 2 or 3,000 men around the shaft of the mine. To proscribe, therefore, the action of the steam-engine of the *Consolidated Mines*, would be to reduce to a state of inactivity a great number of workmen, whose labours are now rendered available; it would be to declare that the copper and tin mines of Cornwall must for ever remain buried under a mass of soil, rock, and liquid, many hundred yards deep. The proposition, brought to this form, would certainly have few defenders; but the form is nothing, whilst the substance remains the same.

"If, from operations which require the greatest development of power, we turn to the examination of different products of industry, whose delicacy of parts and regularity of form have ranked them among the wonders of art, the insufficiency, and even the inferiority of our organs, compared with the combinations of machinery, are equally striking to all. Where, for example, is the skilful spinner who, from a single pound of raw cotton, could produce a thread 150 miles long, as can the mule-jenny."—Translated in *Jameson's Journal.*

PAPER-MAKING MACHINERY.

On April 25th, some very interesting details of "Fourdrinier's Patent" were elicited during a debate in the House of Commons. From a technical description of the invention, by Mr. Mackinnon, it appeared that 1,000 yards, or any given quantity of yards, of paper, could be continuously made by it. The patent had been pirated, and that had led to litigation, in which the patentees' funds had been exhausted before they could establish their rights. They found themselves becoming bankrupts, and thus all the fruits of their invention, on which they had spent 40,000*l.*, were lost to them. The evidence of Mr. Brunel, and of Mr. Lawson, the printer of the *Times* newspaper, were read, to prove the invention of the Fourdriniers to be one of the most splendid discoveries of the present age. Mr. Lawson stated that the conductors of the metropolitan newspapers could never have presented to the world such an immense mass of news and advertisements as was now contained in them, did not this invention enable them to make use of any size required. By the revolution of the great cylinders employed in the process, an extraordinary degree both of quickness and convenience in the production was secured. One of its chief advantages was the prevention of all risk of combination among the workmen, the machine being so easily managed that the least skilful person could attend to it. It had caused a remarkable increase in the revenue: in 1800, when this machine was not in existence, the amount of the paper duty was 195,641*l.*; in 1821, when the machinery was in full operation, the amount of duty was 579,867*l.*; in 1835, it was 833,822*l.* No doubt, part of this increase must be set down to the demand arising from the increased number of publications and readers: still, it was impossible, but for this discovery, that such a quantity of paper could have been consumed. The positive saving to the country, effected by it, had not been less than 8,000,000*l.*; and the increase in the revenue not less than 500,000*l.* a year.

ELECTRO-MAGNETIC MACHINE.

In Gold-street, New York, has been erected an establishment for the construction of machinery to be put in motion by the application of the electro-magnetic fluid. One of these machines, of a large size, consists of a wheel, about 16 or 17 feet in circumference, placed vertically, and surrounded with four large magnets, operating on its outward circumference, with the smaller ones near the centre. Within a few inches stands the galvanic battery, for the generation of the fluid that sets the whole in motion. The battery is in the form of a rectangular tub; in which is placed a series of zinc and copper plates, immersed in a weak solution of sulphuric acid. The battery is attached to the large electric wheel by a series of metallic conductors, and the operation of the wheel is most striking. In a certain portion of the machinery, the fluid, interrupted in its movements, emits, with a snapping noise like a percussion cap, vivid and most brilliant flashes of light, equal in intensity to lightning. On putting the finger into the centre of the flash, only the ordinary sensation is produced; but, on the application of a piece of steel, the material is melted into red and orange sparks, and gradually corroded under the action of the fluid. The wheel is equal in power to that of two able-bodied Irishmen, in giving movement to any kind of machinery. Another machine was equal to four or five men; and preparations are making to construct machines of any extent of power.

This new application of the electric fluid has removed all the difficulties which Cook and Davenport met with in their experiments.* An electric machine to drive a double cylinder printing machine will cost about 300 dollars; and the expense of working it will be only 25 cents. a day, consisting of sulphuric acid to supply the battery; exclusive of the metallic plates of zinc and copper, which only require to be changed at long intervals. The Company who have perfected this great invention have expended 12,000 dollars in experiments, during one year: it has been certified by some of the first engineers of America and Europe; and the application of the power of the electric fluid to mechanics and motion is now complete. The danger from fire is entirely obviated; for, although the fluid, during the operation, flashes forth in a succession of big brilliant drops of liquid fire, to all appearance, yet a piece of paper, even gunpowder, may be applied to the flashes without their ignition. Steel, iron, or other metals are, however, ignited; but no non-conducting substances.—*Abridged from Bennet's New York Herald.*

BRICK AND TILE-MAKING MACHINERY.

Mr. Cottam has patented a machine, whereby may be completed 7,000 bricks a day of sixteen hours, the machine working off 20,000 in a day of ten hours length; and of draining-tiles 10,000, where only 700 could be heretofore worked under similar circumstances, namely, the difference of time, and by hand. The secret of the machine is *mouse cloth*, to prevent the clay sticking to the rollers, a cloth with a nap cut like velvet: its productions are far superior to those of the common method, especially with regard to solidity, and absorption of moisture. For extensive operations, the new machine is combined with the well-known pug-mill, and is worked with water, horse, or steam power.—*Literary Gazette.*

THE MARQUIS OF TWEEDDALE'S DRAIN-TILES MACHINE.

This machine will make 10,000 drain-tiles a day, one man and two boys to attend it, and 20,000 of flat tiles for the drain tile to lie upon; but if the tiles are broad, for roofing, it will make 12,000 a day. These draining tiles are fifteen inches long, so that three machines would make in one season (of thirty weeks) as many tiles as would lay a drain

* For an engraving and description of Davenport's Electro-Magnetic Machine, see Year-Book of Facts, 1839, p. 30.

from London to York. Now, a man and two assistants will only make 1,000 drain-tiles in a day, and these only one foot long, which is 1,000 feet per day; so that if the drain be laid at the distance of twenty-five feet, it will make in one day sufficient tiles for six acres. The advantages are—1st, the tile is much stronger from being compressed, and less pervious to water; it is not only compressed, but it is smoothed over, which gives it a surface as though it were glazed. They are capable of being made of a much stiffer clay than usual; and in nine cases out of ten the clay may be used directly on being dug, if passed through tbe crushers, being much drier. Clay unfit for bricks and tiles, by the common method, is available by the machinery. The expense of draining will be paid in three years, but not unfrequently in one.—*Farmers' Magazine.*

ERICSSON'S FILE-CUTTING MACHINE,

WHICH was patented about three years since, but then deemed impracticable, now turns out files of a superior and more regular cut than the average of those made by hand, and in much greater number in the same time. The principal beauty of the machine consists in the simplicity of its movements, and the skilful application of the principles of mechanics in modifying the stroke according to the varying thickness of the steel; striking lightly at the point, and increasing in strength as the thicker parts of the file come under its action.

NEW AMERICAN FLAX MACHINERY.

IN the United States has been invented machinery for the preparation of flax for spinning, after the manner in which cotton is now spun. A large company, in Delaware, is now engaged in the manufacture of the "short staple" produced by the new invention. The advantages alleged are: — 1.—That there is no loss of fibre, as no tow is to be taken out, all the lint being used up: whereas, by the old plan of hackling, finger-spinning, &c., there was a loss of perhaps half the original weight. 2.—That the expense of labour on the whole process of cloth making is reduced to *one-tenth* of what it was. 3.—That the expense of bleaching in the flax, as now, is much less than in the old plan, and the process less injurious to the texture.—*Letter in the Athenæum.*

IMPROVEMENT IN WOOL COMBING.

A VERY important invention in the woollen manufacture has been patented in Great Britain. It consists in heating the carding-engines and combs by steam, which has the beneficial effect of allowing the wool to be stretched or extended as it is operated on by the cards or combs, without breaking the fibre. By this process, out of every one hundred pounds of undressed wool, ninety-five of the best wool may be obtained; whilst, by the method now in use, sixty-five is the best result. The invention is of French origin.—*Mechanics' Magazine.*

CLOTH MAKING, WITHOUT SPINNING OR WEAVING.

A MACHINE has been invented, by an American, for the making of broad or narrow woollen cloths without spinning or weaving; and, from our acquaintance with the staple manufacture of this district (Leeds), after an inspection of patterns of this cloth, we should say there is every probability of this fabric superseding the usual mode of making cloth by spinning and weaving. The machines are patented in this and every other manufacturing nation. Should this machine succeed to any thing near the expectation of the patentees, its abridgment of labour, as well manual as by machinery, will be very great. It is calculated that one set of machinery, not costing more than six hundred pounds, will be capable of producing six hundred yards of woollen cloth, thirty-six inches in width, per day of twelve hours.—*Leeds Mercury.*

COMPRESSED PEAT.

LORD Willoughby de Eresby, in a pamphlet of a dozen pages, has described the various machines which he has used for the compression of peat, till he has brought the last to working perfection, and patented the same.

In the above pamphlet, the author observes: "in the selection of peat for compression, care must be taken to obtain a black peat, free from fibre." Peat of the proper appearance resembles blackened butter, and is the only sort which will repay the expense of preparation. The peat should be dug of the usual size, viz., eight inches by three, and three deep, and of uniform shape, by a spade of peculiar construction. All attempts to compress peat in large masses have failed. Before compression, the peats must be placed to dry for five or six days under sheds, in the same manner as bricks and tiles; and, after compression, must remain under cover until perfectly free from moisture, when they will be fit for use. All attempts to dry the peat by artificial heat have failed. The peat, when properly compressed, is reduced about one-third in size, is hard and compact, and nearly black in colour; it varies slightly in density, sometimes floating, at others sinking in water. It will be found an excellent substitute for coal. In a steam-engine at St. John's foundry, Perth, the peat outlasted an equal weight of coal, in the proportion of sixteen per cent., the engine being worked at its ordinary rate. It gives off in abundance gas, which burns with a clear white light. The peat may also be prepared by charring, in the same manner as ordinary charcoal, by which its size is reduced about one-half. When thus charred, the slowness and difficulty with which it burns renders it an extremely valuable fuel in many processes of the arts: it is likewise free from sulphur, and leaves few ashes. For the working of steel, especially, this freedom from sulphur renders it greatly superior to charcoal. It has been applied to this purpose by Messrs. Philp and Whicker, of St. James's-street, who have used it in forging razors and surgical instruments. The charred peat has also been employed in working other metals, and in soldering thin brass, with great success. It is likewise as serviceable in the kitchen as common charcoal, and occasions no unpleasant smell or taste.

The quantity of peats which may be thus compressed in one day may be estimated at 27,000, or 45 per minute, under a pressure of

400 pounds on the square inch, with a high-pressure steam-engine of six-horse power.

WIDE VELVET WEAVING.

THE Society of Arts have conferred upon two Spitalfields weavers, named Hanshard and Cole, rewards for their invention of a mode of weaving wide velvets. It appears that about two years ago, a velvet shawl, two yards square, was imported from France, and seen by these weavers, both of whom devised the same means of performing the same work; though Hanshard was first successful. The difficulty in weaving a wide velvet was in the width of the fabric being greater than the stretch of a man's arm, so that he could not pass the wire containing the silk across it, from its thinness and flexibility. To obviate this, Hanshard put the wire into a small pointed brass tube, which held the wire stiff, so that it could be passed across. In working, however, the end of the tube was liable to catch in the fabric, and break the thread; and this difficulty was overcome by putting a pointed cap upon the end of the tube, after the wire had been inserted. Cole followed Hanshard in the invention of the tube; but the cap was solely Hanshard's.—*Mechanics' Magazine; abridged.*

BREECH-LOADING GUNS.

A NEW breech-loading gun, invented in Paris, by MM. Lepage, has been patented in England, in the name of M. Demodion. In this gun, the breech part of the barrel opens by raising a lever in the situation of the breech pin, which carries a part of the breech. When this lever, which turns upon joints in side-plates, is raised, the breech end of the barrel is removed, and the cartridge may be introduced: which being done, the lever is shut down upon the small of the gun, which closes the end of the barrel, and it is made fast by a spring catch in the end of the butt. The cartridge is made up in the usual form, and the copper cap, containing the detonating composition, is inserted into the back end of it. On shutting down the breech pin lever, a solid piece of steel, as a small anvil, is brought close against the side of the detonating cap; an up-striking hammer, impelled by a strong spring placed against the guard, when let off by the trigger, strikes the side of the detonating cap with sufficient force to crush it against the anvil, and thereby discharge the piece. A beautiful specimen musket of this kind has been deposited in the Tower of London,—a present from the French Government.

The Franklin Institute of Pennsylvania have reported that Mr. Jenks, of Massachusets, has invented an improvement, consisting of a piston or plunger, fitting in a chamber in the breech of the piece, which is drawn back by a lever, and several pieces of metal, so as to permit the ball and charge of powder to be put into the chamber through an opening in the upper part of the breech. After this, the lever being depressed, the piston is forced forward by the pieces of metal, which constitute, by their position, a species of toggle-joint. When the charge has thus been forced home, the joints are a little beyond the straight line, from the breech to the extreme point of action. It follows that the piston cannot be forced back by the discharge of the piece, and requires no fastening of any kind. The committee of the Institute consider this invention to be safe and simple, free from the objections which usually accompany breech-loading, and possessing all its advantages.—*Mechanics' Magazine; abridged.*

THE WHEEL RIFLE.

MR. WILKINSON, of Pall-mall, has patented this invention, the novelty of which consists of a wheel, containing seven complete charges, revolving on a centre; which, when discharged, can be replaced in an instant, by other wheels carried in the belt, so as to keep up a continuous firing. As rapidly as the command, "load, cock, fire," can be uttered, can this rifle be discharged, several hundred times, without missing fire, or requiring to be cleaned.—*Literary Gazette.*

NEW RIFLES.

RIFLES of an entirely new construction have been introduced into the British army. They are provided with percussion locks; the barrels have two grooves in the bore, which descend to the touch-hole in a spiral direction. The balls have a projection round them to fit the groove, and each may be described as having the appearance of "Saturn and his belt."

IMPROVED TUNING FORK.

AN improved tuning fork has been invented and constructed by a mechanic of Aberdeen. It is not like the common fork, restricted to the production of one note, but may be made to produce any note within the compass of two or three octavos. This is accomplished by extremely simple means. A stop is made to slide stiffly between the prongs, the effect of which is to check the vibrations in the portion of the prong included between the stop and the hand. The effective length of the prongs being thus diminished, the rapidity of the vibrations is increased, and the pitch consequently raised. In the fork now before us, the fundamental, that is, the note yielded without the use of the stop, is D below the first line of the staff; and the positions of the stop for the production of the other notes of the stop, up to C (beyond which the inventor has not yet carried his divisions), are, of course in the first instance, ascertained by trial. It appears that this improved tuning-fork has only to be known to come into general use.—*Aberdeen Journal.*

THE HARMONIPHON.

THIS new musical instrument is the invention of M. Paris, of Dijon. It resembles the concertina, but is, in some respects, superior to that instrument. The sound is produced by the vibration of thin metallic plates, and it is played by keys, resembling those of the piano-forte; but the air which acts upon the vibrating plates, instead of proceeding from bellows within the instrument, is blown by the mouth, through an elastic tube. Thus, while the fingers on the keys merely mark the different notes of the scale, the expression lies in the mouth, as in the oboe, or clarionet. The harmoniphon is made in three varieties: the first is of the compass of the oboe, the second of the corne Inglese, and the third, of a larger size than the others, combines both these instruments, and has a compass of three octaves. It is calculated to be of great utility in provincial orchestras, where it will be an excellent substitute for the oboe.—*Polytechnic Journal.*

NEW ACOUSTIC INSTRUMENT.

An instrument has been invented in Paris for assisting hearing, and called "the Soniferous Coronal." It goes over the head, whence the sound is collected, and conveyed by small tubes into each meatus. It has this advantage over cornets, that it occasions no noise in the ears. The instrument has been used at the Royal Dispensary by Mr. Curtis, who has improved it by covering the tips of the tubes with caoutchouc.

GLAZING WINDOWS.

Sir John Robison has experienced the evil of the ordinary way of putting in panes of glass, having the convex side outwards. When the action of the wind was strong, as during storms, and its pressure was nearly 8 lbs. on a square foot, the convex side was forced in and rendered concave, and in doing so, was broken. It has been proposed that experiments on this subject should be made by the Society of Arts for Scotland; and Sir John has recommended that the panes should be put in windows with the convex side inwards.—*Jameson's Journal.*

SUBSTITUTE FOR DOOR-SPRINGS.

On March 27, Mr. J. Gilchrist submitted to the Society of Arts for Scotland, an ingenious method of applying lever power instead of springs, for the purposes of shutting, and keeping shut, doors opening either way; and by which the greatest power is obtained just where it is wanted, viz., when the door is shut.—*Jameson's Journal.*

MIKROTYPOPUROGENEION.

Take a page, or any other definite portion of printed paper, cut it into two pieces, note the *size* of the type, and place one piece aside as the muster or test. Thrust the other piece between the bars of a lighted grate, or ignite it in any other manner which may be preferred; place it gently on the hearth, and let it burn away, till entirely consumed. Take up the paper so charred carefully, and, holding it to a good light, the size of the print, which is perfectly legible, will be found to have become considerably reduced, while the sharpness, or purity of the impression, will have been singularly increased.—*Literary Gazette.*

PANTOCRATIC MICROSCOPE.

This beautiful instrument is the invention of Professor A. Fisher, of Moscow. With it, the observer can, by simple and almost imperceptible movements, vary the magnifying power from 270 to 550, without in any degree obscuring the object, the degree of enlargement being registered on the body of the instrument. This is important;

because it saves the operator the necessity of shifting the different parts of the microscope, when he desires to study the same object under different degrees of magnitude.—*Athenæum.*

PAINTING MACHINE.

M. LEIPMANN, of Berlin, has invented a machine for copying paintings in oil with perfect exactness. This discovery is stated to be the result of ten years incessant study; M. Leipmann having been a regular attendant at the museum at Berlin, where he selected a portrait by Rembrandt as the object of his experiment. Fixing single features and parts of this picture in his memory, by hours of daily and incessant observation, he contrived to reproduce them at home, with perfect fidelity, and by the aid of a machine, in what manner is not known. The discovery, however, is so complete, that he has produced, in the presence of the directors of the museum, 110 copies of the painting in question. These copies are said to be perfect, and to retain the most delicate shades of the original picture, confessedly one of the most difficult to copy in the usual way. The price of the copies is but a louis d'or each.—*Foreign Quarterly Review.*

NEW WEAVING-BAR FOR ENGRAVERS.

MR. E. SANG, of Edinburgh, has invented a weaving bar for engravers' ruling machines; the chief novelty of which is the substitution of a cylindric weaving-bar, easily removable, for the usual flat one, by the turning of which may be produced changes in the pattern, and a combination of levers for altering the depths of the undulations.—*Trans. Soc. Arts, Scotland.*

ENGRAVING ON MARBLE.

MR. C. PAGE, of Pimlico, has discovered an improved means of engraving on marble, by covering the surface with a coat of cement before the chisel is used. The cement effectually prevents the marble from chipping, and when the coating is removed the letters remain as perfect as if cut in copper.—*Civil Engineer and Architect's Journal.*

THE EVER-POINTED PENCIL.

AN improvement has been made on that useful article, "the everlasting pencil," that will recommend it again to many who have disused it, owing to the disadvantages remedied by Mr. Riddle's patent. Instead of the slide working in a groove, an internal spiral action is substituted, which, by simply turning the upper part of the case, protrudes or withdraws the point. This not only obviates the inconveniences of the projecting slide to the fingers, but, by dispensing with the slit that weakened the tube, strengthens the case, and preserves its external beauty, that was injured by the traversing of the slide; and prevents the accumulation of dirt, that injured the movement. Mr. Riddle's spiral pencil is at once durable and elegant; in short, this handy invention is now perfected.—*Spectator.*

SELF-ACTING EXTINGUISHER.

THIS ingenious contrivance is the invention of Mr. Jones, who has named it "The Photolyphon, or self-acting extinguisher;" which being slid on to a candle at any distance, beyond which the candle is required not to burn, snuffs it out as soon as the upper part is consumed.—*Spectator*.

WIRE-SEWN BOOTS AND SHOES.

M. SELLIER, of Paris, has patented the right of using brass wire for attaching the upper leather to the welt of shoes and boots. He urges that this metallic thread allows neither moisture nor dust to enter the shoe, and farther that it does not rip. The sewing is performed with as much ease as with waxed thread, and is not more expensive.—*From the French ; Mechanics' Magazine.*

PATENT ELASTIC BOOTS AND SHOES.

MR. J. DOWIE, in some observations read to the Edinburgh Society of Arts, on March 15, states that he has succeeded in a valuable improvement, by making those parts immediately under, or on each side of, the principal arch of the foot, of an elastic material, composed of caoutchouc and animal skin; so manufactured as to bestow on the fabric the elasticity of the caoutchouc, while it retains the tenacity and durability of leather. The introduction of this elastic substance allows considerable changes to be made in the form of the boots and shoes, and gives the wearers the free use of their feet and ankle-joints in walking, to a much greater extent than any hitherto in use.

The patent elastic boots and shoes possess the following advantages:—They are light, elastic, and durable, and admit of being made of the shape and form of the foot when at rest; while their capacity to alter their figure admits of their adaptation to the ever-varyingmotions of the foot. They may be made lighter than the ordinary kind, yet no unequal pressure is felt; indeed, support is given to the foot. They are rendered light by the absence of that weight of rigid leather, placed in the sole immediately under the arch of the foot.—*Abridged from Jameson's Journal.*

IMPROVED SHAKER.

A SHAKER for a thrashing mill, on a new principle, has been invented by Mr. Sherman, of Kirkpatrick, Durham, Kircudbright. It is of the same width as the machine, and about four feet and a half long. The sides, which run from the machine outwards, are three-quarter boards, about four inches deep, with a flat bar across, at each end; to which the bottom, which is parallel rods, is nailed. The rods are five-eighths of an inch thick, five-eighths apart, and about an inch deep. The end of the shaker is attached to the machine by hinges, and the straw is thrown on it by the rake. The other end is suspended by wooden rods from a spindle of inch iron, cranked at each end—an inch and a quarter giving the shaker two inches and a half in throw. The shaker has a small bolt on each side, which runs into a hole in the rods, by which the shaker is connected with the crank. There are several holes in the rods, by which is given more or less declivity, that mostly used being one foot in three. The bolts are about one foot from the extreme end. The motion is raised from the spur-wheel by a pinion on a shaft, on which also there is a pulley, and another on the end of the crank-spindle turned with the rope over them; giving about the same motion as *fanners* require, the fanners being driven from the same shaft. The bottom of the rake might be taken out, and the shaker introduced, in which case room might be saved; and in many cases, the shaker might be driven by a pinion on the crank-spindle attached to the spur-wheel. Should there not be room for the shaker between the rake and fanners, a small rake, just sufficient to relieve the drum of the straw, with a little more motion, might serve. By putting a close bottom on the shaker, two or two and a half inches below the rods, the grain would run into fanners almost without hoppers. By this improvement, the shaker can be made longer, giving the straw more time, the crank can be enlarged, giving it more shake; or more motion can be given, which will shake it oftener.—*Dumfries Times; abridged.*

NEW THRASHING MACHINES.

A NEW hand-thrashing machine has been introduced into the county of Sussex, which will thrash about four quarters of wheat daily; two men turning it, and two others feeding it, and clearing the straw away. This machine, instead of diminishing manual labour, increases it, by giving employment to those labourers who cannot use the flail; and who, in wet days, could not be profitably occupied. It is very portable, and costs but a few pounds.—*Sussex Express.*

TRAVELLING PLATFORM.

THE Society of Arts have rewarded the inventor of a simple and ingenious contrivance, termed "a travelling platform." It consists of two boards, twelve inches long, by four wide, strongly hinged together; the joints of the hinges are uppermost, and a groove in the centre allows a rope to pass between the boards freely when they are brought into a horizontal position. Each board is to be strapped securely to the climber's feet; a belt also passes round his waist, having attached to it a ring, through which also the rope passes. When a person thus equipped wishes to ascend the rope, he hangs by his hands, draws up his knees, keeping the hinged board in a flat position, and it slips up the rope: immediately his weight is put under the boards, the edges on the under side, (the contrary to that on which the hinge joints are placed,) collapse, and firmly clutch the rope; whilst he raises himself, his weight acting on the lever formed by the width of the board, its thickness, and the hinge joints. A workman may thus be supported in many situations where a ladder could not be conveniently used, without the fatigue of keeping the muscles of the legs in constant exertion in clasping the rope, and with the facility of moving upwards or downwards, without extraneous assistance.—*Mechanics' Magazine.*

PNEUMATIC FILTER.

MR. PALMER, of Newgate-street, has lately made a useful application of pneumatic pressure, in the filtering of liquids. His apparatus consists of a tin vessel, divided by a rim at about the middle; where also is a moveable division, formed of a plate of zinc, perforated with holes. This plate supports the superincumbent pressure; and above it is laid a filtering fabric of paper, calico, flannel, leather, felt, or other material. An air-pump is attached to the upper part of the lower division; and a heavy loose brass ring is placed above the edges of the filtering fabric, to keep the rim close to the side of the filter. The filtered liquor is drawn off by a cock. To use this filter, having prepared the fabric, fill the upper portion of the vessel with liquid, exhaust the air from the lower part by means of the air-pump, when the liquid will quickly pass through, leaving the filter nearly dry; then unscrew the air-pump, and draw off the fine liquid.—*Mechanics' Magazine; abridged.*

TRANSPARENT WATCH.

A WATCH has been presented to the Academy of Sciences at Paris, constructed principally of rock crystal. It was made by M. Rebellier, and is small in size: the works are visible; the two teethed wheels which carry the hands are rock crystal, and the other wheels are of metal. All the screws are fixed in crystal, and all the axles turn on rubies. The escapement is of sapphire, the balance wheel of rock crystal, and its springs of gold. This watch is an excellent time-keeper, which is attributed, by the maker, to the feeble expansion of the rock crystal in the balance-wheel, &c.—*Mechanics' Magazine; abridged.*

COOKING CLOCK.

MR. LOUDON describes *an egg-clock*, which rings a bell, or sets off an alarum, at any number of minutes required: it is formed by a dial like that of a watch, but larger, surmounted by an alarum-bell, and with five divisions, representing five minutes on the dial. This being fixed up over the kitchen fireplace, the index is moved to the number of minutes the egg is to be boiled; and, during the boiling, the cook may be otherwise employed till the alarum goes off. The act of moving the index, or pointer, backwards, winds up the clock. The principle may be applied to a larger dial, so as to mark the time requisite for cooking articles generally; and Mr. Loudon has accordingly caused such an apparatus to be made. Hence the ordinary work of the kitchen may go on without the interruption of watching, &c.

SUPERB PALACE GATES.

MR. DEAN, of Bolton, has cast two pairs of beautiful gates and palisading, for the Imperial Palace, at Constantinople. The height of these gates, with the central ornaments over them, is 35 feet; the latter being richly gilt. The height of the gates is 22 feet, and the width of the gateway 12 feet, exclusive of the hanging pilasters of pendent vine-leaves, which are 6 feet 3 inches wide. The gates will be supported by marble columns, surmounted with elegant vases; and the palisades, 23 feet long on each side, are to be fixed in marble basements. The weight of the two pairs of gates is 40 tons; and the cost upwards of 20,000*l.* when fixed; patterns, 900*l.*; packing-cases, 150*l.*—*Mechanics' Magazine.*

NEW METHOD OF PERFORATING GLASS.

PUT a drop of spirit of turpentine on the spot where the hole is to be made, and in the middle of this drop a small piece of camphor. The hole can then be made without difficulty, by means of a well tempered borer, or triangular file. Solid turpentine answers as well as the spirit and camphor.—*Annales des Mines.*

VARIETIES OF IRON.

DR. SCHAUFHAENTL has detailed to the British Association a series of important experiments, which severally go to prove that the purest carbon, at the highest temperature, retained hydrogen, and occasionally azote, and that what was considered to be pure carbon was, in reality, approaching to a carburet. He described a method of obtaining graphite by running puddling slag in a fluid state, or silicates of iron and manganese, over fragments of pit coal; which, on being cooled, left the graphite in thin layers on the surface of the slag. The molecules of iron, according to these experiments, are arranged in the grey cast-iron in the most regular form, having all their surfaces in one plane; the most equal distribution of molecules appeared in hardened steel; collecting in fascicular aggregation in soft steel, and being loose and longitudinally arranged in wrought iron. Pure iron could not be welded; the welding power depends on its alloy with the carburet of silicon; the good and various qualities of all the wrought irons depended on the alloys of pure iron with other metallic bodies; and the presence of most of the electro-negative metals had been generally overlooked in the existing analysis of iron. The presence of arsenic in Swedish steel, when forged red hot, could be ascertained by its smell; as well as in the Low Moor iron. The usual solution of iron under analysis, in order to separate those metals from the iron, must be, for the necessary correction, divided into two parts—one to be treated with a current of sulphuretted hydrogen, the other part dropped into sulphydrate of ammonia, and carefully digested. A small quantity of silica was more difficult to separate from a large quantity of iron than generally seemed to be believed; and the real amount of carbon could only be ascertained by Berzelius's method of burning iron in a current of oxygen, or mixed with chlorate of potash and chromate of lead in a glass tube, used first by Berzelius for analysis of organic bodies. The author maintained that steel was an entirely mechanical production of the forge hammer, which tore the molecules of certain species of white cast-iron out of their original position, into which the forces of attraction, in respect to the centres, as well as to the position of the molecules, had arranged those molecules by the slow action of heat. Steel, as it came out of the converting furnace or the crucible, was nothing more or less than white cast-iron; of which Indian steel, called wootz, was the fairest specimen. The author finally gave an analysis of two specimens of cast-iron and one of steel. The first specimen was French grey iron, from Vienne, department de l'Isère, obtained from a mixture of pea-iron-ore with red hematite, by means of coal from Rive de Gier and heated air, specific gravity, 6·898. The second specimen was Welsh iron, from the tin-plate manufactory of the

Maesteg Iron-works, near Neath, in South Wales; obtained from a mixture of clay iron-stone and Cumberland red ore, by means of coke and heated air. It was silvery white, without signs of crystallization: specific gravity, 7·467. The third specimen was a fragment of a razor forged in the author's presence, in the workshop of Mr. Rodgers, of Sheffield, of the specific gravity of 7·92.

	Grey French Iron.	White Welsh Iron.	Steel.
Silicon	4·86430	1·00867	0·52043
Aluminum	1·00738	0·08571	0·00000
Manganese	0·75130	traces.	1·92000
Arsenic	0·00000	0·00000	0·93400
Antimony	0·00000	1·59710	0.12100
Tin	0·00000	0·00000	traces.
Phosphorus	0·54000	0·08553	0·00000
Sulphur	0·17740	0·32018	1·00200
Azote	0·00000	0·76371	0·18310
Carbon	3·38000	4·30000	1·42800
Iron	89·00740	91·52282	93·79765
Loss	00·27222	00·31428	0·09382
	100·00000	100·00000	100·00000

Several gentlemen, among whom were some connected with the iron trade, expressed a high sense of the value of this communication; from which it appeared that the peculiarities of Swedish iron, in a great degree, depended on the presence of arsenic, and those of Russian iron on the presence of phosphorus.—*Literary Gazette.*

SMELTING IRON.

M. TEPLOFF, mining engineer of Russia, states that, in the Uralian mountains, where many iron mines are worked, fourteen pounds of iron are obtained by a consumption of the same quantity of fuel, the quantity and rapidity of the air which enters into the combustion being properly regulated; whereas, only from four to six pounds of iron are obtained when the blowers are badly managed. In an experiment, it was found that 100 cubic feet of air, under a pressure of two inches of mercury, have produced the same effect as 200 cubic inches of air, under a pressure of one inch; but with this difference — that, in the latter case, the consumption of fuel was double that required for the former. M. Teploff farther states that a furnace had produced 22,000 pounds of iron in 24 hours, with a consumption of only 16,000 pounds of fuel; whilst, before the proper regulation of the blast, double the quantity was required to produce an equal quantity of iron. According to the same engineer, the results thus obtained are more economical than those by the hot blast. — *Recueil de la Société Polytechnique.*

IMPROVED METHOD OF CASE-HARDENING IRON.

THIS discovery consists of the employment of ferrocyanodide of potash (prussiate of potash), the crystals of which are reduced to a coarse powder; the iron articles being well cleaned off, heat them at the forge to a bright red, take them out and strew some of the prussiate of potash over them, which will immediately fuse and spread upon the surface; return the article to the fire, and restore the heat, when sudden quenching in cold water will give the required hardness. By this means, case-hardening is more effectually completed in two minutes than it could be in two hours by the old method; besides which, part of any iron instrument may be case-hardened, while the remaining portion continues in its original state.—*Mr. Baddeley; Mechanics' Magazine; abridged.*

TENACITY OF IRON.

MR. TELFORD made an interesting series of experiments on the strength of wire, an account of which is found in Mr. Barlow's treatise on the strength and stress of timber, page 254.

The same gentleman made various experiments on bars and bolts of iron, detailed in the same work.

He also attended to the successive elongations of bars under different weights, and noted the amount of recoil or contraction when relieved from strain.

Captain S. Brown has also furnished to Mr. Barlow a series of highly interesting experiments on the strength and elongations of iron bolts.

The mean result of three experiments by Capt. Brown on cast iron was 18,564 pounds to the square inch.

Mr. Hodgkinson has published, in the third report of the British Association, three results of experiments on the same material, which make its strength 17,136 pounds per square inch, while the three experiments of the committee of the *Frank. Inst.*, which were considered fair, indicated in the bars a strength of 20,834 pounds.

Mr. Brunton and Mr. Brunel, have each engaged in this interesting department of inquiry, and given the results of experiments on a large scale, which will be found in the work of Mr. Barlow already cited.

Mr. E. Martin, formerly of the Polytechnic school, has given in the *Annales des Mines*, Vol. V., a series of experiments executed in France, under the orders of M. Barbé, on round rods of iron, eighteen or nineteen feet long, and two inches or more in diameter, made with a view, in part, to determine the recoil when released from strain, and the actual amount of elongation under each weight to which it was subjected.

In the same volume of the *Annales des Mines*, M. Vicat has a paper referring to, and controverting some of the positions of Mr. Martin, but not affecting the statements respecting the experimental operations just referred to.

In the same work, Vol. VI., are contained some interesting statements by M. Payen, respecting the manufacture of wire, in which the relative ductility before and after annealling, is established.

In the *Annales de Chimie et de Physique*, for Sept. 1833, is a valuable paper by M. Vicat, showing the influence of time on the gradual extension of wires under different weights. Each of his experiments occupied nearly three years.

The quantity by which iron extends under different degrees of tension, and on the recoil when relieved from strain, has been examined by Professor Barlow. See Journal *Frank. Inst.*, Vol. XVI. p. 124, &c. The relation between the effect of straining and elongating a bar by mechanical means, and that of expanding it by heat, is also noticed.—*Note to the Report of the Franklin Institute on the Strength of Steam Boiler Materials.*

STRENGTH OF CAST IRON.

THE Council of the Institution of Civil Engineers have awarded a Telford premium to Mr. Bramah, for his series of experiments, on the strength of cast-iron. These experiments, undertaken with the view of verifying the principles assumed in the work of Tredgold, on cast-iron, surpass every other series in number and accuracy, since two similar specimens of each beam were subjected to trial. The principles sought to be established by these experiments are that, within the elastic limit, the forces of compression and extension are equal; and that, consequently, a triangular beam, provided it be not loaded beyond that limit, will have the same amount of deflection, whether the base or apex be uppermost; and a flanged beam the same deflection, whether the flange be at top or bottom. This communication was accompanied by some observations by Mr. A. H. Renton, pointing out the agreement which subsists between the experiments and results of the formulæ of Tredgold.—*Mechanics' Magazine.*

PURIFICATION OF COPPER.

MR. L. THOMPSON, of Lambeth, has received a gold medal from the Society of Arts, for a new method of purifying copper, which has long been a desideratum. This method is so simple as to require no particular management on the part of the workmen.

Take of impure copper . . . . 100 parts;
Copper scales . . . . 10 parts;
Ground bottle glass, or any other flux, 10 parts;

heat the whole together in a covered crucible, and keep the copper in a state of fusion for twenty minutes or half an hour, at the end of which time it will be found at the bottom of the crucible, perfectly pure. The quantity of copper scales must vary in proportion to the supposed impurity of the copper to be operated on; but the proportions here given will be found to answer very well for the average kind of English copper. The explanation of this process is sufficiently simple: the impurities contained in the copper, consisting of iron, lead, arsenic, &c., combine with the oxygen contained in the copper scales, and form oxides or acids, which are dissolved by the flux, or fly off in a gaseous form, leaving the purified copper, together with that reduced from the scales, at the bottom of the crucible; consequently, the copper obtained exceeds that put into the crucible, the gain generally averaging from one to one and a half per cent. In this way, Mr. Thompson obtained perfectly pure copper, from brass, bell-metal, gun-metal, and several other alloys containing from four up to fifty per cent., of iron, lead, antimony, tin, bismuth, arsenic, &c.—*Transactions of the Society of Arts.*

PRINTERS' METAL.

A TYPE-FOUNDER of Clermont, named Colson, has patented a new material for printing types, which is harder, capable of more resistance, yet less expensive, than the ordinary composition of lead and antimony. Colson asserts that types of his manufacture will serve for punches in striking matrices, and that they will last ten years without being more worn than the usual composition is in one year.—*Foreign Monthly Review.*

MINES OF CORNWALL AND DEVON.

IF we estimate the value of the metals annually raised in Great Britain and Ireland at about 10,597,000*l.*, and consider that of this sum the iron amounts to 8,000,000*l.*, the value of the remaining metals would be 2,597,000*l.*, of which Cornwall and Devon would furnish about 1,340,000*l.*, or more than one-half, leaving 1,257,000*l.* for the value of all the metals, with the exception of iron, raised in other parts of the United Kingdom. The two great metallic products of the district are copper and tin: of the former it yields one third, and of the latter, nine-tenths of the whole supply of copper and tin furnished by the British Islands, and all the countries of the continent of Europe.—*Report of the Geological Society.*

NEW LIGHT FOR LIGHTHOUSES.

A SYSTEM of illumination for lighthouses has been invented by a serjeant-major in the Austrian artillery, named Selckonsky. The apparatus consists of a parabolic mirror, sixty-two by thirty inches, with a twelve-inch focus; the light being produced by a new kind of wax candle. The invention has been tried at Trieste, where it illuminated the whole of the port and neighbourhood equal to the moon at full; and at the distance of six hundred yards, the finest writing could be read.—*Mechanics' Magazine; abridged.*

ECONOMY IN GAS LIGHTING.

ON March 27, Sir John Robison submitted to the Society of Arts, for Scotland, a paper pointing out the most economical mode of burning gas, by peculiar construction of the burners, and the proper size and fitting of the chimneys, and the disuse of obscured shades. It is found to be more economical, with any burner, to burn the gas to the full height it can attain without smoking. If a small quantity of light be wanted, it is better to use a smaller burner than to reduce the flame of a large one. The best effect of an Argand burner is attained when the holes are all of one size, so that the flame should be of an equal height all round. The paper also describes the method of burning gas in street lamps, pointed out by Sir John Robison, and now very generally used in Edinburgh, so as to prevent the moisture from being condensed on the inside of the globes, and rendering the light obscure.—*Jameson's Journal.*

DR. URE'S EXPERIMENTS UPON THE COST OF THE LIGHT AFFORDED BY DIFFERENT LAMPS AND CANDLES.

THE author, having instituted a series of experiments to determine the advantages of Mr. Parker's new hot oil lamp, adopts as the standard of comparison the French mechanical lamp, in which the oil is raised by machinery, so as continually to overflow at the bottom of the burning wick. The relative illumination was determined by the well known method of the equal intensity of shadows, and verified by that adopted by Professor Wheatstone; namely, by the relative brightness of the opposite sides of a revolving ball.

One peculiar feature of the new lamp is its bell-mouthed glass chimney, above which is a chimney of iron, with a parted diaphragm for the purpose of causing a certain portion of the heat of the flame to reverberate against the interior cylindric cavity of the oil cistern. The bell mouth is formed in a mould, and is far better suited for producing a steady flame than the rectangularly constructed chimney of the mechanical lamp. The intensity of the shadows from the mechanical lamp, and the hot oil lamp, of a wire a few inches long and of the thickness of a crow quill, was equal at a distance of ten and eleven feet respectively; their relative illuminations being as the squares of these are, as 100 and 121 respectively; and the consumption of the best sperm oil was 15·2 and 11·6 grains per minute: the relative cost of illumination for this oil would thus appear to be fifty per cent. in favour of the new lamp. On trying southern whale oil, the cost of illumination appeared to be about one-third that of the mechanical, and one-half that of the hot oil, lamp with sperm oil. The author tried many other substances, and, comparing the various illuminations, concludes that the hot oil lamp with southern whale oil affords an economy of light nearly 12 times greater than stearine or German wax candles, $7\frac{1}{2}$ times greater than tallow mould, 11 times greater than cocoa nut, $8\frac{1}{2}$ times greater than Palmer's, $17\frac{1}{2}$ times greater than spermaceti, and 18 times greater than wax, candles.

The author had also compared the illumination produced by one of the Fresnel lamps deposited at the Trinity House. The lamp consists of four concentric circular wicks, placed in a horizontal plane; the innermost being $\frac{7}{8}$ths of an inch, and the outermost $3\frac{1}{2}$ inches in diameter. The intensity of the shadows from this and from the mechanical lamp were equal at a distance of 13 feet 3 inches, and 4 feet 6 inches respectively; taking the squares of these, the Fresnel lamp gives about 9 times the light of the mechanical, which latter may be assumed as equal to that of 11 average wax candles. On comparing one of the best Argand lamps with the mechanical, the former was to the latter as 10 to 11; so that the illumination of the Fresnel lamp, instead of being, as has been asserted, equal to 40 Argand lamps is not equal to more than 9·6 of those lamps. In the Bude light, a small stream of oxygen is sent up through a small tube within the burning wick, which is $\frac{5}{8}$ths of an inch in diameter, and the flame about $\frac{3}{8}$ths of an inch. The illuminating power is equal to about 30 wax candles. Dr. Ure also examined the illuminating power of different kinds of wax candles, and found that the light from a long-three and a short-three was the same, or 1·11th of the mechanical lamp; also the light emitted from one of the six-to-the-pound was very little less, being 1·12th of that of the mechanical lamp. The consumption of wax in a long or short-three may be taken at 126 grains per hour, and in a short-six at 125 grains per hour.—*Transactions of the Institution of Civil Engineers.*

THE BUDE LIGHT.

THE Bude Light, the invention of Mr. Goldsworthy Gurney, is produced by introducing oxygen into the interior of the flame. An ordinary flame is hollow; the exterior part only being ignited by the atmosphere, the interior part is unburnt, containing vapour of oil and carburetted hydrogen, which forms the interior of an ordinary oil flame, having a cylinder or cone of flame round it. The new light is produced by oxygen, admitted into the bubble or interior of flame. The oxygen strikes the nascent carbon and vapour of oil as it is distilled, and produces an intense light. The difference between that and an Argand lamp is, that one has oxygen in the interior, and the other has common air. The Argand burner consists of two flames, one within the other: the common lamp has but one, and this constitutes the difference between the common ordinary lamp and the Argand burner. In the Bude Light, the outside or atmospheric flame acts as a retort in distilling the combustible matter. The light produced is from the vivid ignition and more perfect combustion of the carbon; such light being in direct ratio with the quantity of disengaged unburnt charcoal.

Mr. Gurney has made several experiments at the Trinity House, with a view to ascertain the quantity, intensity, comparative expense, practicability, and certainty of duration, of the Bude Light; and, taking an Argand burner, one-eighth of an inch in diameter, burnt with atmospheric air, as a standard of light, a Bude Light of one-quarter of an inch in diameter produces a light equal to two such Argand burners.

The expense of the Bude Light, as compared with the Argand burner, taking it as a standard of light, is more in proportion, as thirteen to twelve; as compared with wax-lights at 1*s*. 9*d*. per lb., it is about one half the expense.

The oxygen is of easy production, from manganese found abundantly in Devonshire, Cornwall, Warwickshire, and Cumberland; and, taking the present price of manganese at from 8*l*. to 9*l*. per ton, including the wear and tear of apparatus, expense of fuel and attendance, the oxygen may be produced at about 2*d*. per cubic foot.

The quantity of oil burnt by the Bude Light is not more than one quarter of that consumed in the Argand burner.

There is no difficulty in managing and preparing the gas, and in lighting and superintending the burners, which may be done by an ordinary man, without much previous instruction. It is managed at the Trinity House by a carpenter, who was taken from the premises: the second day he managed the light.

The Bude burner, with good sperm oil, will burn six hours without trimming; if it be impure oil, it will not last so long; and the wick only requires to be trimmed like an ordinary Argand burner.

This invention is called "the Bude Light," in reference to Mr. Gurney's residence in Cornwall, where the experiments were made: his name was associated with the lime light, which he published in 1823, and the above was named "the Bude Light" at the Trinity House, by way of distinction.

Such is the substance of the evidence given by Mr. Gurney before the Select Committee appointed to superintend his experiment of lighting the House of Commons with the Bude Light. The Committee have also examined several very eminent scientific and practical men, who have investigated the nature and properties of light.

Professor Faraday, having examined the Bude Light, reports that the lamp furnished by Mr. Gurney to the Trinity House burns with remarkable steadiness for eight hours together, not requiring so much attention as an ordinary Argand lamp for the same time. There is no fear from this lamp, except from the great heat which it produces, which can be easily guarded against by carrying off the hot air from the burner. Professor Faraday considers there to be no danger from explosion; and, should any of the oxygen escape, it would not be deleterious, but rather the contrary. The expense of the Bude Light, compared with that of oil, is 8¼*d.* by the Argand, and 10½*d.* by the Bude Light, including every expense, save men's wages. Professor Faraday has reported the new light to be so good and constant that he has recommended it to the Trinity Board for their light-houses, and one has been erected at Oxfordness. The Bude Light is very manageable, it being easy to make a single burner give a variation of light from one to two and a half. The adjustment of the supply of fuel (manganese), and then of the oxygen, enables you to have a great command over the lamp: that burned by Professor Faraday gave a light of twenty Argands, for twelve hours, consuming in that time six pints and four-tenths of oil, and sixty-four cubic feet of oxygen, which is the best proportion; but the light can be increased to almost any intensity.

Sir David Brewster proposed the Bude Light, many years ago, to the Commissioners for Northern Lighthouses, and afterwards recommended it to the public in the *Edinburgh Review*.

Dr. Ure considers it an admirable light for the House of Commons. "It is extremely simple in its mechanism, seems to be very easily managed, and is so very brilliant that it may be removed to a much greater distance than wax-lights could be. It could be taken entirely out of the house, and would throw down its light perfectly, without polluting the air with the products of combustion."

Dr. Lardner considers that with the Bude Light in large apartments, like the House of Commons, you can illuminate them effectually, diffusing light in sufficient quantity and splendour, without producing any injurious effects upon the air, or without interfering with its temperature or ventilation; for the polluted air arising from the combustion of the luminaries may be conducted through proper tubes into the atmosphere, without entering the house at all.

The evidence of Sir George Cayley, Bart., and others, concur in the general principle of the great superiority of the Bude over any other mode of illuminating large buildings like the Houses of Parliament; and the session of 1840 is expected to bring the plan into perfect operation.

GURNEY'S OXY-OIL LAMP.

On Feb. 15, Professor Faraday read to the Royal Institution the following paper upon this invention:—On this occasion, Mr. Faraday observed, he came forward to give a brief but fair description of a lamp most eminently and best calculated for the central one of a lighthouse arrangement, quite disinterestedly, solely on his own account, and not on that of Mr. Gurney. He would only compare it with the Argand as a standard, and with the French one of Fresnel, because intended for the same purpose. He first illustrated the circumstances, or particular conditions, under which light is obtained; the ignition of vaporous particles, or of solid particles, by either throwing off a mass of matter in ignition of great mobility in an aerial state of substance, and by charcoal, glowing in oxygen gas, incapable of acquiring the gaseous state, now and then beautifully scintillating, emitting solid glowing particles. The oxy-hydrogen gas lamp for microscopes, the combustion of the two gases igniting the particles of lime, was cited as an example of a light produced by a solid glowing surface: and of flame-lamps, or the combustion of vaporous substance, were mentioned those of tallow, oil, gas, &c. Mr. Gurney's is of the latter class, being a farther adaptation of the principles of the Argand lamp; pure oxygen gas being, beautifully in effect, and ingeniously in arrangement, substituted for common air to feed the inner surface of the flame. Beautiful in effect, not only because of its brilliancy and intensity, but also because the most powerful mass of light is concentrated in the given space required for the particular purpose to which the oxy-oil lamp is at present intended to be applied; and ingenious in arrangement, because after three years' indefatigable exertion and continued perseverance, to remove the rigidly and philosophically practical objections of Mr. Faraday, to whom the original lamp had been submitted by the Trinity Board for an opinion, Mr. Gurney has succeeded in constructing an apparatus calling for the warmest approbation of that severe scrutineer; and because, which has great weight to our mind, the form of the feeder whence oxygen flows to the centre of the flame has been reduced to the simple elementary form of the flame itself, the core, as it were, of natural, or rather of common artificial flame: many forms had been tried, and as often were the orifices in a short time closed by a deposit of carbon. Means had been adopted, occasionally, to sweep away these deposits, as with a finger, but the desired end was not satisfactorily gained. At length, the form of the deposit was observed invariably to assume a peculiar conical shape: this shape was discovered to be identical with what we have termed the elementary form of flame: to the like form was the feeder fashioned, and then no longer any deposit, no longer any necessity for sweepers or fingers—the lamp was complete and perfect. The form may be more familiarly exemplified as the Will-o'-the-Wisp-like light at the end of a burning stick or chip, dancing up and down, departing and yet loth to leave. Such, then, is the form of the vent through which the oxygen passes, and upon which no deposit accumulates.

But we have a mass before us to prove the superiority of Gurney's oxy-oil lamp, to which we must confine ourselves, omitting much interesting matter. The desideratum for lighthouse improvement has long been an intense light within a space of about three and a half inches in diameter and one and a half in height, in order that its rays may be thrown by a reflector, or directed by a refractor to the greatest possible distance in the greatest possible quantity; and in order to avoid diffusion, the size of the image being in proportion to the size of the light. The elementary lamp, the size of an ordinary Argand, is the common arrangement of cotton wick and oil, with the oxygen introduced into the centre of the flame, as before described; by which the flame is singularly compressed, and the brilliancy greatly exalted to two and a half times above the common Argand. The oxy-oil lamp consists of seventeen of these elementary lamps, arranged in a circle of three and a half inches diameter, the given dimensions; and has been calculated, thus heaped together, to afford a mass of light equal to thirty-seven and a half Argands. The flame under ordinary circumstances, without the oxygen, and from common lamps so arranged, would rise to some height, and produce a longer light than desired; and also burn with a reddish hue, from imperfectly burnt carbonaceous particles. Such appears to be the principal defect of Fresnel's lamp, constructed also on the principle of the Argand, with three, four, or any number, (four is the number used,) of concentric wicks, with space between, so that air might be supplied to the interior and exterior surface of each flame. But in Gurney's, as soon as the oxygen is turned on, the light is reduced from the height of about three to one and a half inches, and its brillancy exalted from reddish to the purest, brightest white, to the most intense flame lamp ever beheld. Thus has been obtained the desired improvement in the central light for guiding the wayfaring mariner; but there is one other consideration to be noted previously to pronouncing a final decision upon its practical utility and benefit,—and that is, the expense. The cost of oxygen is a very great addition to the expense of oil, &c. The value of a pint of oil is about ten-pence, which is calculated to burn, say for one hour; the oxygen required with that measure, and for that time, in the oxy-lamp, would be ten cubical feet, and its value would be twenty-pence. Here is an addition of double this cost of the oil; and the light cannot be obtained for less than 2*s*. 6½*d*. (we take Mr. Faraday's figures, which, of course, include the charge for cotton). But compare this with the other arrangements. To produce the same light for the same time, not now at all taking into consideration the form or dimensions of the light—which question has been previously settled—it would require thirty-seven and a half Argands, which would consume two and a half pints of oil, and cost 2*s*. 2*d*. The same light in Fresnel's would incur a charge of 3*s*. 11*d*. Thus it is shown, satisfactorily, that, in every respect, Gurney's oxy-oil lamp is superior, for lighthouse purposes, to any other hitherto invented. One curious fact is, that an addition of two-thirds to the cost of the material consumed in the other arrangements should only cause the expense of the oxy-oil lamp to be very little more than that of the Argand, and considerably less than the French, to produce the same light in the same time. The cause of this, Mr. Faraday states to be the suppression of the consumption of the oil by the oxygen to nearly two-thirds. This we conceive to mean, that the oxygen causes every particle of the oil to do its work by rendering the combustion more complete, and thus effecting a saving of two-thirds of the quantity consumed to produce the same result.—*Literary Gazette.*

PARKER'S NEW LAMP.

MR. S. PARKER has patented a lamp, in which the oil is heated by the flame of the lamp, to the temperature of from 200 to 250 degrees before arriving at the wick. The light emitted by this lamp, when supplied with the viscid and very cheap southern whale oil, surpasses, in purity and whiteness, the light of the best mechanical lamp, though it be fed with the best vegetable or even sperm oil. This superiority is, in part, due to the form of the chimney, and to the oil being maintained uniformly at the level of the bottom of the flame; but it must also be ascribed, in a certain measure, to the high temperature and fluency of the oil, by which it enters more readily into complete combustion than cold and viscid oil could possibly do. The preparatory heating seems to act on the same principle here as it does in the smelting of iron by the hot blast. Rape seed, for example, is so viscid as to burn with difficulty in lamps of the ordinary construction; but in the hot oil lamp of Parker it affords a very vivid light.—*Dr. Ure; Proceedings of the British Association.*

HEMMING'S SHADOWLESS LAMP.

THE reservoir of the oil is in a cylinder in the body of the stand, in which moves a piston resting on the surface of the oil, and having a hollow piston-rod, formed of sliding tubes, through which the oil is forced up for the supply of the lamp by the pressure of a spiral spring upon the piston. Provision is made for the overflowing of the oil, and for raising the piston when a fresh supply of oil is required. The spring is pretty equable in its pressure; but as equilibrium, with the hydrostatic pressure, is effected only by a variation in the height of the column of oil, the falling force of the spring, as it expands, is compensated by a descent of the oil.—*Mechanics' Magazine.*

FRENCH SAFETY-LAMP.

THIS lamp was invented by Baron du Mesnil in 1834, and has been adopted by the French Government, after a favourable report on it by

M. Ch. Combes. The lamp consists of a body of flint-glass, defended by twelve iron bars. The air is admitted by two conical tubes inserted at the bottom, which are capped with wire gauze, and enter by the side of the flame. The latter rises into a chimney, which has, over its top, an arched piece of metal; the chimney, however, being quite open: consequently, a strong current is constantly passing up the chimney. When carburetted hydrogen gas passes in, the fact is discovered by numerous small explosions, and the whole glass-work is thrown into vibrations, which emit a loud and shrill sound, to be heard at a very considerable distance.

Professor Graham states the novelty in the above lamp to be the open chimney. He considers that the Davy lamp was left almost perfect by that philosopher, and that all accidents proceed from carelessness. [We are happy to see the fame of Davy thus vindicated; for there has been too evident a disposition to asperse it, of late.] Professor Graham alludes to the deleterious effects of after-damp, or carbonic acid, left in the atmosphere of a mine after an explosion, which is believed to occasion often greater loss of life among the miners than the original explosion, and has often prevented assistance being rendered in accidents. In many cases, the oxygen of the air was not exhausted by the explosion; although, from the presence of five or ten per cent. of carbonic acid, it was rendered irrespirable. The atmosphere might, therefore, be rendered respirable by withdrawing this carbonic acid; and he suggests a method by which this might be effected. He has found that a mixture of dry slaked lime and pounded Glauber's salts, in equal proportions, has a singular avidity for carbonic acid; and that air may be purified completely from that deleterious gas, by inhaling it through a cushion of not more than an inch in thickness, filled with that mixture, which could be done without difficulty. He suggests the use of an article of this kind by persons who descends into a mine to assist sufferers after an explosion; indeed, wherever the safety-lamp is necessary, and the occurrence of an explosion possible, the possession of this *lime-filter* will be an additional source of security.—*Proc. British Association; Athenæum.*

NEW MODE OF ILLUMINATING CLOCKS.

THE Horse Guards clock was illuminated for the first time, on July 16, by means of the "Bude Light," which falls on the face; thus differing from the ordinary transparent clocks, to which it is much superior, both in clearness and beauty. It gives the clock face an appearance of being shone upon by a very powerful moonlight.

ECONOMY IN CANDLES.

IF you are without a rushlight, and would burn a candle all night, unless you use the following precaution, it is ten to one an ordinary candle will gutter away in an hour or two, sometimes to the endangering the safety of the house. This may be avoided by placing as much common salt, finely powdered, as will reach from the tallow to the bottom of the black part of the wick of a partly burnt candle, when, if the same be lit, it will burn very slowly, yielding sufficient light for a bedchamber ; the salt will gradually sink as the tallow is consumed, the melted tallow being drawn through the salt, and consumed in the wick.—*Economist.*

IMITATIVE WAX CANDLES.

TAKE equal parts of gum benzoin and resin mastic ; put each into a separate vessel of glass or lead, add spirit of wine, and heat them gently till the resinous parts are dissolved. Let each of the solutions remain awhile at rest, and then mix them. Before using this varnish, heat it to eighty or ninety degrees Fahrenheit; dip into it a candle from five to ten seconds, and dry it carefully. By this means, common candles may be made to resemble wax lights.—*Mechanics' Magazine ; abridged.*

GAS FOR HEATING.

SIR JOHN ROBISON recommends to be used in various processes of the arts, heating by the flame of a mixture of gas and common air passing through a tube having a diaphragm of wire-gauze; this manner being particularly adapted for the hardening and tempering of fine edge-tools and other articles of steel. No scaling or oxidation takes place, and the finest edge is preserved.—*Jameson's Journal.*

PATENT FUEL.

MR. ORAM has patented certain " Improvements in the manufacture of fuel," which consist in compounding small coal or dust with other materials, as mud, alluvial deposits, marl, clay, or any other earth containing vegetable matter; water; and several other substances, as mineral tar, coal-tar, gas-tar, mineral pitch, vegetable pitch, resin, asphaltum, or any other bituminous matter; chalk, or lime ; sawdust; anthracite or stone coal, coke or coke dust, and breeze.—*Mechanics' Magazine ; abridged.*

RESIN FUEL.

AT a late meeting of the Institute of Civil Engineers, Mr. C. W. Williams presented specimens of peat, from the first state, as taken from the bog, to the last, when compressed and converted into a hard coke; and of his new resin fuel, or artificial coal, which is composed of resin and turf coke. This resin fuel is used advantageously in long steam voyages, with a proper proportion of coal, as it enables the fireman to maintain the requisite pressure of steam with great regularity, and also to raise steam more rapidly on any emergency. It is not adapted for use as a fuel by itself; but when about 2¼ cwt. of this fuel is used with 20 cwt. of coal, by throwing it in front of the fire with each charge of fresh coal, a much better combustion of the coal takes place, and the effect is equal to that which would be produced by 27 cwt. of coal. Thus, 2½ cwt. of this fuel, so employed, is equivalent to 7 cwt. of coal. The cost is from thirty-five shillings to forty shillings per ton. The transatlantic steamers have carried from forty to sixty tons of it ; and, besides the advantages attending its use, there was a considerable saving in room.

In the discussion which the above statement elicited, it could not be doubted that nine pounds of coke will do as much in any department of the arts as twelve pounds of coal; for, on adding to coal a peat and hydro-carbon far more inflammable than coal, the result is equivalent to that which is produced by all the carbon, hydrogen, and oxygen, in many times the quantity of coal. It was remarked, that the circumstances under which fuel was employed ought to be considered. It could not be believed, that the absolute quantity of heat from the coke of a ton of coals is the same as of the ton of coals; for, in that case, all the heat of a coke oven would go for nothing; and this has been sometimes beneficially employed.—*Civil Engineer & Architect's Journal.*

GEARY'S PATENT COAL.

MR. S. GEARY, the architect, has patented an artificial coal, consisting of the following ingredients: first, about half of small coal, or coal-dust; second, pitch, or bituminous matter obtained from coal or other mineral, or vegetable tar; third, coal or coke cinders, powdered peat, or bark of trees, sawdust, or tan; fourthly, powdered clay, freestone, chalk, plaster, earth, sand, or other earthy matter; and lastly, about a fortieth part of sulphuric or other acid. The whole being melted and mixed, is run into moulds, or if required for steam-boat fuel, the mixture is allowed to cool, and is then pressed or beaten into blocks. The novel ingredient in this composition is the acid. Professor Brand and Mr. Squire, chemist, have reported favourably of this invention, either for domestic uses or steam-engines. Its rate of consumption in a steamer, in comparison with common coal, is as three and a half to five; its price, eighteen shillings per ton.—*Ibid.*

STIRLING'S ARTIFICIAL FUEL.

A NEW description of fuel for steam purposes has been invented by Mr. Stirling, of Limehouse. In an experiment to prove its superiority over even the best Welsh coals, at the furnaces of Messrs. W. Fairbairn, engineers, Bankside, under a high-pressure engine, during eleven hours, the saving in fuel alone was not less than twenty per cent. in comparison with that description of the best coals usually burnt in the above works; and the space occupied by Mr. Stirling's fuel was alone one-third less than that usually set aside for coals.—*Mechanics' Magazine.*

NEW STOVE.

ON Feb. 27, was communicated to the Society of Arts for Scotland, the description of a stove of a new construction, said to have all the properties of the Arnott stove, (radiation, equalization, and regulation of heat, and economy of fuel,) without the tendency in weakly drawing vents, to return smoke or carbonic acid gas into the room. This stove can, in an instant, be converted into a cheerful open fire, and, at the expense of about a penny per day, render an apartment perfectly comfortable, night and day. The stove is the invention of Mr. W. Kirkwood, of Edinburgh. In certain circumstances, such as where there is an ill-drawing vent, Arnott's stove is said not to have answered well, and to be dangerous to the health and even the life of individuals;

which defects Mr. Kirkwood's stove is calculated to remedy.—*Jameson's Journal.*

PNEUMATIC STOVE.

MR. J. JEFFREYS has exhibited before the British Association, a beautiful model of his pneumatic grate, the main feature of which is—the bold manner in which it is projected into the room. Instead of the fire being *immured*, and three-fourths of the heat lost up the chimney, it stands so far forward as vastly to increase the field of its radiation. This important point of bringing forward the fire, without risking the entrance of smoke into the room, is effected by the peculiar arrangement of parts behind the grate. The various currents of air, flowing towards the chimney, converge in the chimney throat, and by retarding each other's passage upward, often turn the smoke into the room. To use Mr. Jeffreys' term, the air there suffers "congestion;" the plane of greatest pressure of these currents, is a horizontal one in the chimney throat. In the pneumatic grate, the course of these converging currents is changed to a parallel direction, by their being made to pass through the intervals, between several flat and hollow tubes, which are ranged vertically behind the fire. In passing through these narrow alleys, they acquire a parallel course, and make their way up the chimney without mutual interference. The pressure is here transferred to the vertical plane, formed by the front of these tubes. In addition to the advantage of bringing the grate forward, a copious body of warm fresh air is brought into the room, by making the vertical tubes communicate with an air-box below. The smoke current, passing through their intervals, warms the air inside without any danger of over-heating it, and it is discharged into the room over the chimney-piece. — *Literary Gazette.*

ARNOTT'S STOVE.

THE greatest improvement in this stove has been made by Mr. Jeakes, Great Russell-street, who, by placing it within a case and introducing fresh air, has rendered it effective in ventilating as well as heating. Dr. Arnott himself has also made some improvements in his stove; but though we have examined those recently erected under his direction in the custom-house, they do not appear, even on the large scale on which they are formed in the long room of that building, to give sufficient ventilation. In short, this is only to be accomplished by the principle adopted by Mr. Strutt, in his cockle stove, by which fresh air, heated to a proper temperature, is continually introduced. This has been done by Mr. Strutt, by smoke, on a large scale; and, by Mr. Jeakes, in his improvement on Arnott's stove, by smoke on a small scale; while, by Mr. Manby and others, it is effected by hot water. In the case of small houses this plan is unsuitable; and perhaps Arnott's stove, as improved by Jeakes, will be found preferable: but, in all large houses, a cockle or hot water apparatus placed on the cellar floor, and the heated air admitted into the hall and staircase, are all that is required for comfort.—*Mr. Loudon; Gardeners' Magazine.*

PATENT CHUNK STOVE.

THIS invention, by Mr. R. Prosser, of Birmingham, consists of three parts—a base, or stand; a portable furnace, or fire-pot; and an envelope, or case. The base is a circular plate of cast iron, standing on three feet, with three concentric rings on its upper surface, having an aperture in its centre, and a valve adjusted to such aperture, to regulate the admission of the external or atmospheric air; and a tube or flue for the escape of the gases of combustion. The portable furnace consists of a conical bucket of sheet iron, having an iron grate inside, supported at about one-sixth part of the depth of the bucket by three studs or brackets, projecting about half an inch on the inside; which prevent the grate from being displaced by the weight of the fuel, and forming underneath the grate an ash-pit or receptacle for the ashes or dust caused by the combustion of the fuel employed. In the centre of the ash-pit is placed a funnel or chimney, covered at the top to prevent the ashes falling through, and perforated on all its sides to admit the external air. From the centre of the lid of the bucket a short tube projects, covered with a valve which closes by its own weight; when the envelope is removed from the base, and when the stove is in use, a lever opens the valve by pressure against the envelope. This envelope, or case, is a cylinder of sheet iron, closely covered at the top, and adjusted at the bottom to the upper surface of the base of the stove. When it is desired to use the stove, it is requisite that the tube for the escape of the gases of combustion should be let into a chimney, common air flue, or be connected with the atmosphere in any other suitable manner, so as to cause a draught of air through the stove; a sufficient quantity of sand is then to be strewed between the rings on the upper surface of the base of the stove to prevent the escape of smoke or vapour; fuel is next to be put into the fire basket or furnace, which is to be placed over the aperture in the base; and the envelope is lastly to be adjusted over the furnace and on the base of the stove. When the stove is thus in use, the external air for maintaining combustion is admitted through the aperture in the base of the stove and up the funnel at the bottom of the furnace into the ash pit, formed between the grate which supports the fuel and the bottom of the fire bucket; the air then passes through the fuel, and the gases evolved during combustion are carried upwards through the tube in the lid of the fire bucket, and thence downwards between the outside of the furnace, and the inside of the envelope into the flue communicating with the external air. Or, in case of there being no flue, the gases may be exhausted by mechanical means, and delivered into the atmosphere at any suitable place. The valve in the lid of the furnace is kept open, (as it must be to allow of combustion going on,) by the lever pressing against the inside of the envelope. The ring in the centre of the base of the stove is for the purpose of preventing the sand falling through the aperture; the next ring is to receive the fire-bucket; and the outer ring to receive the rim of the envelope. The patentee has obviated the great objection against close stoves, of their becoming too *hot*, in consequence of their contact with burning fuel. In the construction of his stove, no part of the envelope is in contact with the fuel, but is situate at such a distance from the furnace as prevents its being over-heated, and cannot, therefore, contaminate the air of any apartment. On removing the top of the envelope, the furnace can be withdrawn and replaced, for the purpose of removing ashes, and recharging the apparatus with fuel.

Directions for use:—"Remove the outer case of the stove; withdraw the fire-pail, and take off the cover; place the pail upon one corner of the kitchen fender, or upon something that will admit of the air drawing in at the bottom; throw into the pail, to the depth of five or six inches, burning coke or cinders from the kitchen fire that have ceased to smoke; in a few minutes afterwards, when it is seen that they are well kindled, fill the pail with coke or cinders: put on the cover, replace the pail on the bottom of the stove, and put on the outer case, giving it a slight to and fro motion, so as to fit it into the sand, with which the whole of the bottom must always be kept covered half an inch deep; see that the valve-rod is drawn out. Should the heat be too great, partly close the valve; to extinguish the fire, entirely close the valve. The stove will not require attention oftener than once every twenty-four hours; at the expiration of which time remove the pail, empty it of the ashes, not any of which can fall out during the consumption of the fuel, or the removal of the pail; replace the grate in the pail upon the brackets provided for it, and charge it with fuel as before."

The following experiments with this stove are from the *Mechanics' Magazine*:—The experimenter's office is 20 feet by 18 feet, and 9 feet high; it has three external walls (that is, walls which are exposed to the weather), and three windows; the area of glass, in the three, is =72 square feet. The fire-bucket, or furnace, contains, when full, 16 lbs. weight of coke.

Temperature of room.	Temperature of external air at same time.
61½ degrees.	57 degrees.
66½	52
68½	44
65	40

The above are the results of experiments on four different days.

FIRES IN CHIMNEYS PREVENTED.

THE principle of Davy's safety-lamp has been successfully applied to prevent fires in chimneys, by M. Maratueh, in France. He has found by experiments, that, if three frames of wire-work are placed near the base of the chimney, one above the other, about one foot apart, no flame will pass through them, while the draught of the chimney will not be impaired, and, consequently, no fires can ever happen in the chimney. As most of the soot lodges on the uppermost wire, but little on the second, and none on the third, he suggests that with a brush applied once a day to the lowest or two lowermost, the chimney will never want sweeping.—*Railway Magazine.*

MILLER'S PATENT FIRE-BARS.

A PATENT has been taken out for a new fire-bar, which is suited not only to the common steam-engine furnaces, but can with equal facility be applied to the furnaces of marine engines, and the locomotive engines of railways, &c. The principle of the invention consists in moving each alternate bar longitudinally in one direction, whilst the intermediate bars are moved in the opposite one. This movement, aided by the channelled surface of the bars, breaks up the clinkers the instant they are formed, or prevents their formation, and thus keeps the air-way perfectly free. The advantages secured are very considerable; for not only, by the perfect freedom from all obstruction of the air-way, is the combustion of the fuel and its heating power considerably increased, but coal of an inferior quality can be used without the usual effect of choking up the grate. By the vigorous combustion which this grate insures, it prevents large masses of coal from passing away unconsumed in the form of smoke, and consequently must effect a considerable saving in fuel. The ingenious patentee is the chief engineer of the extensive works of Messrs. Thomson, Brothers and Sons, Primrose, near Clitheroe.—*Manchester Guardian.*

VENTILATION OF THE NEW COURT, OLD BAILEY.

MR. PERKINS has caused capacious subterranean chambers to be formed, in which are placed coils of hot water pipes, and others containing cold air; which are so arranged, that by turning a valve, the warmed fresh air is admitted through apertures made in the floor and wainscoting of the court, so that a comfortable temperature may be preserved, whether the court be more or less crowded. The foul air, which naturally generates in a crowded court, is drawn off by a shaft under the prisoners' dock, as well as from the gallery and ceiling; which, communicating with large curves on the roof, the foul air, makes a thorough exit, and fresh air, either warm or cold, can be supplied in such quantities as may be requisite. The above is an important improvement upon the old method of ventilating the courts by canvass bags, and warming them with braziers filled with charcoal.—*Morning Advertiser; abridged.*

PREPARED NIGHT SOIL MANURE.

MM. PAYEN and Buren, chemists, in France, have produced a powder, the mixture of which with night soil, almost immediately frees it from offensive smell. It is then prepared, and when ready for use, resembles fine black mould, so dry and powdery as to be passed through the drill, and deposited with the seed. The efficiency of this manure has been most satisfactorily proved in France. A quantity sufficient for manuring two acres can be packed in a sugar hogshead, at an expense of from thirty shillings to twenty-six shillings an acre.—*Farmers' Magazine; abridged.*

NEW EXPERIMENTAL RESEARCHES ON CAOUTCHOUC. BY DR. URE.

THE specific gravity of the best *Para* caoutchouc, taken in dilute alcohol, is	0·941567
The specific gravity of the best Assam is	0·942972
Sincapore	0·936650
Penang	0·919178

In the process of making the "elastic tissues," the threads of caoutchouc are first deprived of all their elasticity, to prepare them for receiving a sheath upon the braiding machine. For this purpose, they are stretched by hand, in the act of winding upon the reel, to seven or eight times their natural length, and left two or three weeks in a state of tension upon the reels. Thread thus *inelasticated* has a specific gravity of no less than 0·948732; but, when it has its elasticity restored and its length reduced to its pristine state, by rubbing between the warm palms of the hands, the specific gravity of the same piece of thread is reduced to 0·925939. This phenomenon is akin to that exhibited in the process of wire-drawing, where the iron or brass gets condensed, hard, and brittle, while it disengages much heat; which the caoutchouc thread also does in a degree intolerable to unpractised fingers.

Dr. Ure has subjected to chemical examination two samples of caoutchouc juice; the first from M. Sievier, and the second from Mr. Beale, the engineer.

That of M. Sievier is greyish brown, that of Mr. Beale is of milky grey colour; the deviation from whiteness in each case being due to the presence of aloetic matter, which accompanies the caoutchouc in the secretion by the tree. The former is of the consistence of thin cream, and has a specific gravity of 1·04125; and yields, by exposure upon a porcelain capsule in a thin layer, for a few days, or by boiling for a few minutes, with a little water, twenty per cent. of solid caoutchouc. The latter, though it has the consistence of pretty rich cream, has a specific gravity of only 1·0175. It yields no less than thirty-seven per cent. of white, solid and very elastic caoutchouc.

It is interesting to observe how readily and compactly the separate little clots or threads of caoutchouc coalesce into one spongy mass in the progress of the ebullition, particularly if the emulsive mixture be stirred; but the addition of water is necessary to prevent the coagulated caoutchouc from sticking to the sides or bottom of the vessel and becoming burnt. To convert the spongy mass into good caoutchouc, it is exposed to moderate pressure between the folds of a towel. By this process, the whole of the aloetic extract, and other vegetable matters, which concrete into the substance of the balls and junks of caoutchouc prepared in Assam and Java, and contaminate it, are entirely separated, and an article nearly white and inodorous is obtained. Some of the cakes of the American caoutchouc exhale, when cut, the fetour of rotten cheese; a smell which adheres to the threads made of it, after every process of purification.

Neither of the above two samples of caoutchouc juice affords any appearance of coagulum when mixed in any proportions with alcohol of 0·825 specific gravity; wherefore, Dr. Ure infers that albumen is not a necessary constituent of the juice, as Mr. Faraday inferred from his experiments published in the 21st vol. of the Journal of the Royal Institution.

The odour of M. Sievier's sample is slightly acescent; that of Mr. Beale's, which is by far the richer and purer, has no disagreeable smell whatever. The taste of the latter is at first bland and very slight; but eventually very bitter, from the aloetic impression upon the tongue. The taste of the former is bitter from the first, in consequence of the great excess of aloes which it contains. When the brown solution which remains in the capsule, after the caoutchouc has been separated in a spongy state by ebullition from 100 grains of the richer juice, is passed through a filter and evaporated, it leaves 4 grains of concrete aloes.

Both of these emulsive juices mix readily with water, alcohol, and pyrolixic spirit, though they do not become at all clearer; they will not mix with *caoutchoucine* (the distilled spirit of caoutchouc), or with petroleum-naphtha, but remain at the bottom of these liquids as distinct as mercury does from water. Soda caustic lye does not dissolve the juice; nitric acid, (double aqua fortis), converts it into a red curdy magma. The filtered aloetic liquid is not affected by the nitrates of baryta and silver: it affords with oxalate of ammonia minute traces of lime.—*London Journal of Arts.*

PREPARATION OF CAOUTCHOUC.

On Jan. 19, was read before the Royal Asiatic Society, a paper on the preparation of caoutchouc, by Mr. E. Solly; who stated that in the different processes now employed, the extraneous soluble matters contained in the sap were allowed to remain with the caoutchouc, and become incorporated with it as it solidified. He considered that these impurities probably exerted great influence on the strength, elasticity, and consequent value, of the article; and described the kinds of caoutchouc in which these impurities were most abundant, and those in which their deteriorating influence was the greatest. The want of perfect adhesion between the layers of caoutchouc which composed the India rubber bottles, he attributed to the presence of a very thin layer of these impurities between them. Mr. Solly then detailed some experiments undertaken by him on the recent sap, with a view to its improved preparation, and concluded by some remarks on the importation of the sap of the caoutchouc tree into this country; the probable causes of the repeated failure of almost all attempts, and on the means most likely to succeed, in obtaining so desirable an object.—*Literary Gazette.*

FILTRATION OF HYDROGEN GAS THROUGH CAOUTCHOUC.

M. Cagniard Latour, in experimenting on the permeability of membraneous envelopes, ascertained that a small balloon of caoutchouc, well prepared, appears to be completely tight to atmospheric air; *i. e.*, once inflated, it remained so indefinitely; but that the same bottle filled with another gas—for example, oxygen, azote, carbonic acid or hydrogen, became very sensibly loose or uninflated. With hydrogen the collapse was more prompt than with the others.

In seeking for the cause of this rapid escape of hydrogen, he filled a balloon with it, and kept it for some time under water. In a few hours, the surface was nearly covered with small gaseous bubbles, and the size of the balloon appeared sensibly diminished. On examining the gas after its escape, it appeared to be the same as before, nor had the caoutchouc membrane undergone any alteration. The result seemed to be simply the effect of mechanical filtration.

The thickness of the balloon membrane in these experiments was only the fourth of a millimetre (= 1-1000th of an inch), but the author had tried others much thicker, and found that when suitably compressed, hydrogen would escape through them.—*Franklin Journal; Mechanics' Magazine.*

NEW METHOD OF WORKING CAOUTCHOUC.

THE employment of either spirits of turpentine, the volatile oil of caoutchouc, balsam of copaiva, and the oils obtained from gas-works, as solvents of India rubber, have the disadvantage of being expensive, and produce a varnish which dries but with difficulty. For some time past ammonia has been advantageously used; the gum-elastic being cut into shreds, covered with caustic ammonia, and left for several months. The ammonia becomes brown, and the gum assumes a brilliant and silky appearance, resembling a fresh nerve. The caoutchouc swells but is still elastic; though, when drawn out, it breaks more easily than raw caoutchouc. In treating this swelled caoutchouc with spirits of turpentine, it is easily converted by agitation into an emulsion; in a short time it swims on the surface like butter on milk, after which it acts like varnish. But a much smaller quantity of spirits of turpentine is sufficient to dissolve it than when it has not been softened by ammonia.—*Annales des Mines.*

NON-ADHESIVE CAOUTCHOUC.

IN the report of the Committee on the late exhibition of domestic manufactures, held at the Frank. Inst., Philadel., it is observed "these articles (gum elastic goods) manufactured at Roxbury, near Boston, recommend themselves strongly to favourable notice. They consist of gum elastic attenuated into thin sheets, and these sheets, in some specimens, cemented apparently by simple pressure to the printed surface of calicoes, chintzes, engravings, maps, &c., and in others made themselves the ground upon which various coloured patterns are imprinted. The peculiar merit of these goods is their retaining, in a perfect degree, all the original qualities of the gum elastic; its elasticity, its toughness, freedom from odour, and absence of all adhesiveness; the latter feature giving to this manufacture a decided superiority over any other preparation of the gum hitherto attempted. The attention of the committee was particularly attracted to the beauty and evenness of texture of a shawl, consisting wholly of the gum elastic, upon which a very tasteful pattern has been impressed.—*Mechanics' Magazine.*

NICKELS'S NEW INDIA RUBBER FABRIC.

WATERPROOF cloth has been made with combinations of caoutchouc with silk, cotton, wool, and other fibrous materials, felted, woven, platted, plated, and in other ways. The present patentee adopts the plan of placing a series of strands or threads side by side, parallel to each other, and combining them into a fabric, by attaching them together by means of a solution of caoutchouc. Shellac, or other resinous varnishes, may be used, instead of a solution of caoutchouc, to cause the longitudinal or parallel strands or threads to adhere together.

Machinery for performing these operations is described by the patentee, as an exemplification of the way in which the invention is to be carried into effect. The thread is wound round a cylinder, the diameter of which shall be equal to the size of the sheet to be produced. The India rubber solution, or resinous varnish, is then spread over the helical coil of threads, the superfluous stuff removed, and when it has dried or set, the sheet has to be removed from the drum. A sheet of figured or ornamental silk, or other fabric, may be superadded to the sheet of caoutchouc cloth thus produced, which may be used for ladies' cloaks and other like purposes.—*Mechanics' Magazine.*

INDIA RUBBER BOAT.

THERE has been launched on the Neva an India rubber boat. It is made of sail-cloth, impregnated with caoutchouc. It may be rolled up, and in the space of ten minutes can be filled with air by means of four little cocks, by which inflation it assumes the form of a boat. During its trial on the river, it held three persons and excited much attention, as well by the readiness of its movements as by its very graceful appearance.

PURPOSES TO WHICH BITUMEN IS APPLICABLE.

1. THE structure and repair of public and private roads, promenades, footpaths, terraces, &c. 2. Railways. 3. Fortifications. 4. Bridges. 5. Canals. 6. Churches, and all buildings of public resort. 7. Colleges, and all similar establishments. 8. Piers, wharfs, the flooring of dry and other docks. 9. Prisons and convict-cells. 10. Markets and bazaars. 11. Slaughter-houses. 12. Bonding warehouses and granaries, where the exclusion of rats and vermin is most desirable. 13. Tombstones, monuments, and vaults, in cemetries. 14. The flooring of public buildings, halls, &c., where the absolute exclusion of moisture is required. 15. Baths, aqueducts, reservoirs, tanks, pipes, sewers, drains, &c. 16. The facing of all masonry and timber-work. 17. Sea and river walls. 18. Roofs, cellars, vaults, underground kitchens, &c. 19. Fastening balustrades, balconies, &c. 20. Paving the bottoms of ships, to preserve them from worms, &c. 21. Water-proofing cloth, tarpaulins, and all textile substances. 22. Varnishes for pictures and other purposes. 23. Medical uses. 24. Coating ironwork of almost every description, which will prevent its oxidation from air and water.—*From a Theoretical and Practical Essay on Bitumen; Railway Magazine.*

USES OF BONE.

IN two lectures delivered to the Society of Arts, Mr. Aikin has shown that bone contains a considerable quantity of valuable nutriment; that, in its entire state, it is applicable to a variety of useful purposes; that the worker in steel employs it for case-hardening small and delicate articles; that, in proportion to its weight, it is the most valuable and active of all manures; that, in the absence of other combustibles, it may be and is largely used as fuel in the plains of Tartary and South America; that, by its decomposition in close vessels, it produces hartshorn, ammonia, and animal charcoal; and that when burnt to ashes, it becomes useful to the assayer, furnishes a valuable polishing powder, and is the material from which is produced phosphorus, the most combustible of all solids.

DURABILITY OF STONE.

THE Report recently made to the Commissioners of her Majesty's Treasury, by Messrs. Barry, De La Beche, W. Smith, and Charles H. Smith, on the Sandstones, Limestones, and Oolites of Britain, forms, with the numerous tables and results of experiments by Messrs. Daniell and Wheatstone appended to it, one of the most valuable contributions to architectural science that has been made in modern times. One hundred and three quarries are described, ninety-six buildings in England referred to, many chemical analyses of the stones given, and a great number of experiments related, showing, among other points, the cohesive power of each stone, and the amount of disintegration apparent, when subjected to Brard's process. *Mr. G. Godwin, jun.; in the Civil Engineer and Architect's Journal.**

Of this document we have only space to quote the following:—

Of the necessity and importance of the inquiry upon which we have been engaged, the lamentable effects of decomposition observable in the greater part of the limestone employed at Oxford, in the magnesian limestone of the minster, churches, and other public buildings at York, and in the sandstones of which the churches and other public buildings in Derby and Newcastle are constructed, afford, among numerous other examples, incontestable and striking evidence. The unequal state of preservation of many buildings, often produced by the varied quality of the stone employed in them, although it may have been taken from the same quarry, shows the propriety of a minute examination of the quarries themselves, in order to acquire a proper knowledge of the particular beds from whence the different beds have been obtained. An inspection of quarries is also desirable for the purpose of ascertaining their power of supply, the probable extent of any given bed, and many other matters of practical importance.

It frequently happens that the best stone in quarries is often neglected, or only in part worked, from the cost of baring and removing those beds with which it may be associated, and, in consequence, the inferior material is in such cases supplied, especially when a large supply is required in a short space of time and at an insufficient price, which is often the case with respect to works undertaken by contract.

As an economical supply of stone in particular localities would often appear to depend on accidental circumstances, such as the cost of quarrying, the degree of facility in transport, and the prejudice that generally exists in favour of a material which has been long in use; and as the means of transport have of late years been greatly increased, it becomes essential to ascertain whether better materials than those which have been employed in any given place may not be obtained from other, although more distant, localities, upon equally advantageous terms.

With respect to the decomposition of stones employed for building purposes we would observe that it is effected according to the chemical and mechanical conditions to which such stones are exposed. As regards the sandstones that are usually employed for such purposes, and which are generally composed of either quartz or silicious grains cemented by silicious, argillaceous, calcareous, or other matter, their decomposition is effected according to

* The publication of this document has occasioned a Mr. John Mallcott to observe, in the *Times*, "that all stone made use of in the immediate neighbourhood of its own quarries is more likely to endure that atmosphere than if it be removed therefrom, though only 30 or 40 miles."

the nature of the cementing substance, the grains being comparatively indestructible. With respect to limestones composed of carbonate of lime, or the carbonates of lime and magnesia, either nearly pure or mixed with variable proportions of foreign matter, their decomposition depends, other things being equal, upon the mode in which their component parts are aggregated, those which are most crystalline being found to be most durable, while those which partake least of that character suffer most from exposure to atmospheric influences.

Sandstones, from the mode of their formation, are very frequently laminated, more especially when micaceous, the plates of mica being generally deposited in planes parallel to their beds. Hence, if such stone be placed in buildings, with the planes of lamination in a vertical position, it will decompose in flakes according to the thickness of the laminæ, whereas, if it be placed so that the planes of lamination be horizontal, that is, most commonly upon its natural bed, the amount of decomposition will be comparatively immaterial.

Limestones, such at least as are usually employed for building purposes, are not liable to the kind of lamination observable in sandstones; nevertheless varieties exist, especially those commonly termed "shelly," which have a coarse laminated structure, generally parallel to the planes of their beds, and therefore, the same precaution in placing such stone in buildings so that the planes of lamination be horizontal, is as necessary as with the sandstones above noticed.

The varieties of limestones termed Oolites, being composed of oviform bodies, cemented by calcareous matter of a varied character, will, of necessity, suffer unequal decomposition, unless such oviform bodies and the cement be equally coherent and of the same chemical composition. The limestones which are usually termed "shelly," from being chiefly formed of either broken or perfect fossil shells, cemented by calcareous matter, suffer decomposition in an unequal manner, in consequence of the shells, which, being for the most part crystalline, offer the greatest amount of resistance to the decomposing effects of the atmosphere.

The effects of the chemical and mechanical causes of the decomposition of stone in buildings are found to be greatly modified according as such buildings may be situate in town or country. The state of the atmosphere in populous and smoky towns produces a greater amount of decomposition in buildings so situate, all other conditions being equal, than in those placed in the open country, where many of the aeriform products which arise from such towns, and are injurious to buildings, are not to be found.

The chemical action of the atmosphere produces a change in the entire matter of the limestones, and in the cementing substance of the sandstones, according to the amount of surface exposed to it. The mechanical action due to atmospheric causes occasions either a removal or a disruption of the exposed particles; the former by means of powerful winds and driving rains, and the latter by the congelation of water forced into, or absorbed by the external portions of the stone. These effects are reciprocal, chemical action rendering the stone liable to be more easily affected by mechanical action, which latter, by constantly presenting new surfaces, accelerates the disintegrating effects of the former.

Buildings in this climate are generally found to suffer the greatest amount of decomposition on their southern, south-western, and western fronts, arising, doubtless, from the prevalence of winds and rains from those quarters; hence it is desirable that stones of great durability should at least be employed in fronts with such aspects.

Buildings situate in the country appear to possess a great advantage over those in populous and smoky towns, owing to lichens, with which they almost invariably become covered in such situations, and which, when firmly established over their entire surface, seem to exercise a protective influence against the ordinary causes of the decomposition of the stone upon which they grow.

As an instance of the difference in degree of durability in the same material subjected to the effects of the atmosphere in town and country, we may notice the several frusta of columns and other blocks of stone that were quarried at the time of the erection of St. Paul's Cathedral in London, and which are now lying in the island of Portland, near the quarries from whence they were obtained. The blocks are invariably found to be covered with lichens, and although they have been exposed to all the vicissitudes of a marine atmosphere for more than 150 years, they still exhibit beneath the lichens their original form, even to the marks of the chisel employed upon them, whilst the stone which was taken from the same quarries (selected no doubt with equal if not greater care than the blocks alluded to), and placed in the cathedral itself, is, in those parts which are exposed to the south and south-west winds, found in some instances to be fast mouldering away.

Colour is of more importance in the selection of a stone for a building to be situated in a populous and smoky town than for one to be placed in an open country, where all edifices usually become covered, as above stated, with lichens; for although in such towns those fronts which are not exposed to the prevailing winds and rains will soon become blackened, the remainder of the building will constantly exhibit a tint depending upon the natural colour of the material employed.

It should be stated that the object of the above investigation was the selection of stone for building the Houses of Parliament; and the Report concludes as follows:—

In conclusion, having weighed to the best of our judgment the evidence in favour of the various building stones which have been brought under our consideration, and freely admitting that many sandstones as well as limestones possess very great advantages as building materials, we feel bound to state, that for durability, as instanced in Southwell Church, &c., and the results of experiments, as detailed in the accompanying tables; for crystalline character, combined with a close approach to the equivalent proportions of carbonate of lime and carbonate of magnesia; for uniformity of structure; facility and economy in conversion; and for advantage of colour, the magnesian limestone, or dolomite of Bolsover Moor and its neighbourhood is, in our opinion, the most fit and proper material to be employed in the proposed new Houses of Parliament. [The precise locality is the Steatly quarries, a short distance from Worksop, Notts.]

HENNESSEY'S ELASTIC LIFE-BOAT.

This boat is perfectly elastic, except about three-fourths of its keel, which are secured by three bars, or plates of copper or iron; one plate being on each side of the keel, and one on the flat at bottom. These plates give strength and stability, and prevent the upsetting of the boat. The stem and other parts of the keel are secured by thinner plates, in joints, so as to give great strength to these parts, but still preserve their elasticity. The timbers, which are very slight, are of oak, tarred and covered with light strong canvass, with a casing over that of thin whalebone, taken out of a stove or boiling pan, then sewed like a rope with spun yarn, and the outside finished with leather, or improved canvass, sewed on. They are removable at pleasure, when the boat is stowed away. The covering, or skin, in place of plank, (an essential part of the invention), is a kind of cloth, of great strength and durability, and perfectly waterproof. There are two skins of this fabric, graduating in thickness from the outside inwards, so as to prevent any possibility of water or damp coming through. The materials of this cloth are saturated in the loom, by a chemical process, which preserves it from wet, and the action of the atmosphere, heat, mildew, or rot. This boat, from its strength and elasticity, is capable of sustaining concussions that would destroy life-boats of the usual construction: it was invented in 1823; but the originator has only just brought it into use.—*Liverpool Newspaper.*

TRACTIVE POWER OF THE HORSE.

Mr. E. Morris, C. E., has communicated to the *Franklin Journal* some important experiments upon this question; whence he takes the mean day's work of horses, fully loaded, to be as follows:—

Of *picked* horses at 16½ millions of pounds, lifted one foot high per day.

Of *common* horses at 12 9-10 millions of pounds, lifted one foot high per day.

This average result for *common* horses is equal to a continued pull of near 135¾ lb. at the rate of 2¼ miles per hour for 8 hours out of 24.

The pace of 2¼ miles per hour, or more accurately 200 feet per minute, has been ascertained by the writer from numerous experiments upon horses hauling earth for embankments in single carts, as well as from some trials with teams of four, to be that at which horses in common use on public works generally proceed when working with full effect.

Note.—The horses employed upon public works, averaging the working time of the whole year, are worked nearly 10 out of 24 hours; for which time, however, they would only be able to exert a mean force equal to a steady pull of about 108½ lb., and the pace and the dynamical result would both be found to remain nearly as above.

TRACTION OF CARRIAGES.

In 40 experiments undertaken in France, on an extensive scale, by M. Morin, with different loads and wheels, varying from 2¾ feet diameter to 6½ feet, he found that the tractive force was directly as the load, and inversely as the diameter of the wheels. In 97 observations, with different breadths of tire, where the road was a little soft, the force of traction diminished as the breadth of the tires increased. But, when the breadth reached 8½ inches, it had no sensible influence. The harder the roads, the less the breadth of the tire affected the draft. On paved roads, it was insensible; and on hard roads of broken stone not new laid, the tires had no effect at 3¾ inches broad, and beyond. This breadth, therefore, is the limit on good roads. It also appears by 168 experiments, that the traction increases proportionally to the increase of velocity. This, however, is less sensible on springs than without. A carriage well hung on springs can carry at a trot a load as great as that on a carriage without springs, with no greater injury to the road. This M. Morin proved in several instances.—*Railway Magazine.*

GRAY'S PATENT SAFETY COACH.

In August last, a trial of this ingenious improvement was made at the Hippodrome, near Kensington. Every person who has visited the Hippodrome will recollect the ruggedness of the ground which is below the rails round "the hill." In some parts the rills themselves are a foot or 18 inches deep. Over these the coach was driven at every pace, from the walk to the gallop, sometimes with a full load of passengers, at others with but three or four, and at all times with the same result. However low the wheels on the one side were running, it produced no prejudicial effect on the coach, which always found its level. Indeed, it could not well be otherwise, seeing that the body, or rather the coach, is hung on springs which work longitudinally from the roof instead of laterally; from immediately above the perch, or under the coach itself. But the main point of the invention is, that these longitudinal springs work on sliding blocks, by which a constantly levelling position is produced. In the course of the experiment the wheels on the "off" side passed over a surface which was more than four feet higher than those which were on the "near" side. This was regarded as a prominent trial of the principle; and therefore Mr. Gray, as charioteer, drove round and round with the horses at full length for at least ten minutes. The result was, that at each evolution, and at each jerk, the ground being particularly uneven, and therefore peculiarly adapted for the experiment, the coach preserved its perfect equilibrium or level. Beyond this it is unnecessary to say more than that all parties, whether as riders or as lookers on, appeared to entertain but one opinion—namely, that the turning over of a coach hung upon this patent principle was an impossibility.—*Times.*

MANU-MOTIVE AND PEDO-MOTIVE CARRIAGES.

MR. REVIS, of Cambridge, has exhibited in the metropolis a manumotive carriage, which he calls "the Aellopodes," the chief novelty of which consists in the name. Mr. Baddeley, (in the *Mechanics' Magazine,*) describes this machine as consisting of two large driving wheels, urged round by two cranks, acted upon by treddles; on each of which the rider's weight is thrown alternately, and causes rotation of the wheels, the direction of the machine being effected by a guide-wheel in the front. This plan has often been tried under various modifications, but not having been found to realize any practical advantage, it has in almost every case been abandoned. It has been asserted that "on common roads" the Aellopodes "may be propelled at the rate of from twenty to thirty miles an hour;" * upon which, observes Mr. Baddeley, "I presume the word 'common' means rail-road, for on nothing else can such a speed be attained; it would be an utter impossibility to maintain half thirty miles an hour on any turnpike road." Mr. Revis has offered his carriage to the Post Office; but Mr. Baddeley trusts "the heads of the Post Office are not so heedless as to entertain such an absurd and extravagant proposition." Now, the Aellopodes is no novelty, for a machine was constructed upon the same principle by a coach-maker of Bristol several years since. Some months previously to Mr. Revis's exhibition, a machine in every respect the counterpart of the "Aellopodes," (except that the driving-wheels were not quite so large), was constructed by Mr. R. Merreweather, and is described in the *Mechanics' Magazine;* No. 823. In the same Journal, No. 818, Mr. Baddeley describes a carriage of the above description, in which the rider sits, and the cranks are worked by hand instead of by treddles—a decided improvement upon Mr. Revis's carriage.

* This assertion we take to be a statement of the inventor.

"The Accelerator" is another of these "hobbies." It consists of four huge wheels connected by a strong perch, to which is suspended a carriage for passengers. Beneath the axle-tree are shafts for two horses, who are described as "*in some degree* suspended by bands passing under their bodies to the axle-tree. Their feet barely touch the ground, as the slightest movement of the horses will set the large wheels in motion." The conductor sits between the fore-wheels, and by means of pulleys "can support or lower the horses at pleasure!"

We are inclined to think with Mr. Baddeley, that all pedo-motive and manu-motive machines, even for gymnastic exercise, have but a limited range of usefulness; "but to imagine they can ever be made extensively or usefully subservient to the purposes of trade or commerce in any shape, is a chimera altogether unworthy of this mechanical age."

ATTEMPT TO STEER A BALLOON AGAINST THE WIND.

M. Eubriot has ascended from Paris in a balloon of a new construction: the gas bag was egg-shaped, lengthwise; the car was a kind of chair, or cabriolet seat, but in lieu of wheels were appended on each side sails like those of windmills; and at each end of the car were two machines like ships' poops. The secret was that the flaps, paddles, or windmill-sails would act against the currents. On the balloon rising, however, despite of the sails, it took the exact direction of the pilot balloon: the balloon dragged away the car, not the car the balloon, as M. Eubriot had expected. The strength of the windlass, with which the sails were worked, was not sufficient to cope with the huge body propelled by the ascending power of the gas.

ICE-BOAT.

Mr. Ballard has presented to the Institution of Civil Engineers, the description of an Ice-Boat. His principle of breaking ice consisting in forcing the ice upwards, instead of forcing it through horizontally, or by pressing it down. For this purpose, a frame, coated with a sheet of iron, is laid over the front of a boat, with an inclination downward from the boat, the lower end being under the ice.

THE TURBINE.

This invention was noticed in the *Year-Book of Facts*, 1839, (p. 68). Mr. Rennie has since communicated to the *Railway Magazine*, No. 41, a valuable paper upon its theory and construction; and the Editor of the above Magazine has inserted a similar paper in the succeeding Number, whence are extracted the following notes: "The secret of this instrument, as the report of the French Academicians states, lies in constructing the curves, and laying out the angles in which the water from the drum shall issue into the canals of the Turbine, so as to enter without shock, and leave them without velocity." From Mr. Rennie's paper we gather that "the Turbine is an instrument into which water enters at the centre, and after passing through the curved canals of a fixed drum, issues obliquely into other curved canals, in a moveable annulus, or ring, closely surrounding the drum, but not so as to rub it, whence it is rejected at the outer circumference. By the action of the water on the canals of the ring, or Turbine, it is driven rapidly round in the opposite direction to that in which the water issues from the canals of the drums, and," as Mr. Rennie says, "carries with it all the machinery." "In conclusion," states the *Railway Magazine:* "it is obvious that the velocity of the Turbine will accommodate itself to any pressure of the fluid, since it is expressly constructed on the principle of revolving, when producing the most useful effect, at the same rate as the water issues out."

WHITELAW'S HYDRAULIC ENGINE.

The advantages of this machine are stated to be very great. In the first place, while, by the common water-wheel, in some circumstances, only a small portion of the water-power can be used, and under the most favourable circumstances not more than 65 per cent., it is calculated that, by this new machine, not less than 95 per cent. of the motive power of the water is rendered available. Secondly, the most trifling rivulet, provided it have a good fall, can be taken advantage of by the new machine; thirdly, the expense of the improved Barker mill is not more than one-fifth of the expense of a water-wheel, to work in the same stream. We are informed that one of the new machines has been put up on the Shaws Water, near Greenock, for about 300*l*, to do the work of a water-wheel which would have cost, according to estimate, 1500*l*.—*Aberdeen Herald.*

MUNTZ'S YELLOW METAL FOR SHEATHING SHIPS' BOTTOMS.

A few years ago, a patent was obtained for a metal wherewith to cover the bottoms of vessels, and which promised all the advantages of copper, with at least equal durability in any climate, and a considerable saving in cost. It is manufactured in sheets in the same manner as copper, and from its resemblance to yellowish tinted brass, has obtained the name of "yellow metal." Several vessels belonging to London and other ports were plated with it some years age; but it was found to be less ductile than copper, and therefore less suitable for being laid round angles, or into the grooves or hollows of the stern-post or rudder, where the rudder-irons (or rather coppers) are attached. In some instances, too, when it required renewing, it was found, on taking it off, that it had, to a considerable degree, perished, having become brittle, and full of pores. Such improvements have, however, since been made, as will, in all probability, ere long, render the use of the "yellow metal" almost universal in our mercantile marine. The metal is now made sufficiently ductile for sheathing; several fine large ships have been bottomed with it; and it is so rapidly gaining ground in the estimation of navigators, that we are informed Messrs. Pascoe Grenfell and Sons, copper dealers, have, during the last half-year, done much more business in the article than in copper itself. The captains who have lately given it a trial, speak of it as showing a durability considerably beyond the copper generally used. The price is 1½d. per lb. cheaper than copper, and it is 6½ per cent. lighter than that metal. It is also found to retain its weight to a greater extent, and the old metal is taken back in exchange by the manufacturers. Another, and perhaps an equally important, use of this metal is, for the bolts or fastenings of ships. From its hardness, when manufactured for this purpose, it is vastly superior to copper; inasmuch as being nearly as hard as iron, it can be driven into a much smaller augur-hole than copper, and to a greater depth in dead wood.

The metal is a combination of copper and zinc, 60 per cent. of the former, and 40 per cent. of the latter. It is submitted to the action of a reverberatory furnace, or melted in pots, from which it is cast in plates or bars. It is subsequently submitted to heat, and when of a "cherry red," is worked in the cylinders or rollers, or drawn out in rods. —*Mining Journal; abridged.*

EXPERIMENTS ON CAST-IRON.

In March, 1837, experiments were commenced by Mr. Freebairn, with a number of bars of Coedtalon iron, cast from the same model, five feet long, and one inch square, which were placed horizontally on props four feet six inches asunder, and had different weights, as two and a half, three, three and a half, and four cwt., laid upon the middle of each; the last weight being within a few pounds of the breaking weight. The intention was to ascertain what effect would arise from each of these weights lying constantly upon the bars. The results are,—1st,—The bars are still bearing the loads, and apparently may do so for many years. 2nd,—The deflections, which are frequently measured, the temperature being observed at the time, are constantly increasing, though in a decreasing ratio,—a fact, which shows that, though cast iron may be safely loaded far beyond what has hitherto been deemed prudent, still it is extremely probable that the bars are advancing, by however slow degrees, to ultimate destruction.—*Proc. Brit. Assoc.*

KYANIZED WOOD.

The effects of Kyanizing Wood, with reference to living plants, seem to be that the Kyanizing liquor is made so strong, that more corrosive sublimate is deposited, not only on the outer surface of the wood, but in its interior pores, than is wanted for entering into combination with the albumen. This superfluous quantity, then, in the case of hot-house rafters or sash-bars, being sublimated by the great heat of the house, escapes even through several coats of paint, contaminates the atmosphere, and injures vegetation for a time, till the source is exhausted. As the wood, when steeped in the Kyanizing fluid, absorbs water in a considerable quantity, as well as mercury, it is easy to conceive that the heat of a hot-house will raise the temperature of the rafters and sash-bars, which are always in the hottest part of the house, to such a degree as to expand the water into vapour, as well as the corrosive sublimate; and, thus pent up, the elastic fluids produced will readily burst through any layers of paint that could be laid on. A safe mode, when Kyanizing for plant structures, would appear to be, to use the Kyanizing liquor much weaker than is generally done; and, after the wood is removed from the tank, to dry it as thoroughly as if it had not been Kyanized, in order to get rid of the water absorbed during that process.—*Gardeners' Magazine.*

BETHELL'S PATENT METHOD OF RENDERING WOOD MORE DURABLE, ETC.

Mr. Bethell has patented improvements in preparing wood, so as to render it more durable, less pervious to water, &c. They are founded upon the supposition that the mere external application of any solution or mixture, either by steeping, painting, &c., never can be effective in preserving wood from decay; as the first symptoms of *dry rot* are generally to be discovered in the heart of the timber, where the sap being untouched by the solution, ferments and putrifies, if the wood be placed in a moist and warm situation.

The view which is here taken has been strongly supported by Sir John Barrow, Secretary to the Admiralty, who, in his *Life of Lord Anson*, lately published, when speaking of the different modes of preserving wood that have hitherto been suggested, observes: "The steeping of large logs of timber in solutions of any kind is *perfectly useless*, the solution penetrates *only skin deep;* whereas the real dry rot commences at the centre, where the fibres, being the oldest, first give way, as is the case in the standing tree."

Mr. Bethell's process consists in a mode of impregnating the wood to the centre, and saturating it throughout its whole substance with various solutions and mixtures, which have the effect of *coagulating the sap*, and thus preventing its putrefaction. But for wood that will be much exposed to the weather, and alternately wet and dry, the mere coagulation of the sap is not sufficient; for although the albumen contained in the sap of wood is the most liable and the first to putrify, yet the ligneous fibre itself, after it has been deprived of all sap, will, when exposed in a warm damp situation, rot and crumble into dust. To preserve wood, therefore, that will be much exposed to the weather, or to a warm and damp situation, it is not only necessary that *the sap should be coagulated, but that the fibres should be protected from moisture.* This Mr. Bethell effects by impregnating the wood throughout its whole substance with oil of creosote mixed with tarry bituminous matters. The creosote coagulates the albumen, and the tarry and bituminous matters fill up every pore, and form an insoluble waterproof covering to the fibres of the wood.—*Patentee's Circular.*

OPTICAL GLASSES OF THE LATE DR. RITCHIE.

At a meeting of the Astronomical Society on June 14, Mr. Simms read a paper upon the optical glasses of the late Dr. Ritchie. Besides several pieces of very excellent crown, Mr. Simms obtained twenty-nine discs of flint-glass, varying in diameter from 4 to 7½ or 8 inches. These were circular, and had evidently been cast in a mould; in thickness they varied exceedingly, some being scarcely thick enough for the purpose for which they were intended; and others at least three times as thick as necessary. Of these twenty-nine discs, only three give promise of being fit for telescopes at all; two of 4, and one of 4½ inches diameter; and even these not likely to prove of the first order. The rest are, literally full of veins and striæ, and the bestowal of any labour on them would obviously end in loss and disappointment. The results of a series of comparative experiments are stated in the following table, in which the refractive and dispersive powers, for the lines of the spectrum distinguished by Fraunhofer as B and G, are pretty accurately given for the four specimens of flint-glass that were examined. The first is of glass, made by Guinand, of Neuchâtel, the second was made several years since, at the Stangate Glass-works, near Lambeth; the third, by Dr. Ritchie; and the fourth, at the Falcon Glass-house, Whitefriars; and it will be seen that in all those qualities which were made the subject of investigation, the flint-glass of Dr. Ritchie differs little, if at all, from that which is generally made in England. The fact, however, of a disc of flint-glass, 7½ inches in diameter, having been made sufficiently perfect for the construction of a good achromatic telescope is unique in the history of glass-making in this country.

Flint Glass.	Specific Gravity.	Index of Refraction for B.	Index of Refraction for G.	Dispersive Power B. to G.
Guinand	3.6459	1.616	1.644	.044
Stangate	3.3747	1.594	1.619	.041
Ritchie	3.2269	1.573	1.597	.041
Falcon	3.1964	1.570	1.593	.039

Mechanics' Magazine.

SELF-ACTING GAS APPARATUS.

MR. HEGINBOTHAM, of Stockport, has patented this improvement. The retort, which is four feet long, produces upwards of 8000 feet in 24 hours, being three times as much as can be made upon the present system, from one retort. The gas has a superior illuminating power, and one third more is extracted from a given quantity of coal.

LIQUID LEATHER.

DR. BERULAND, of Larria, in Germany, is said to have discovered a method of making leather out of certain refuse and waste animal substances. He has established a manufactory near Vienna: no part of the process is explained; but it is stated that the substance is at one stage in a state of fluidity, and may then be cast into boots, shoes, &c.—*Bristol Mirror.*

DYE WOOD.

A METHOD of extracting the colouring matter from wood has been lately employed by a M. Besseyre with much success. He first reduces the woods to very small divisions, and then immediately places them in a closed vessel exposed to a current of steam. When the whole has attained 80 degrees of heat, it is uncovered, and watered with several pints of cold water. By means of a tap below, the condensed liquid is drawn off, and thrown back upon the chips, and this operation is repeated until the dye has acquired sufficient strength; it is then subjected to evaporation over an open fire, and subsequently in a sand-bath, till the extract becomes a mass, which is soluble in warm water. —*Athenæum.*

NEW DYE PLANTS.

On June 1, Mr. Solly read to the Royal Asiatic Society, a Report on some Lichens received from India, and a collection of Indian Lichens from Dr. Royle's herbarium, with a view to ascertain their importance as dye substances. The specimens from Bombay had failed; but those received from Ceylon contained good colouring matter, and some of them, by comparative experiments with the lichens of commerce, yielded as good a dye as kinds worth more than 250*l.* a ton. The lichens from Dr. Royle's collection failed; save one, called *Chulchæleera*, which yields much colouring matter, and has long been used in India by the native dyers.

Mr. Solly also read a paper "On the Bengal Safflower;" stating that the best safflower imported in very limited quantities from China, fetches as high as 30*l.* per cwt., whilst good Bengal safflower averages 7*l.* per cwt.; which difference he attributes to their mode of cultivation and preparation.—*Literary Gazette.*

TO JOIN SHEETS OF PAPER.

To prevent the puckering caused by paste or glue, substitute a thick solution of caoutchouc, which, being applied to the edges of the paper, is allowed a little time to dry and get sticky, before the sheets are joined together.—*Mechanics' Magazine.*

PRESERVATION OF FLOUR.

M. Robineau states that by very strongly compressing flour into rectangular moulds it may be preserved both from damp and insects; the bran being previously separated from it. A cake of flour thus prepared, has been kept in a damp cellar for six weeks, without injury. —*Athenæum.*

NEW METHOD OF CLEANING GLASS.

Powder finely indigo, and dip into it a moistened rag, with which smear the glass, and wipe it off with a dry cloth. Very fine sifted ashes applied by a rag dipped in spirit will also answer well: but Spanish white is apt to roughen the glass.—*From the French; Mechanics' Magazine.*

IMPROVED CRAYONS FOR DRAWING ON GLASS.

Melt together equal quantities of asphaltum and yellow wax; add lamp black, and pour the mixture into moulds for crayons. The glass should be well wiped with leather, and in drawing, be careful not to soil the glass with the fingers. In trimming these crayons, if the edge be bevelled, like scissors, the point may easily be rendered very fine.—*From the French; Mechanics' Magazine.*

NEW MODE OF MARKING LINEN.

Simply cover the linen with a fine coating of pounded white sugar; impress upon it a stamp of iron very much heated, and in two seconds the linen will be slightly but indelibly scorch-marked with the mark.

LABOUR-SAVING SOAP.

TAKE two pounds of soda, two pounds of yellow soap, and two quarts of water. Cut the soap into thin slices, and boil all together for two hours; then strain it through a cloth, let it cool, and it will be fit for use. Directions for using the soap.—Put the clothes in soak the night before you wash, and to every pail of water in which you boil them add one pound of soap. They will need no rubbing; merely rinse them out, and they will be perfectly clean and white.—*New York Paper*.

NEW FRENCH PHOSPHORIC MATCHES.

IT is well known that phosphoric matches have proved so dangerous as to have been almost totally supplanted by the red or chlorate matches. A safer method of making the former has been devised by the Committee of Health and Safety in Paris. The recipe is as follows:—Put a quantity of mucilage of gum arabic into an earthenware vessel, and heat it to the temperature of 100° or 125° Fahrenheit: to four parts of this mucilage add one part of phosphorus: it will instantly melt, and should be stirred and well mixed with the mucilage: add chlorate of potash in powder, nitrate of potash, and a little benzoin, to form a soft paste; into which dip the ends of the match sticks. The mass constitutes what is called *fulminating tinder*.

CURIOUS CLOCK.

Two French gentlemen, MM. Christa and Mejanardi, have submitted to the Academy of Sciences, a clock which tells the hour, day of the week, day of the month, month, year, cycles of the sun and moon, dominical letter, epact, Easter-day, and phases of the moon.—*Railway Magazine*.

ECONOMICAL CLOCKS.

THE Societé d'Encouragement, of Paris, report that the clocks lately invented by M. Wagner, perform with exactitude and safety, at a reduction of from three to four hundred francs, in the price of a large public clock. At Moret is a manufactory of what are called *Jura clocks*, in which all the pieces of the mechanism are made in the large way, and all made to fit; wheels, weights, bells, dial-plates, hands, hammers, &c., are made, set up, and delivered in the market for forty francs; and the clock goes very well; especially when the escapement, as is done by Wagner, has been retouched. This apparatus is transformed by Wagner into a public clock, which strikes the hours on any required bell, by a proportionate hammer. This is done by adding to the Jura clocks the requisite mechanism. Thus, imagine added to the Jura clock, a heavy striking weight, hammer, and fly to moderate the descent of the weight, and a very simple mechanical connexion of these with the clock. The heavy weight is retained by a lever, held in its place by the wheelwork of the common striking part of the clock. As soon as the hammer is set free to strike, the lever is also freed; and by the descent of the heavy weight, the great hammer is set in motion, which consequently strikes as many blows as the hammer of the clock;

the hour being thus struck twice, a large dial-plate and hands must be added to the Jura clock, adapted to its height or distance to be seen.—*Mechanics' Magazine.*

HELIACAL SPRING-JOINT DOOR-HINGE.

MESSRS. HOYT and BULKLEY, of Connecticut, have invented and patented a new hinge to operate as a door-spring; it being so constructed that instead of the middle knuckles of the hinge, a heliacal spring surrounds the joint-pin, the two ends being so attached as to cause the hinge to close by the elastic force of the spring. Two portions of the hinge may be provided with brass wire heliacal springs.

PAPE'S PATENT TABLE PIANO-FORTES.

AMONGST the objects of the fine arts admitted at the late Exhibition of the produce of French Industry, Pianos were, unquestionably, the most remarkable. Sixty-seven masters sent to the exhibition nearly 200 pianos, amongst which were several of an entirely new shape; such as table, gueridon, ovalt, hexagon, and consol. These new instruments are made at the manufactory of M. Pape, piano maker to the king, who also exhibited a square piano, which judges have justly considered as a master-piece of its kind. The latter is veneered with sheets of ivory, part of which is carved and inlaid, and forms a most beautiful mosaic design. M. Pape obtains these ivory sheets by means of spiral machinery of his own invention, which produces from elephant's teeth of an ordinary size sheets of from 12 to 15 feet in length, and 2 feet in width. This invention will, no doubt be appreciated by miniature-painters, to whom this mechanical discovery will be of very great advantage. M. Pape also exhibited an horizontal grand piano of a small size. The most remarkable improvement in this instrument is the sounding-board, which is so disposed that the tension of the string stretches and keeps the sounding-board level. The consequence is, that the sound improves in course of time, whilst in pianos of the ordinary construction, the contrary will happen.—*Musical World.*

CARD-MAKING MACHINE.

AT the model exhibition of the British Association, one of the inventions which attracted most attention was a machine for the fabrication of carding web. It unwinds the wire from the reel, bends it, cuts it, pierces the holes, inserts the tooth, drives it home, and, lastly, gives it, when inserted, the requisite angle, with the same, or rather with greater, precision and accuracy than the most skilled set of human fingers could; and with such great expedition that one machine performs a task which would require the labour of, at least, ten men. An engine of 500-horse power, would drive, it is calculated, one hundred such machines.—*Mechanics' Magazine.*

DRYING OF STUFFS.

MM. PEUZOLT and LEVERGUE have invented an apparatus for rapidly drying stuffs, without fire or pressure. It consists of a double drum, which turns on its axis at the rate of 4,000 times a minute. The stuffs are placed in it as they are taken out of the water; and, by the effect of rotation, the water contained between the threads is carried towards the external covering of the drum, which is bored with holes. Woollen stuffs are thus dried in less than three minutes, when the apparatus is small, and in eight minutes when it is larger. Flax and cotton require a short exposure to the air, after being taken from the drum.

METALLIC PRINTING.

On June 7, Dr. Faraday delivered, at the Royal Institution, a lecture "On Mr. Hullmandell's Mode of producing Designs and Patterns on Metallic surfaces." Great facility is acquired in preparing the plate, by the process of biting out. Advantage is taken of the powers or forces of attraction of aggregation existing in aqueous and oily substances. A watery body, (an ink, of water, gum, and lamp-black), is put upon the metallic surface, previously greased. Immediately, both the substances draw up into spots, or bands, producing a peculiar, irregular, web-like surface. Varnish is then laid over the whole face of the plate thus prepared, a film of which dries over the inky and oily figure, retaining both, until put into water to soak. Then, from the attraction of the water for the watery particles of the ink, the ink retracts, and removes the film of varnish, leaving it only on the oily parts. The figure, the spots, and bands, which had been occupied by the ink, are now left clean, and free to be acted upon by acid. The use of gum-arabic, or gum-tragacanth, in the watery body, curiously varies the web-like figure. By a variation of the process, any ground may be produced in a considerably shorter period, and at much less expense, than by mechanical ruling, by parallel lines, or by any process of hand-engraving; and upon the ground chiefly depends the shading, &c. Improved and complicated patterns, which the great expense of engraving excluded, will in future adorn the fabrics for female dresses, &c.—*Literary Gazette.*

MOULDED LEATHER ORNAMENTS.

Mr. Esquilant has communicated to the Society of Arts, a new method of making ornaments to substitute for carvings in wood, and castings, composition, or metal, or for those of *papier maché.* Metal moulds of the separate pieces are first prepared; leather of the required thickness is then soaked for a day or two, in a solution of resin in oil of turpentine. The leather is then taken out, wiped, and cold-pressed in the mould with sufficient force to give it the intended figure; it hardens as it dries by the evaporation of the essential oil, and, when once dry, retains its form without warping on exposure to damp or draught. The separate pieces are then put together by ties and glue, and are painted, varnished, or gilt. Such is the process for leaves. Fruit is made of a paste of saw-dust powder, glue, resin, and turpentine; to be shaped by hand, or pressed into moulds. For flowers, rolled zinc is shaped by pressing the parts separately in a mould, and then cementing them together. Leather, prepared as above, is equal in hardness to wood or *papier maché;* and is so tough as not to chip by a blow. The cost is moderate.

IMPROVED PAPER-HANGINGS.

MESSRS. EVANS, of the Alder Mills, near Tamworth, have patented the application of a pneumatic pump in the compression of the moisture from the pulp, by which means the substance is almost instantaneously converted into paper. By this improvement, they are enabled to manufacture a continuous sheet of paper, 6 feet in width and nearly 2,000 yards in length, every hour. This paper, as it is taken off the reel, is fit for immediate use, and is conveyed on rollers to another part of the mill, where the printing machinery is erected, and through which it is passed with great rapidity, receiving the impression of the pattern with extreme precision and beauty of finish; the only hand-labour employed being that of one man, who superintends the machinery, and four girls employed in rolling up the paper in pieces of the required length. To complete their invention, Messrs. Evans calculate on being able to print, glaze, and emboss the most complicated and delicate patterns in paper-hangings, in every variety of shade and colour, as rapidly as the paper can be manufactured.—*Midland Counties Herald; abridged.*

SEA-REED PAPER.

AT a meeting of the Boston Society of Natural History, the President read a Report on the specimens of paper and pasteboard manufactured from the Beach grass, and presented by its inventor, Mr. Sanderson, of Dorchester. The plant is the *Arunda arenaria*. Lin. It is placed in the genus *Calamagrostis* by Withering and De Candolle; *Ammophila* by Hort and Hooker; *Psamma* by Palissot de Beauvais, Torrey, Eaton and Beck; *Phalaris* by Nuttall. It is called sea-reed or mat-reed, in England, and is found on all the shores from Iceland to Barbary; and all the Atlantic shores from Greenland as far south as New Jersey, at least. Its principal use therefore has been a negative one, connected with the very terms of its existence. It effectually secures the shifting sands on which it grows; and for that purpose large sums are annually appropriated by Government, that by its cultivation important harbours may be preserved. The paper is smooth, soft, and pleasant to write upon, and takes ink well. It is firm and very strong, and may be whitened readily. The pasteboard appears to be specially valuable.—*Silliman's Journal.*

FEATHER FLOWERS.

AT a late meeting of the Horticultural Society were exhibited a rose-tree, and some other specimens of artificial flowers, made by Mrs. Randolph, of Bridge-street, Westminster. The flowers are entirely composed of the feathers of various birds, from the common goose to the humming bird and parrot. With these, the blossoms, the bud, and the leaf, are so perfectly imitated, that you cannot distinguish the artificial from the real rose, jasmine, wallflower, pink, camellia, &c.; and as no dye or colouring matter is used, but simply the feathers, occasionally clipped into the necessary forms and moulded into shape by the hand, it is evident that even the most delicate of them must be of a lasting and unalterable quality. In this way, examples of the rarest floral productions may be made and preserved in cabinets of natural history; and we commend the invention and its accurate and beautiful application as a novelty in this country, and well deserving of notice.—*Literary Gazette.*

PROJECTILE EXPERIMENTS.

By recent experiments made at Metz, it has been ascertained that a 16-pounder impels its ball, with the ordinary charge of powder, 506 yards in the first second of time; and that by increasing the charge, it may be projected 817 yards within the same short space of time.—*Naval and Military Gazette.*

INK FILTER.

A very ingenious contrivance for filtering ink for immediate use has been produced by Mr. Perry, the steel pen manfacturer. It is an inkstand, in which is a strainer of very fine material, for purifying the ink, which is propelled into a receiving funnel by means of an air-pump. The whole occupies little more space than a common ink-glass. It also possesses the advantage, from being air-tight, of preserving ink for almost any period of time.—*Times.*

REPAIR OF WESTMINSTER BRIDGE.

Mr. W. Cubitt, the contractor for this work, in eight months from its commencement, completed the coffer-dam round the 13 and 14 feet piers, on the Westminster side; and, notwithstanding all that had been said about the impossibility of keeping the water out, and that it would require 150 men to pump constantly in a dam round *one* pier only, it appears that *one* man only, working two or three hours during the day, was sufficient in the dam round *two* piers.

When the mud which had accumulated during the execution of the dam, and the coverings of gravel, were removed, the caisons were found in a perfect state, the wood (fir) even retaining its resinous smell; their construction agrees very nearly with the description given by M. Labeyle. The sill is formed of the whole timber, extending longitudinally under the pier, and framed at each end, so as to run parallel with the cutwaters. Upon this the grating is placed: it is composed of timbers 10 in. × 10 in.; its outer frame is of the same shape as the sills, but 7 inches less in width all round, thus forming an offset or footing; the transverse timbers upon which the pier rests are one foot apart, and finely morticed and trenailed into the frame, and trenailed into the sill. Round the pier, a curb of planking 6 in. thick and 2 ft. 8 in. broad, was fastened to the grating; which has since been removed, to make way for the stone-work.

In securing the foundations, the sheet piling which surrounds the caisson is beech 12 in. thick, and 15 ft. long; the waling is 8 in. thick by 12 in.; every third pile is bolted to the wale with a $1\frac{1}{4}$ inch screw bolt, the head counter sunk into a cast iron washer; the wale is bolted to the caisson by $1\frac{1}{2}$ inch tie-bolts, 6 ft. long, let into the timber; the inner end has a cast-iron carriage, bedded; the angles of the waling are secured with wrought-iron straps. The space between the sheet piling and the caisson, and also between the timbers of the grating, is filled in with brick-work, thus forming a solid bed for the pavement, which is of roche Portland stone, 6 ft. in depth of bed, and 18 in. in height next the pier, bevelled off to 12 in. next the piles.—*Civ. Eng. and Arch. Journal.*

THE CALEDONIAN CANAL.

THE Caledonian Canal, which was begun by Mr. Telford in 1803, and opened in 1822, is, doubtless, the grandest specimen of inland navigation in the world. Its total length is 60 miles, 23 of which are formed artificially; the other 37 being the united lengths of the three fresh-water lakes, namely, Loch-Ness, Loch-Oich, and Loch-Lochy, through which it passes. As it was intended to admit the largest vessels that trade between Liverpool and the Baltic, West Indiamen of an average size, and frigates of 32 guns when fully equipped, its dimensions were necessarily great. The width of the water surface was to be no less than 120 feet, the bottom width 50, and the depth 20; while the locks, which are 28 in number, are no less than 170 or 180 feet long, and 40 wide. Although the three fresh-water lakes just mentioned greatly facilitated the formation of the canal, yet the construction of eight junctions with the canal itself occasioned much labour and expense, as well as great embarrassment to the engineer. These junctions, as well as the shallows at Loch-Oich, have been cleared by dredging-machines to the depth of 15 feet; but, in various places, a farther excavation of five feet is necessary to obtain the intended depth of 20 feet. The estimated expense of this great national work was 474,531*l.*, while the actual expenditure was nearly double that sum. (The expenditure from the 20th Oct. 1803, to the 15th August, 1831, was 990,559*l.*) Still, the work was far from completed. The new grants of money which were necessary to carry on the canal, brought the subject annually before the public, some of whom began to stigmatize the undertaking as a job; when, to pacify the discontent, Mr. Telford unwillingly consented to finish the works on a less perfect and substantial scale, and to open the canal for the passage of shipping in that unfinished, unsafe, and unsatisfactory condition in which it at present remains. The last parliamentary grant was made in 1824, and hence the remaining works were executed in a hurried and imperfect manner; so that the depth of water, in place of being made 20 feet, to admit that class of vessels for which it is intended, is not greater than 11 or 12 feet in some parts after a continued drought.

At the close of 1837, however, the Commissioners of the Caledonian Canal reported to the Lords Commissioners of Her Majesty's Treasury that an accident had taken place at the loch at Fort Augustus, which considerably impeded the navigation, and that immediate impending danger from inundation was to be apprehended from the state of the works in other parts of the line of the canal. The Treasury then deputed Mr. Walker to examine the works along the whole line, which he commenced surveying early in 1838, with the assistance of Mr. Gibb and of Mr. George May, the resident engineer and superintendent of the canal. Mr. Walker's Report to the House of Commons was ordered to be printed in July, 1838. Its general result is that 24,287*l.* is required for immediate repairs; that 104,490*l.* will be necessary for completing and perfecting the locks, and giving the canal a depth of 17 feet; that 13,200*l.* will be needed for five steam tug-boats—three upon the canal, and one at each end of it; and that, allowing ten per cent., or 1,320*l.* for contingencies, 143,837*l.* will be sufficient for completing this great work, and fitting it for the reception of vessels of 38 feet beam, and 17 feet draught of water. "With these improvements and additions," says Mr. Walker, "the passage from Fort George to the Sound of Mull might generally be depended on to be made in *five* days, and certainly, even in foul weather, within a week."

The subject of this grant will, it is hoped, occupy the attention of Parliament in the current session; notwithstanding it has been rumoured that rather than advance so large a sum as 150,000*l.*, Government will allow this great national work to fall into decay; although its important objects and the interests of the country demand otherwise.—Compiled from a paper in the *Edinburgh Review*—"On the Life and Works of Thomas Telford."

ARITHMETIC BALANCE OR NEW CALCULATING MACHINE.

WHEN we project a road, canal, or railway, it is not sufficient to calculcate, by processes more or less expeditious, the volumes of cuttings and embankments: the mean distance of the cutting to the embankment is an important element of the expense, which must be exactly determined. To obtain the value of this mean distance, we multiply the partial cubes of the cutting by the respective distances to which they are carried, and divide the sum of all the products thus obtained by the total cube of the cutting. But this series of operations is very long and tedious, and subject to error.

M. Lalane, engineer, has shown that to determine the mean distance of transport, *without calculation*, it suffices to suspend upon one of the arms of a lever, in equilibrium about its centre, weights proportional to the volumes to be carried, at distances proportional to the distances of transport; and to find at what distance from the centre we must suspend, on the other arm, a weight equal to the sum of the weights which are on the former. M. Lalane has presented to the Academy of Sciences a machine constructed upon this principle, at the expense of the French Government, by that skilful optician, M. Erust. We may represent it under the form of a common balance, whose beam is not furnished with scales, but has a breadth of several centimetres, parallel to the axis of suspension. The two arms of the beam are divided into equal parts on each side of the centre; and one of them, in the direction of its length, into equal intervals, by aid of small laminæ perpendicular to the beam, between which are to be placed the weights in the form of plates; the total weight, which is suspended at the other arm, being contained in a small moveable scale. In its actual state, the apparatus has 150 divisions on each side the axis upon a length of thirty centimetres. Each of them answers to an interval of four metres, so that upon the machine we may indicate distances of transport as far as 6,000 metres, a quantity which is never exceeded in the construction of an ordinary route. The scale of weights is a demi-centigramme for a cube metre.

Repeated trials have shown the time necessary for the research of a mean distance by means of this machine, to be at most but a quarter of the time required by the ordinary method. For the formulæ of calculations and other details, the reader is referred to the *Polytechnic Journal,* No. 4, p. 296-7; whence the above have been extracted.

SOCIETY OF ARTS.

THE distribution of rewards adjudged by the Society, took place on June 17; and the following are those connected with mechanics and other practical arts:—

1. To Mr. T. J. Cooper, Polytechnic Institution, Regent Street, for his method of preparing Paper for Photographic Drawings: the Silver Medal.

2. Mr. R. Redman, 43, Great Wild Street, Lincoln's Inn Fields, for his method of making Transfers from Copper-plate Printing to Zinc or Stone: the Silver Isis Medal and 5*l*.

3. Mr. H. C. Page, 5, Commercial Road, Pimlico, for his method of Lettering on Polished Marble: the Silver Medal.

4. Mr. W. H. Thornwaite, 3, James Place, Hoxton, for his Apparatus for the use of Divers: the Silver Medal.

5. Mr, W. Jones, 6, Horse-shoe Court, Ludgate Hill, for his Travelling Platform: the Silver Isis Medal.

6. Mr. M. Jennings, Portsmouth Street, Lincoln's Inn Fields, for his Night Signals for Steamers: the Silver Isis Medal.

7. Mr. J. Hopkins, King's Row, Horselydown, for his Scale Lever: the Silver Isis Medal.

8. Mr. B. Holmes, clerk of the works at Chatsworth, for his Spring-bolt plate: the Silver Isis Medal.

9. Mr. J. Gray, 25, Old Burlington-street, for his improved instrument for extracting teeth: the Silver Medal.

10. Mr. A. Wivell, 20, Cardington-street, Hampstead-road, for his fire escape: the Silver Medal.

11. Mr. Joseph Jeay, 6, Oxford-market, for his method of determining the lengths and bevels of timbers in a hip roof: the Silver Isis Medal.

12. Mr. C. Hanshard, 35, Sebright-street, Bethnal-green: the Silver Isis Medal and 5*l*.

13. Mr. James Cole, 38, Sebright-street, Bethnal-green, 5*l*.

14. Mr. J. Sodo, 6, Sebright-street, Bethnal-green, 5*l*.

For their respective Shares in the invention and improvement of the tube used in weaving wide silk velvet.

15. Mr. J. Dove, 9, Surat-place, Green-street, Bethnal-green Road, for his improved machine for weaving silk tissue: the Silver Isis Medal and 10*l*.

Among the prizes in the department of the Fine Arts, was the Acton Medal, awarded to Mr. Robert Billings, Manor House, Kentish Town, for his analysis of the great Eastern window of Carlisle Cathedral. This prize was the first of the kind, being conferred in pursuance of the object of a fund of 500*l*. lately presented by the widow of a gentleman who had long been a member of the Society. The window above named, though now in a state of decay, bears evidence of extreme artistical beauty, and the successful efforts of Mr. Billings has been so to analyse its composition as that, if the window should now fall to pieces, an exactly similar one could again be constructed, The window proves the profound mathematical knowledge of the ancient architects.

Natural Philosophy.

DIFFERENCE OF LONGITUDE BETWEEN GREENWICH AND NEW YORK.

Mr. Dent has communicated to the *Athenæum* the second instance of the successful transport of Chronometers from London to New York, for the purpose of determining the Longitude of these two cities. The first took place in July and August last, and when compared with the result given by M. Daussy, in the *Connaissance des Tems*, the difference was 2·63 sec. This observation was made in the first voyage of the *British Queen* across the Atlantic. On her second trip, Mr. Dent sent a second set of four chronometers from London to New York, with the following result:

	h.	m.	sec.
By this second experiment the difference of longitude between the Observatory at Greenwich and the City Hall, New York, is	4	56	0·24 W.
According to M. Daussy it is	4	56	0·72
Difference of the two observations	0	0	0·48

The difference of the two observations does not, therefore, amount to half a second! This fact is the more gratifying when we consider that these estimates were made independently of each other, "by different observers in different years, and in vessels propelled by different agents."

During the first voyage, all the chronometers showed a difference between the mean travelling rate and the mean stationary rate, which had the remarkable character of being always on the same side, viz., the *losing* rates were always *increased*, and the *gaining* rates always *diminished*. The same curious fact again occurred in the second voyage. From this circumstance, the longitude of New York was given by each chronometer scarcely enough to the westward in the outward bound voyage, and too much in the homeward one.

The great rapidity and accuracy with which this important branch of nautical inquiry may be pursued over the whole surface of the globe, as the agency of steam shall be extended, is now considered demonstrated. The instances under consideration show that observations may be made connecting very distant countries, and their several results compared in a few weeks—a circumstance of great consequence—for with the diminution of duration in a voyage, proceeds, in a higher degree, the diminution of all the chances and causes of error in chronometrical experiments at sea. Within the space of ninety-nine days, we have seen the *British Queen* carry chronometers *four* times across the Atlantic, and give ample time during each of her visits to New York for the necessary observations of rates, &c. All objections founded on the idea that the motion of a steam-vessel would affect injuriously the more delicate movement of a chronometer are also groundless.

FRODSHAM'S COMPENSATING PENDULUM.

It is an ordinary pendulum, with a steel rod: over this Mr. Frodsham slips a zinc tube, which passes through a brass bob, and rests on the adjusting screw at the lower end of the rod, the bob being fastened at the centre by two connecting rods of steel to the tube, at the point at which the expansion of the tube is the same as that of the rod; so that, as the steel rod expands downwards, and is lengthened by heat, the zinc tube expands upwards in the same degree: and, therefore, if the lengths of the rod and the tube be rightly proportioned, the pendulum may be regarded as of invariable length. But, as it is seldom found that different specimens of the same metal have precisely the same expansibility, Mr. Frodsham proposes to have several small pieces, or rings of different lengths, cut from the same tube, as correcting pieces, which are to be slipped on or withdrawn, until the length of tube is found that will compensate the pendulum for change of temperature. Mr. Frodsham states, that the hole in the bob through which the zinc tube passes is larger than the tube, but there are brass fillets at both ends, with a hole in each exactly fitting the tube: these fillets are perforated with several small holes to admit the air, so that any change of temperature may not be prevented from affecting the part of the tube which is within the bob. The zinc tube is larger than the steel-rod of the pendulum; fillets being also placed at each end of the tube, with a hole in each, just large enough to let the rod pass through. The tube is pierced with small holes throughout its whole length, to allow the air access to the rod. In the suspending part of the pendulum, Mr. Frodsham directs attention to what he calls an isochronal piece: it is a brass tube about five inches in length, with a slit about an inch in length at the bottom, to form a spring, so as to slide rather stiffly on the rod. At the upper end of the tube is a clasp, which, by means of two screws, is made to embrace the suspending spring; so that after the pendulum has been adjusted to the length for time, the acting part of the suspending spring may be varied.—*Proceedings of the British Association; Athenæum.*

ON WAVES.

The Committee on Waves, (Sir John Robison and Mr. Russell,) have reported to the British Association their researches during the past year; which have confirmed or corrected the results formerly obtained by them, and extended their acquaintance with several interesting phenomena. Concerning the first object of their attention—the nature and laws of certain kinds of waves, Mr. Russell, in a previous report, then stated that all the phenomena of the great wave of translation had been obtained, and that he himself had found no difficulty in obtaining the laws of this wave from the equations of M. Laplace, on the hypotheses, that the motion of the water particles was not infinitely small, and the oscillations infinitely small, as had been hitherto supposed, but that they had the magnitude and nature actually found in these experiments. During the past year, considerable progress has been made in this examination, and confirmations of the truth of these views had been obtained by the labours of Prof. Kolland; who has also introduced the hypothesis of the particles of water having the motions observed in the experiments, viz., a motion of permanent translation in a given course; and although his results did not perfectly accord with the experiments, they had much closer approximated to them than previous examinations, and were to be regarded as additions to the theory of the motion of waves.

In the second subject of inquiry—the connexion between the motion of waves, and the resistance of fluids to the motion of floating bodies—the phenomenon of vessels at high velocities riding the waves, had been exhibited to an extent never before witnessed. The wave, divided in two by the prow of the vessel, had risen on each side to a height far greater than that even of the vessel itself; it expanded on each side of the vessel from stem to stern, in a broad, unbroken sheet of water, bearing along the vessel, as it were, between a pair of extended gossamer wings, giving, at extremely high velocities, a resistance much less than had hitherto been observed.

The third point concerned the form of a solid of least resistance,* which had been inferred by Mr. Russell, from theory, to be that which he called the Wave Form of vessel. This form was, that the lines of anterior displacement should correspond to the outline of the great wave of translation anteriorly, and to the outline of the posterior wave of replacement, towards the stern; and the truth of the hypothesis had been confirmed by the experiments of the preceding year. A very remarkable practical confirmation is the fact of a large steam-vessel, of 660 tons, with an engine of 220 horse power, being built on the wave principle, and turning out the fastest vessel in Great Britain. This vessel, built as a pleasure yacht, combined the qualities of sailing fast, and carrying a large cargo: she was named *the Fire King*, and belonged to Mr. Ashton Smith, of Wales.

The last point of inquiry had been the nature and laws of the tidal wave, as propagated along our shores, and up the estuaries of our great rivers. But the nature of its propagation along our shores, after it ceased to be affected by the celestial influence, formed a terrestrial mechanism, with which we were still very imperfectly acquainted. On this subject, the Committee had made simultaneous observations, at thirteen different stations along the Frith of Forth. It was found that there were four tides a day in the Forth instead of two—four high waters and four low waters. Mr. Russell exhibited drawings of these tidal waves, and gave, what he conceived the explanation of a phenomenon, which is, he thinks, much more common than hitherto supposed. It is well known that the tidal wave which brings high water from the Atlantic to the south-western shores of Great Britain, becomes divided into two parts, one of which passes upwards through the English Channel, and the other passing round the west and north of Ireland and Great Britain, brings high water to the east coast of Scotland and to the Frith of Forth. Now, it appears not to have been

* See Year-Book of Facts, 1839, p. 84.

recollected, that the other wave, after coming up through the English Channel, and bringing, along with the former, high water to London, must pass on northwards, and, in doing so, will enter the Frith of Forth considerably earlier than the northern wave passing southwards. This southern wave, smaller, but earlier than the other, appears to enter the Frith, and may be traced at every station. It is followed up, however, very rapidly, by the great northern wave, and the former moving more slowly than the other, according to the law of the great wave of translation, is overtaken by it at the higher parts of the Frith; and being both greatly exaggerated by the form of that channel, produces the two tides of the Frith of Forth. Mr. Russell expressed his opinion, that the tides in the upper part of the Frith of Forth would be found to rise as high above the mean level of the sea as the tides of the Bristol Channel. The observations on this subject were not, however, completed, but would be finished in the course of next (the present) year.—*Athenæum; abridged.*

THEORY OF WAVES.

In illustration of the theory of waves, it is recorded in the *Literary Gazette*, that, on Sept. 1, a fir-tree, about 40 feet in length, and covered with great barnacles, was drifted across Dover harbour, and towed into that port. It is supposed to have been parted from the American shore by a hurricane, and borne on its long voyage across the Atlantic by the tidal action of the wave.

HEIGHT OF WAVES.

M. Alme, from experiments made in the Bay of Algiers, from December, 1838, to July, 1839, during the continuance of heavy north and north-east winds, concludes: 1st, that the motion of the sea produced by the agitation of the waves, may be sensible 40 yards in depth; 2ndly, that the motion of the bottom is oscillatory; and, 3rdly, that the extent of this oscillation varies slowly from the bottom to the surface.

INFLUENCE OF ATMOSPHERIC PRESSURE ON THE TIDES.

Mr. Walker, Assistant-Master-Attendant in H. M. Dockyard, Devonport, (who has long devoted much time to tidal phenomena,) considering that half-tide levels on oceanic shores, such as those of a large part of Cornwall and Devon, give the equilibrium level of the sea, proposes the following method for obtaining it; which, whatever opinion may be ascertained of the general value of half-tide levels, affords very considerable facility in ascertaining that level at any given place. The mode is: when the barometer stands at its mean annual height, and the air is calm and still, set up a tide-pole, (or select a rock), in some sheltered corner on the coast. Mark upon it the high and low water levels, and half-way between these points will be found the mean level of the sea. Under the above condition, a single observation will give the mean level very nearly; but numerous observations are necessary when great accuracy is required.

Mr. Walker has observed, with respect to the influence of the pressure of the atmosphere upon the tidal waters on the shores of Cornwall and Devon, that a fall of 1 inch of the mercury in the barometer, corresponds with a rise of 16 inches in the level of the sea, more than would otherwise happen at the same time, under the other general conditions; a rise in the barometer of one inch marking a corresponding fall in the sea-level of 16 inches. This he has found to be the usual rate of such alterations in level; but *very sudden* changes in the pressure of the atmosphere are accompanied by elevations and depressions equal to 20 inches of sea-water for 1 inch of mercury in the barometer. Regarding the whole pressure of the atmosphere over the globe as a constant quality, all local changes in its weight merely transfer a part of the whole pressure from one part to another; whence he concludes that the subjacent water only flows into, or is displaced from, those areas; where, for the time, the atmospheric pressure is either less or greater than its mean state in accordance with the laws which would govern the conditions of two fluids situated in the manner of the atmosphere and sea. We might account for the difference observed by Mr. Walker, in the amount of depression or elevation of sea-level produced by sudden changes in atmospheric pressure, by considering that a sudden impulse given to the particles of water, either by suddenly increased or diminished weight in the atmosphere, would cause a perpendicular rise or fall in the manner of a wave beyond the height or depth strictly due to the mere change of weight itself.*

As regards the influence of the winds on the mean level upon the south coast of Cornwall and Devon, Mr. Walker observes that east and west winds scarcely affect it; but that southerly winds raise the sea above it from 1 to 10 inches, and off-shore winds depress the water beneath it as much, according to their force. From occurrences, which we have not space to quote, it would appear that, around the shores of the above districts, when the winds which traverse it have considerable force, the levels which would obtain in calms are considerably disturbed; consequently, minor effects of the same kind are caused by less powerful winds, according to their velocity. To obtain, therefore, true heights in this district above the sea, which should correspond, above a level in both channels, supposing such a level to exist, calm weather is essential for accuracy.--From a valuable *Report on the Geology of Cornwall, Devon, and West Somerset; by H. F. De la Beche.*

* A circumstance connected with this subject, of considerable practical value, has been noticed by Mr. Walker. He has found that changes in the height of the water's surface resulting from changes in the pressure of the atmosphere, are often noticed on a good tide-gauge, *before* the barometer gives notice of any change. Perhaps, something may be due, in those cases observed by Mr. Walker, to the friction of the mercury in the barometer tube. The practical value of the observation is, however, not the less, be the cause of the phenomenon what it may; for if tide-gauges at important dockyards show that a sudden change of level has taken place, indicative of suddenly decreased atmospheric weight, before the barometer has given notice of the same change, all that time which elapses between the notices given by the tide-gauge and the barometer is so much gained; and those engaged with shipping know the value of even a few minutes before the burst of an approaching hurricane.

RESEARCHES ON THE TIDES.

PROF. THE REV. W. WHEWELL has communicated to the Royal Society his Tenth * Series of Researches—"On the Laws of Low Water at the Port of Plymouth, and on the Permanency of Mean Water," —In this valuable memoir, the indefatigable author investigates the question, how far the *mean water*, that is, the height of the tide midway between high and low water, is permanent during the changes which high and low water undergo. That it is so approximately at Plymouth, having been already ascertained by short series of observations, it was desirable to determine the real amount of this permanency by induction from longer series of observations. A period of six years was chosen for that purpose; and the method of discussing these observations was the same, with slight modifications, as in former researches. The height of low water, cleared from the effects of lunar parallax, and very nearly so from those of lunar declination, and compared with the height of high water, similarly cleared, enabled the author to ascertain whether the mean water also was affected by the semi-menstrual inequality. The results of the calculation show that the height of mean water is, within two or three inches, constant from year to year; and that, for each fortnight, it has a semi-menstrual inequality amounting to six or seven inches;—the height being greatest when the transit is at 6 h., and least when at 11 h.;—the immediate cause of this inequality being, that the semi-menstrual inequality of low water is greater than that of high water: this inequality, however, is probably modified by local circumstances. These researches have also verified the theoretical deduction, that, the height both of low and of high water being affected by the moon's declination, their mean height partakes of the variations in this latter element, in successive years, consequent on the change of position of the moon's orbit. At Plymouth, the increase in mean low water amounts to about two inches for each degree of increase in the declination. In the high water, this change is less marked. The parallax correction of the height of low water is obtained from all years alike, by taking the residue of each observation, which remains when the semi-menstrual inequality is taken away, and arranging these residues for each hour of transit, according to the parallax. The declination correction is obtained in a manner analogous to the parallax correction, from each year's observations, with some correction for the variation in the mean declination of the moon in each year.

The Eleventh Series—"On certain Tide Observations made in the Indian Seas," contains the results of the examination, by the author, of certain series of tide observations made at several places in the Indian Seas, which were forwarded to the Admiralty by the Hon. East India Company. These localities were Cochin, Corringa River, Surat roads in the Gulf of Cambay, Gogah, on the opposite side of the same Gulf, and Bassadore, in the Island of Kissmis, in the Persian Gulf.

The Reports made to the British Association, "On Tides," are briefly as follow:—

* For the Ninth Series, see Year-Book of Facts, 1839, p. 85.

Prof. Whewell communicated some tide observations, forwarded to him by the Russian Admiral Lütke. These observations supplied, —first, the tide hours of various places on the coasts of Lapland, the White Sea, and the Frozen Sea, and the coasts of Nova Zemlia. These observations enable us to follow the progress of the tide mean farther than had hitherto been done. Mr. Whewell's map of Co-tidal Lines (the second approximation contained in the Phil. Trans. 1836,) follows the tide only as far as the North Cape of Norway, eastward of Nova Zemlia. Professor Whewell stated that he was informed by Admiral Lütke, that in the Frozen Sea, east of Nova Zemlia, there is little or no perceptible tide. The observations communicated by Admiral Lütke offered various other results, and especially the existence of the *diurnal inequality* in the seas explored by Russian navigators; as on the coast of Kamtschatka, and the west coast of North America.

Prof. Whewell made some observations on Capt. Fitzroy's views of the tides. In the account of the voyage of H. M. S. *Adventure* and *Beagle*, just published, there is an article in the Appendix, containing remarks on the tides. Capt. Fitzroy observes that facts had led him to doubt several of the assertions made in Mr. Whewell's memoir, published in the *Philosophical Transactions*, 1833, and entitled, "Essay towards a First Approximation to a Map of Co-tidal Lines." * (Appendix, p. 279.) Prof. Whewell stated that he conceived that *doubts*, such as Capt. Fitzroy's, are reasonable, till the assertions are fully substantiated by facts. Capt. Fitzroy has farther offered an hypothesis of the nature of the tidal motion of the waters of wide oceans, different from the hypothesis of a progressive wave, which is the basis of Prof. Whewell's researches. Capt. Fitzroy conceives that, in the Atlantic and the Pacific, the waters oscillate laterally between the eastern and western shores of these oceans, and thus produce the tides. This supposition would explain such facts as these, that the tide takes place along the whole west coast of South America at the same time; and the supposition might be so modified as to account for the absence of tides in the central part of the ocean. Prof. Whewell stated that he was not at all disposed to deny that such a mode of oscillation of the waters of the ocean is possible. Whether such a motion be consistent with the forces exerted by the sun and moon, is a problem of hydro-dynamics hitherto unsolved, and probably very difficult. No demonstrative reason, however, has yet been pub-

* Among the points which I could not establish in my own mind, by appeals to facts, were :—" The tides of the Atlantic are, at least in their main features, of a derivative kind, and are propagated from south to north;" "that the tide wave travels from the Cape of Good Hope to the bottom of the Gulf of Guinea, in something less than four hours; that the tide wave travels along this coast (America) from north to south, employing about twelve hours in its motion, from Acapulco to the Strait of Magalhaens;" "from the comparative narrowness of the passage, to the north (of Australia), it is almost certain that these tides must come from the southern end of the continent." "The derivative tide, which enters great oceans, (North and South Pacific,) from the south-east, is diffused over so wide a space that its amount is greatly reduced."

lished, to show that such a motion of the ocean waters may not approach more nearly to their actual motion than the equilibrium theory, as usually applied, does. When the actual phenomena of the tides of the Atlantic and Pacific have been fully explored, if it appear that they are of the kind supposed by Capt. Fitzroy, it will be very necessary to call upon mathematicians to attempt the solution of the hydro-dynamical problem, either in a rigorous or in an approximate shape.

The Report on the application of the sum assigned for Tide Calculations to Prof. Whewell, in a letter from T. G. Bunt, Esq., Bristol, was then read. (See *Athenæum*, No. 618.)

In a discussion which followed, Mr. Russell observed that with respect to there being no tide in the neighbourhood of the South Sea Islands, he thought it probable that the coral reefs in that part of the ocean would have the effect of obliterating the tide wave; for he found, during his researches on waves, that the slightest obstacles placed at the bottom of the trough were sufficient almost instantly to obliterate the wave; insomuch that, whereas, at first, he had to wait after one experiment sometimes fifteen or twenty minutes, before the surface was still enough to commence the next, after he discovered the fact, he could at pleasure quiet the surface, by elevating certain pieces of wood, which in general lay even with the bottom. Now, the coral reefs were just such obstacles; and the tide wave being clearly proved a wave of translation, he had little doubt in his own mind that they would furnish an explanation of the fact.

NEW THEORY OF THE TIDES.

On Nov. 11, Prof. Whewell explained to the Cambridge Philosophical Society, more at length, the new theory of the Tides referred to in the above Report; and which, to a certain extent, coincides with the views of Capt. Fitzroy. The new theory is as follows: The tide of each large ocean may be considered as nearly independent of the tides of other waters. The central area of each ocean is occupied by a *lunar* wave, which oscillates, keeping time with the moon's returns, and having its motion kept up by the moon's attraction acting at each return. From the skirts of this oscillating central area, tides are carried on all sides by *free* waves, the velocity of which depends upon the depth and local circumstances of the sea; and thus the *littoral* tides may travel in any direction, while the *oceanic* tides near the centre of the oscillating area may be small, or may vanish altogether.

This theory was confirmed by a reference to tide observations on the eastern and western sides of the Pacific, and by mathematical calculations, tending to show that such a motion is mechanically possible. It was remarked that single observations can be of small use in deciding upon such a theory; and that it can be judged of only when we have the *systems* of co-tidal lines which belong to the shores of the Pacific. With this view, it is very desirable to obtain numerous and connected observations on the tides on the eastern shores of Australia, the Indian Archipelago, the Philippine Isles, the Loo Choo Isles, and Japan.—*Philosophical Magazine.*

SALINE MATTER AND TEMPERATURE OF SEA-WATER.

Capt. Wauchope, R. N., having enclosed a thermometer in a glass tube in the centre of an instrument, let down the same 653 fathoms, in lat. 0° 33′ N., long. 8° 16′ E., when the temperature of the water proved to be 43°; that of the surface-water being 78¾°. During this experiment, there was no current. Mr. Kemp having examined these two specimens of sea-water, found the deep water to contain more saline matter than the surface-water, being about 4½ per cent., the water at the surface having about 3½ per cent. It contains also sulphate, and a small quantity of carbonate, of lime. The surface-water differs chemically from the deep water only in having no sulphate or carbonate of lime in its constitution:

Specific gravity of the deep water = 1·30
Do. of the surface = 1·23½.

A trace of iodine and bromine was found in both.

From four other experiments, Capt. Wauchope thinks it not improbable that, at some given depth, the temperature of the sea may be found at 40° all over the world. The instrument employed by the captain for letting down the thermometer consists of a series of cases, one within the other, having valves opening up so as to allow the water to pass through on descending, but which close on hauling the instrument up.—*Selected from Jameson's Journal.*

EXPERIMENTS TO ASCERTAIN THE DEPTH OF THE SEA BY ECHO.

Prof. C. Bonnycastle, of the University of Virginia, conceiving that an audible echo might be returned from the bottom of the sea, and the depth be thus ascertained from the known velocity of sound in water, has made several experiments to determine this inquiry. His apparatus consisted, first, of a petard or chamber of cast-iron 2½ inches in diameter, and 5¼ inches long, with suitable arrangements for firing gunpowder in it under water; secondly, of a tin tube, 8 feet long and 1¼ inches in diameter, terminated at one end by a conical trumpet-mouth, of which the diameter at the base was 20 inches, and the height of the axis 10 inches; thirdly, of a very sensible instrument for measuring small intervals of time, and which was capable of indicating the 60th part of a second. Besides these, an apparatus for hearing was roughly made on board the vessel, in imitation of that used by Colladon in the Lake of Geneva: it consisted of a stove-pipe, 4½ inches in diameter, closed at one end, and capable of being plunged four feet into the water. The ship's bell was also unhung, and an arrangement made for ringing it under water. For the details of the experiments the reader is referred to the proceedings of the American Philosophical Society, (*Philosophical Magazine*, No. 92, pp. 538—540); the conclusion thence being, either that an echo cannot be heard from the bottom of the sea, or that some more effectual means of producing it must be employed.

THEORY OF CRYSTALLOGRAPHY.

PROF. NECKER has presented to the Royal Society a new hypothesis, on which he explains how common salt, alum, sulphate of iron, &c., crystallize in pure water in the most simple forms; the reciprocal attraction of their molecules being controlled and diminished by the affinity exerted on them by the molecules of the water; whilst, if some of these molecules of water are neutralized by mixture with another soluble principle, they cease to act as an obstacle to the crystallization of the body, which then takes forms more complicated, and approaching nearer to that of the normal solid with a curved surface.

M. Necker considers that his new views require, for their complete development, many ulterior details, new experiments, and new facts; but that the tendency which the crystals of all systems present, to progress towards the curved surface form appropriate to each system, by the complication of their forces, is a fundamental fact of the first importance; and that an advance has been made by showing the bearing of the important experiments of MM. Beudant and Beudant, and by having brought the theory of crystallography nearer to those views which the progress of chemistry and of physics has led us to adopt, relative to the form of the elementary molecules of bodies.—*Proc. Royal Soc.; Philosophical Magazine.*

ATOMIC WEIGHTS OF ELEMENTARY BODIES.

A VERY important paper has been read to the British Association, by Dr. Clarke, of Aberdeen; entitled, "On the Limits within which the Atomic Weights of Elementary Bodies have been ascertained." In this communication, the author adduced numerous arguments proving the equivalent of carbon to be 75·6 on the oxygen scale, instead of 76·44, the number assumed by Berzelius. The atomic weight of carbon on the hydrogen scale, taking oxygen as 8, will thus, according to Dr. Clarke, be 6·048. On account of the great importance and value of this paper, it will be published entire in the *Transactions* of the Association.—*Literary Gazette.*

RELATION OF ATOMS IN ORGANIC COMPOUNDS.

MR. EXLEY has reported to the British Association an account of his continued experiments to determine the truth of his theory of organic compounds. He holds that there are three kinds of atoms: viz., tenacious atoms of common matter, of which fifty-five different sorts have been chemically ascertained; electrical atoms, observed in electrical phenomena, with less absolute forces, but larger spheres of repulsion, than the tenacious atoms; and ethereal atoms, with forces still more diminished, and with spheres of repulsion still more enlarged.—*Literary Gazette.*

FLUIDITY OF LIQUIDS.

DR. URE has arrived at the following results by a series of new experiments, the *modus operandi* of which is as follows: the liquid was put into a glass funnel, terminated at its beak with a glass tube of uniform bore, about one-eighth of an inch in diameter, and three inches long. The funnel was supported in a chemical stand, and discharged its contents, on withdrawing a wooden pin from the beak, into a glass goblet placed beneath; by the side of which a chronometer was placed, to indicate, in seconds, the time of efflux. The volume of liquid used in each case was the same, viz., 2000 grain measures, at 65° Fahr. The times of efflux with liquids of the same specific gravity and bulk, in the same vessel, vary with the viscidity of the liquids, and serve to measure it:

	Fahr.		Sec.
2000 grain-measures of water, at	60°,	ran off in	14.
....	68	....	13.
....	164	....	12.

2000 grain-measures of	Fahr.	Spec. grav.	Sec.
Oil of turpentine	65°	0·874	14
Pyroxylic spirit	—	0·830	14½
Alcohol	—	0·830	16
Nitric acid	—	1·340	13[illegible]
Sulphuric acid	—	1·840	21½
Ditto	262	—	15
Saturated solution of sea salt	65	1·200	13
Sperm oil	—	0·890	45½
Fine rape-seed oil	—	0·920	100
Fine pale seal oil	—	0·925	66
Fine South Sea whale oil	—	0·920	66
Sperm oil	254	—	15
Rape-seed oil	254	—	17
South Sea oil	250	—	17

Literary Gazette.

ANALYSIS OF WOOD.

THE experiments of M. Payen have led him to conclude that the ligneous body so universally existing in phenogamous vegetables is not an immediate principle of vegetation, but that it is composed of two parts, chemically distinct. Having obtained the cellular tissue in its earliest state, from various ovula, and the radicles, or radicle fibres, of several plants, he only found in it various combinations of carbon, hydrogen, and oxygen: consequently, it is not truly ligneous; but the thickening substance in the interior of the fibrous cells is operated on by agents which have no effect on the elementary tissue, such as soda, potash, and azotic acid. Remarkable differences take place in the composition of the woods, according to their species, and, in the same species, according to climates. Hence, the proportion of carbon relative to that of oxygen, and the predominance of hydrogen over oxygen in the strongest woods. In combustion, an excess of hydrogen tends to the production of heat, and offers a reason for preferring what are called heavy woods; with the exception of the birch, which owes its superiority to a principle named betuline.—*Athenæum.*

RESPIRATION OF SEEDS.

MM. EDWARDS and COLIN have read to the French Academy of Sciences the details of some new experiments upon the respiration of seeds. All previous researches have been made in the air, or, if made in water, the phenomena which occurred in the liquid have been limited to the explication of what took place in the air; the gas disengaged in the liquid has not been examined, and, consequently, its proportion has not been determined.

The experimenters filled a bell-shaped bottle with a narrow neck, holding 6 or 8 pints of water, into which they put forty garden beans of large size, without fissure in the husk, or any other defect whatever. To this bottle was adapted a bent tube, also filled with water, which was finally introduced into a receiver full of the same liquid. Thus the beans were in contact only with the water, and the air which it contained, air which, under the circumstances, could not be removed; and this was one of those important circumstances which led to all the success of the experiments. Air-bubbles were disengaged from the beans—at first, very minute, but in twenty-four hours very conspicuous; which evolution of gas had not previously been pointed out, and scarcely seemed to agree with the received ideas upon germination; still less with the supposition that this disengagement proceeded from air contained in the beans; which idea soon became wholly improbable from the unceasing continuance of the disengagement of the gas, and to such an extent that it could not possibly be attributed to this cause.

For the details of these experiments the reader is referred to *Jameson's Journal*, No. 53. The results are: 1st, that the water is decomposed, a fact quite foreign to the popular theory of the present day; 2nd, that the oxygen of the decomposed portion unites with the carbon of the seed, and forms carbonic acid gas; 3rd, that this carbonic acid disengages itself from the seed, in whole or in part; and, 4th, the other portion of the decomposed water, the hydrogen, is absorbed by the seed, in whole or in part. It also results from the facts which the experimenters have propounded, that respiration is not, as it has hitherto been considered, solely a function of excretion; but it at the same time exhibits, according as they have demonstrated, a fundamental fact concerning the nutrition and development of the embryo by the absorption of hydrogen.

ABSORPTION OF AZOTE BY PLANTS DURING VEGETATION.

M. BOUSSINGAULT has determined, by numerous experiments, made with great care, that, while shooting, wheat and trefoil neither increase nor diminish the portion of azote which analysis shows them to contain; and that, during germination, these grains lose carbon, hydrogen, and oxygen; and that each of these elements, as well as the proportions in which the loss occurs, varies at different stages of germination. It appears also, that, during the cultivation of trefoil in soil absolutely deprived of manure, and under the influence of air and water only, this plant acquires carbon, hydrogen, oxygen, and a quantity of azote, appreciable by analysis: wheat, cultivated exactly in the same circumstances, also takes from the air and water, carbon, hydrogen, and oxygen; but analysis does not prove that it has either lost or gained azote.—*Annales de Chim. et de Phys.; Philos. Mag.*

INFLUENCE OF NATIVE MAGNESIA ON VEGETATION.

THE presence of magnesia was regarded as a cause of barrenness in lands, until the investigations of Bergmann proved magnesia to form one of the principal constituents of fertile soils. This inquiry has been taken up by Prof. Giobert, who concludes from various experiments: 1st, that native carbonated magnesia is not injurious to the various functions of vegetables; 2nd, that on account of the solubility of magnesia in an excess of carbonic acid, this earth can exercise an action analogous to that of lime; that a magnesian soil may become fertile when the necessary manure is employed. From these facts naturally proceeds the conclusion, that if the magnesia was dissolved in an excess of carbonic acid and water, and had entered, like the lime, into the composition of the sap, it ought to be found in the plants with the potash, lime, oxide of iron, &c. M. Abbene has ascertained this by the analysis of the ashes of plants which had grown in magnesiferous mixtures. He has also confirmed the experiments of Prof. Giobert as regards the influence of magnesia on vegetation being analogous to that of lime; and he likewise concludes that when lime and magnesia exist in arable lands, the former is absorbed in preference by the plants, on account of its greater affinity for carbonic acid. The barrenness of magnesian lands is not referable to magnesia, but to the cohesive state of their parts, to the want of manure, of clay, or of other composts, to the large quantity of oxide of iron, &c. Barren magnesian soils may be fertilized by means of calciferous substances, as rubbish, chalk, ashes, marl, &c., provided the other conditions be fulfilled.—*Journal de Pharmacie; Philos. Mag.*

POROSITY OF COTTON.

Fill a glass tumbler completely with some spirit, so that a few more drops would cause it to overflow. This done, you will find no difficulty in introducing into the tumbler, so filled, *a whole handful of raw cotton.* This experiment was suggested by the accidental recovery of some wet cotton from a boat which had been some time sunk in a river in America; when it was found that after the water was squeezed from the cotton, the vessel which had contained it remained nearly as full as before the cotton was removed.

Spirits answer better than water for this experiment, from the rapidity with which they are absorbed by the cotton. Several theories have been started in explanation of this result: such as, that the filaments of cotton occupied the vacancies between the globules of water; or that, by its capillary attraction, the cotton subdivided the globules, and caused them to occupy a less space, &c.; but, Mr. Trantwine, who has communicated this experiment to the *Franklin Journal*, accounts for the effect more satisfactorily, by supposing the fluid to insinuate itself between the filaments of cotton, and thus permit the latter to occupy no more space than is due to their actual solidity.

INFLUENCE OF COLD UPON CAPILLARY CIRCULATION.

M. Poisseville has observed, the temperature being 68° Fahr., that, by putting particles of ice in a receptacle, as, for instance, a bull-head, the circulation in the capillary tubes become slower and slower, the globules separate, becoming deranged in form as they make way across these vessels, and they resume their primitive form when passing through vessels of a larger calibre. By a longer stay in water at a temperature of from 33° to 35°, the circulation ceases in the greater number of capillary tubes; and if we then measure the calibre of these vessels, we find their diameters varying from 0·18 to 0·20 of a millemetre from before the ice was applied; but, if we maintain this low temperature by means of a new supply of ice, at the end of a certain time, the globules of the capillary vessels in which the circulation had entirely ceased, will undergo a slight disturbance through the influence of the contractions of the heart. These oscillations of the globules, acquiring a greater and greater amplitude, very soon change into a progressive motion; so that in about three-quarters of an hour, the circulation is as thick in this ambient medium, which is some degrees above zero, as in the open atmosphere. Certain vessels, which, previous to the action of the ice, gave way to two or three globules in front, offer no more than one single rank of globules, which move on their axes. The latter vessels, as also those of a greater calibre, appear not to have changed their volume; but the capillary vessels, which then produce a circulation as quick as in the normal state, have a greater diameter, by double and treble their volume. Some few other capillaries, which were quiescent, had not, however, augmented their calibre; but, by removing the ice, the circulation is soon re-established, and, in a few hours, all the capillaries resume their primitive volume.

Thus, under the influence of the contractions of the heart, the *living* tubes would acquire, by a prolonged action of cold, a more considerable volume.

Having preserved, during the winter, some bull-heads, which were collected in July, M. Poisseville found that in January their capillary vessels were augmented in volume; the place occupied by them being at 2° above zero.

The same experiments made upon the bladder of very young mullets presented much greater difficulties. In fact, from the moment that the ice comes in contact with the capillaries of these animals, and continues but for a few minutes, the circulation ceases, and finally is entirely destroyed. But, if we replace the ice by some water at 50° or 53°, (the circumambient temperature being at 77°, the circulation becomes gradually re-established, but only in those capillary vessels which have, as in the preceding case, augmented their volume.

From the above facts, M. Poisseville concludes that the points of the tegumentary surface of the human body which are usually exposed, as that of the face, neck, hands, &c., and, consequently, submitted to a mean temperature less than that of the body itself, have their capillary vessels of a larger volume than those of other portions of the skin.—*Translated in the Polytechnic Journal.*

CAUSE OF ANIMAL HEAT.

Dr. I. M. Winn, whilst making experiments with caoutchouc, was forcibly struck with the property it possesses of evolving heat when suddenly stretched; and was thence led to infer the probability of other bodies being similarly endowed. The elastic coat of arteries, especially, from the mechanical resemblance it bears to caoutchouc, appeared to be one of the substances most likely to exhibit this calefactory principle; and in the event of this being the case, it would not be unreasonable to conclude that the incessant contraction and dilatations of the arteries during life, must prove an efficient source of animal heat. Dr. Winn has verified this conjecture in an experiment with part of the aorta of a bullock, having pulled which to and fro for about a minute, he found the same caused the mercury in a thermometer to rise two degrees; and on removing the thermometer, the heat immediately began to diminish.

The following physiological and pathological facts appear to Dr. Winn to corroborate the view he has taken of the mechanical source of heat: 1st, the minute distribution of the arteries to every part of the system insures a general and equal distribution of heat; 2nd, the ossification of the arteries in old age, by diminishing their elasticity, is a probable cause of the diminution of animal heat at the close of life; 3rd, the increased warmth of the body from exercise appears to be more readily explicable upon the principle of increased force in the arteries, rather than that of increased vigour in the function of the lungs, inasmuch as the immediate effect of exercise is evidently to embarrass the breathing, as shown by the hurried respiration; 4th, in many diseases of the lungs, where its functions are all at fault at a time when the arteries are beating with increased violence, the heat of the body is found to be above the usual standard; 5th, medicines which diminish the contractibility and elasticity of the arteries, almost invariably reduce the heat of the body; 6th, the heat of local inflammations, in cases where the constitution does not sympathize to any extent, cannot be easily referred to any other source, as the arteries immediately in the neighbourhood of the affected part are throbbing with violence at a time when its capillaries (which are supposed to play so large a share in the chemical theory of heat), are generally considered to be entirely arrested. Many facts of a similar nature could be enumerated; but enough, Dr. Winn thinks, have been stated to establish the truth of the theory in question. — *Communication to the Philosophical Magazine.*

MOTION OF THE BLOOD.

On April 25, was read before the Royal Society, a paper "On the motion of the Blood," by J. Carson, M.D. After referring to his paper contained in the *Philosophical Transactions* for 1820, relative to the influence of the elasticity of the lungs as a power contributing to the effectual expansion of the heart, and promoting the motion of the blood in the veins, the author endeavours, from a review of the circumstances under which the veins are placed, to show the inconclusiveness of the objections which have been urged by various physiologists against his and the late Sir David Barry's theory of suction: namely, that the sides of a pliant vessel, when a force of suction is applied, will collapse and arrest the farther transmission of fluid through that channel. The considerations which he deems adequate to give efficacy to the power of suction in the veins of a living animal are:—1st, the position of the veins by which, though pliant vessels, they acquire in some degree the properties of rigid tubes; 2nd, the immersion of the venous blood in a medium of a specific gravity at least equal to its own; 3rd, the constant introduction of recrementious matter into the venous system at its capillary extremities, by which the volume of the venous blood is increased, and its motion urged onwards to the heart in distended vessels; and, lastly, the gravity of the fluid itself, creating an outward pressure at all parts of the veins below the highest level of the venous system. The author illustrates his positions by the different quantities of blood which are found to flow from the divided vessels of an ox, according to the different modes in which the animal is slaughtered.—*Proc. Royal Soc.*; *Athenæum.*

RESPIRATION AT DIFFERENT PERIODS OF THE DAY.

Mr. C. T. Coathupe has communicated to the *Philosophical Magazine*, No. 91, a series of elaborate experiments upon this inquiry. A few of the results follow:

1st period	8 a. m. to 9½ a. m.	32 exp.	indicated	4·37	per cent. of carb. acid gas.
2nd do.	10 a. m. to 12 noon	15 do.	do.	3·90	
3rd do.	12 noon to 1 p. m.	7 do.	do.	3·92	
4th do.	2 p. m. to 5½ p. m.	29 do.	do.	4·17	
5th do.	7 p. m. to 8½ p. m.	17 do.	do.	3·63	
6th do.	9 p. m. to midnight	24 do.	do.	4·12	
		124 exp. comprising 8 days.			

Hence we find the carbonic acid gas produced by respiration to be less during active digestion, and to increase with increased abstinence from food; and to vary, in the same individual, at *similar* periods of different days. Excitement of any kind, (whether from the exhilirating stimulus of wine, or from the irritating annoyances which are wont to occur to most persons, actively engaged,) caused a diminution of carbonic acid in the air respired, as compared with the ordinary average of that respired at a similar period of the day, and during ordinary tranquillity. The *total daily average* indicated 4·09 per cent. of carbonic acid gas. The *maximum* observed at any single examination was 7·98 per cent. It was at 8 a.m. February 5th. The *minimum* observed at any single examination was 1·91 per cent. It was 7½ p.m. February 7th.

1st, The average number of respirations made by most adult healthy individuals, (varying from 17 to 23 per minute,) may be stated as 20 per minute.

2nd, The average bulk of air respired at each respiration made by such individuals, (varying from 14 to 18 cubic inches,) may be stated as 16 cubic inches.

3rd, The average daily amount of carbonic acid gas found in the air respired by such individuals, (varying at its extremes from 1·9 to 7·98 per cent.) may be stated as 4 per cent.

Hence 460800 cubic inches, or 266·66 cubic feet of air pass through the lungs of a healthy adult of ordinary stature in 24 hours, of which 10·666 cubic feet will be converted into carbonic acid gas, = 2386·27 grs. or 5·45 ounces avoirdupois, of carbon. This gives 99·6 grs. of carbon per hour, produced by the respiration of one human adult, or 124·328 pounds annually; and, if we multiply this by 26½ millions, (the calculated population of Great Britain and Ireland in the year 1839,) we have 147,070 tons of carbon as the annual product of the respiration of human beings then existing within the circumscribed boundaries of Great Britain and Ireland.

Hence, also, the *maximum* quantity of fresh atmospheric air that can possibly be required by a healthy adult during 24 hours, even supposing that no portion of the air respired could be again inspired, will not exceed 286·666 cubic feet.

THE LAWS OF VITALITY.

On Jan. 10, a paper by Charles Jellicoe, Esq., "On the Laws of Vitality," was read to the Royal Society; in which the author, considering that the variations and discrepancies in the annual decrements of life which are exhibited in the tables of mortality hitherto published would probably disappear, and that these decrements would follow a perfectly regular and uniform law, if the observations on which they are founded were sufficiently numerous,—endeavours to arrive at an approximation to such a law, by proper interpolations in the series of the numbers of persons living at every tenth year of human life. The method he proposes, for the attainment of this object, is that of taking, by proper formulæ, the successive order of differences, until the last order either disappears, or may be assumed equal to zero. With the aid of such differences, of which, by applying these formulæ, he gives the calculation, he constructs tables of the annual decrements founded principally on the results of the experience of the Equitable Assurance Society.—*Proc. Royal Soc.*; *Philos. Magazine.*

CALCULATING YOUTH.

On the 28th June, Master Bassle, who was only thirteen years of age, went through an extraordinary mnemonic performance at Willis's Rooms. Five large sheets of paper, closely printed, with tables of dates, specific gravities, velocities, planetary distances, &c., were distributed among the visitors, and every one was allowed to ask Master Bassle a question relating to these tables; to which he received a correct answer. He would also name the day of the week on which any day of any month had fallen in any particular year. He could repeat long series of numbers, backwards and forwards, and point out the place of any number in the series; and to prove that his powers were not merely confined to the rows of numbers in the printed tables, he allowed the whole company to form a long series, by contributing each two or three digits in the order in which they sat, and then, after studying this series for a few minutes, to commit it to memory, repeated it entire, both backwards and forwards, from the beginning to the end. These performances are believed to be not the result of any natural mnemonic powers, but of a method acquirable by any person in a course of twelve lessons.—*Abridged from the Times.*

FORMER HIGH TEMPERATURE OF EUROPE.

VEGETABLES appear to have formed the commencement of organic life on the earth. Their debris are the only things met with in the oldest beds deposited by water, and these belong to plants of the most simple structure,—ferns, reeds, and lycopodiums.

Vegetation becomes more and more complicated in the upper formations. Finally, near the surface, it resembles the vegetation of the present continents, but with this very remarkable addition, that certain vegetables which flourish only in the south, such as large palm-trees, are found, in a fossil state, in all latitudes, and even in the midst of the frozen soil of Siberia.

In the ancient world, these northern regions must have possessed, during winter, a temperature at least equal to that which is experienced at present in the parallels where large palm-trees begin to flourish. At Tobolsk, there was the climate of Alicant or Algiers!

We shall discover fresh proof of this mysterious result, from an attentive examination of the dimensions of plants.

There are, at the present day, species of reeds, of ferns, and lycopodiums, as well in Europe as the equinoctial regions; but it is only in warm climates that they are of great dimensions. Thus, a comparison of the dimensions of the same plants is, in fact, to compare, in reference to temperature, the regions where they were produced. Place beside the fossil plants of our coal formations, not the analogous European plants, but such as abound in those regions of South America most celebrated for the richness of their vegetation, and you will find the former incomparably larger than the latter.

The *fossil floras* of France, England, Germany, and Scandinavia, exhibit, for instance, ferns nearly 50 feet high, and with branches 3 feet in diameter, or 9 feet in circumference.

The *Lycopodineæ*, which, at the present time, in cold or temperate regions, are creeping plants, scarcely rising above the surface; which, even at the equator, under the most favourable circumstances, do not rise to more than 3 feet; reached in Europe, in the ancient world, to the height of 80 feet.

One must be blind, not to see, in these enormous dimensions, a new proof of the high temperature formerly possessed by our country before the last irruptions of the ocean.—*Arago's Eloge of Fourier; Jameson's Journal.*

INTERIOR OF THE EARTH.

ON Jan. 17, was read to the Royal Society, a paper, entitled, "On the State of the Interior of the Earth," by W. Hopkins, Esq., *Second Series.* "On the Phenomena of Precession and Nutation, assuming the Fluidity of the Interior of the Earth." * In this Memoir the au-

* An abstract of Mr. Hopkins's First Memoir will be found in the Year-Book of Facts, 1839, p. 103.

thor investigates the amount of the luni-solar Precession and Nutation, assuming the earth to consist of a solid spheroidal shell filled with fluid. This shell is first supposed to be bounded by a determinate inner spheroidal surface, in which the elasticity is equal to that of the outer surface; the change from the solidity of the shell to the fluidity of the included mass being, not gradual, but abrupt. Both the shell and the fluid are also supposed to be homogeneous, and of equal density. The author then gives the statement of the problem which he proposes to investigate; but the investigation itself is wholly analytical, and insusceptible of abridgment. The following, however, are the results to which he is conducted by this laborious process: namely, that on the hypothesis above-stated,—1, The precession will be the same, whatever be the thickness of the shell, as if the whole earth were homogeneous and solid. 2. The lunar nutation will be the same as for the homogeneous spheroid to such a degree of approximation that the difference would be inappreciable to observation. 3. The solar nutation will be sensibly the same as for the homogeneous spheroid, unless the thickness of the shell be very nearly of a certain value, namely, something less than one quarter of the earth's radius; in which case this nutation might become much greater than for the solid spheroid. 4. In addition to the above actions of precession and nutation, the pole of the earth would have a small circular motion, depending entirely on the internal fluidity. The radius of the circle thus described would be the greatest when the thickness of the shell was the least; but the inequality thus produced would not, for the smallest thickness of the shell, exceed a quantity of the same order as the polar nutation, and for any but the most inconsiderable thickness of the shell would be entirely inappreciable to observation.

In the *Third Series* of these Researches, the author extends his inquiry to the case in which both the interior fluid and external shell are considered as heterogeneous. His expectation that the solution of this problem would overturn the hypothesis of the earth's entire solidity was founded on the great difference existing between the direct action of a force on a solid and that on a fluid mass, in its tendency to produce a rotatory motion; for, in fact, the disturbing forces of the sun and moon do not tend to produce directly any motion in the interior fluid, in which the rotatory motion causing precession and nutation is produced indirectly by the effect of the same forces on the position of the solid shell. A modification is thus produced in the effects of the centrifugal force, which exactly compensates for the want of any direct effect from the action of the disturbing surfaces; a compensation which the author considers as scarcely less curious than many others already recognised in the solar system, and by which, amidst many conflicting causes, its harmony and performance are so beautifully preserved.

The solution of the problem obtained by the author destroys the force of an argument, which might have been used against the hypothesis of central fluidity, founded on the presuemd improbability of our being able to account for the phenomena of precession and nutation on this hypothesis, as satisfactorily as on that of internal solidity. The author conceives that he has demonstrated an important fact in the history of the earth, presuming its solidification to have begun at the surface; namely, the permanence of the inclination of its axis of rotation, from the epoch of the first formation of an exterior crust. This permanence has frequently been insisted on, and is highly important as connected with the speculations of the author on the causes of that change of temperature which has probably taken place in the higher latitudes: all previous proofs of this fact having rested on the assumption of the earth's entire solidity; an assumption which, whatever may be the actual state of our planet, can never be admitted as applicable to it at all past epochs of time, at which it may have been the habitation of animate beings.—*Proc. Royal Soc.; Philos. Mag.*

TEMPERATURE OF VEGETABLES.

On Nov. 18, M. Dutrochet read to the French Academy of Sciences a notice containing an account of new researches which he has made with the thermo-electric apparatus, on the Vital Heat of Plants. The results confirm the observations previously communicated by him to the Academy; whence it appears certain, that all plants have a proper and vital heat; that this heat exists especially in the green parts; and, lastly, that it undergoes a quotidian paroxysm, which reaches its maximum during the day, and its minimum during the night. The following are the most striking facts contained in the new communication: — The *Euphorbia lathyris* is a plant, a stalk of which has afforded the greatest degree of vital heat. The hour of the quotidian maximum is always the same for each plant. The hour varies from 10 o'clock in the morning till 3 in the afternoon, in different plants. No proper heat has been detected in ligneous tissue, even in that of recent formation. Complete darkness does not prevent the reproduction of the diurnal paroxysm of the proper heat of the stalks of plants, and that during some days; but it goes on gradually diminishing its intensity, till the complete extinction of this vital heat. —*Jameson's Journal.*

NEW CALORIMETER.

Dr. Ure, when engaged in some experiments for the Measurement of Heat disengaged in Combustion, has found that with the water calorimeters which he employed, the estimate of a portion of the heat could not be depended upon, in consequence of the combustion being kept up by the current of a chimney through which a quantity of the heat passed away. This defect Dr. Ure has entirely removed by means of a copper bath traversed with zig-zag flat pipes, and an enclosed furnace.

PHOTOMETRY.

Dr. Ure having failed in obtaining accurately the intensities of different lights, by a comparison of the relative shadows they project, —has employed the following photometric means. He placed several pieces of paper, prepared with the salts of silver as for photogenic drawings, in the rooms of a house darkened by a high wooden wall, or board, before the windows; and also in those of a neighbouring house not so circumstanced. In a certain time, those exposed to the action of free daylight acquired a certain depth of tint; and by observing the time required to produce the same tint on those papers placed in the darkened room, Dr. Ure was enabled to determine the amount of daylight so diminished. By photogenic impressions, Dr. Ure considers the relative degrees of diurnal illumination in different rooms in any house, in different countries, or on different days in the same house or country, also the extent or strength of daylight in any part of the world,—may be correctly measured and registered.—*Proc. British Assoc.; Literary Gazette.*

INTENSITY OF SOLAR LIGHT.

PROF. DAUBENY has exhibited to the British Association the model of an apparatus for obtaining a numerical estimate of the intensity of solar light, at different periods of the day, and in different parts of the globe. The contrivance consisted of a sheet of photogenic paper, moderately sensible, rolled round a cylinder, which, by means of machinery, would uncoil at a given rate, so as to expose to the direct action of the solar rays, for the space of an hour, a strip of the whole length of the sheet, and of about an inch in diameter. Between the paper and the light was to be interposed a vessel, with plane surfaces of glass at top and bottom, and in breadth corresponding with that of the strip of paper presented. This vessel, being wedge-shaped, was fitted to contain a body of fluid of gradually increasing thickness; so that, if calculated to absorb light, the proportion intercepted would augment in a gradually increasing proportion from one extremity of the vessel to the other. Hence it was presumed that the discoloration arising from the action of the light would proceed along the surface of the paper, to a greater or less extent, accordingly as the intensity of the sun's light enabled it to penetrate through a greater or less thickness of the fluid employed. The results were to be registered, by measuring, each evening, by scale, how many degrees the discoloration had proceeded along the surface of the paper exposed to light, during each successive hour of the preceding day. To render the instrument self-registering, some contrivance for placing the paper always in a similar position with reference to the sun, must, of course, be superadded. Mr. Jackson thought that a heliostat, for throwing the reflected light of the sun upon the instrument, would be objectionable; and suggested, in preference, that the heliostat should rather turn the instrument to the sun; an alteration to which Dr. Daubeny assented. Dr. Daubeny, in conclusion, observed that the indications of the scale were not intended to furnish absolute, but only relative, results.—*Athenæum.*

RESTORATION OF COLOUR BY LIGHT AND AIR.

IN the summer of 1838, there were exhibited in the Haymarket, London, certain Raphael tapestries, the colours of which were restored and altered by exposure to light and air, after they had faded during centuries of exclusion from these mighty agents. The proprietor of the tapestries, Mr. Trull, has since removed them to a finely situated factory in Coventry, where the changes have been as follow: the greens had all become blue, but have now almost returned to their original tints; the robes and full colours, from being dull and heavy, have become very brilliant, especially the gold; the flesh has recovered from pallid white to the high tint, deep shadow, and strong anatomical effect, of Raphael. A renewed freshness now reigns over the whole, and the clearing up of the light in many of the landscape parts is most extraordinary, giving a depth and breadth the cartoons themselves do not *now* convey, particularly in the Keys to St. Peter, St. Paul at Athens, and the Death of Ananias; where extensive landscapes, ranges of buildings, and foliage have sprung up on parts *quite obscured* when in London eight months back; much of which is either worn or torn out of Raphael's patterns at Hampton Court, and painted over, and known only through the means of these Leo tapestries. Again, Mr. Trull observes: "Some colours entirely change; others, in confusion, and apparently gone, by the *mere* effects of light and air, slowly and quietly resume the chief of their original tints. Flesh re-appears, hair on the head starts up; the grand muscular effect and unique power of expression, only found in Raphael and Michael Angelo, are finely developed where a few months back appeared a plain surface.—*Communicated by Prof. Faraday to the Philosophical Magazine.*

NEW SPECULUM.

A NEW, and it may turn out important, instrument has been produced on the following ingenious principle.

Mr. Nasmyth, in offering to the British Association a few remarks "On the Difficulties in the general Use of Metallic Specula for Reflecting Telescopes of very large size, in consequence of their excessive Weight, and of the great nicety required in casting and grinding them," drew attention to an invention of his, viz., a plate-glass pneumatic speculum. The dimensions of the plate were 3 feet 3 inches n diameter, and 3-16ths of an inch thick. It had been placed on a concave cast-iron bed, the edges only of the glass resting on a rim perfectly turned, and fastened in with bees'-wax, which rendered the apparatus air-tight, and was also of a yielding character. By removal of the air from behind the mirror, which Mr. Nasmyth effected with his mouth, sucking it out by a pipe 6 or 8 inches long, the surface of the glass, previously a plane, was pressed by the weight of the external atmosphere considerably out of the level; and by this means the focus of the mirror could be varied to any length. The form which the glass takes is, as it were, a curve of its own making, not exactly a parabola, but more like an ellipsoid. Mr. Nasmyth, however, conceived that the cast-iron back being turned to an exact figure of any kind, the glass might be made by this simple mode to lie flat on the metal, and again, at pleasure, to resume the plane form. He then named some of the advantages which he thought would result from the use of this contrivance for the specula of reflecting telescopes.—*Literary Gazette.*

THE SOLAR SPECTRUM.

On May 16, was read to the Royal Society a paper "On the visibility of certain rays beyond the ordinary red rays of the Solar Spectrum;" by J. S. Cooper, Esq. The author has observed an extension of the red portion of the Solar Spectrum, obtained in the ordinary way, beyond the space it occupies when seen by the naked eye, by viewing it through a piece of deep blue cobalt glass. He finds that the part of the spectrum thus rendered perceptible to the right is crossed by two or more very broad lines or bands; and observes that the space occupied by the most powerful calorific rays coincides with the situation of the red rays thus rendered visible by transmission through a blue medium.—*Proc. Royal Soc.; Philos. Mag.*

LAWS OF IRRADIATION.

M. Plateau has presented to the Brussels Academy a Memoir on Irradiation, with a view of putting an end to the uncertainties which still exist among astronomers and physicists as to the very existence of Irradiation; and of examining the causes of the phenomenon, its influence in astronomical observations, and the laws to which it is subject. The principal results of these inquiries may be thus summed up:

1. Ocular Irradiation is perceptible at all distances, from any separation whatever to the shortest distance of distinct vision.
2. Ocular Irradiation increases with the continued contemplation of the object.
3. Two ocular Irradiations near one another, which tend to act in a contrary direction, and to encroach upon each other, mutually destroy each other, and the more completely the nearer they are made to approach.
4. Ocular Irradiation varies considerably in different persons.
5. In observations made through astronomical glasses, that part of the total error which arises from ocular irradiation is dependent upon the enlargement of the brightness of the image, and on the greater or less degree of sensibility of the eye of the observer for irradiation.
6. This part of the total error necessarily vanishes in observations where a double image micrometer is used.
7. The part of the total error attributable to the aberrations of the glass necessarily varies with different instruments; but for the same glass it may be considered as constant, that is to say, independent of the magnifying enlargement.
8. The irradiation in glasses, or the total error arising from the ocular irradiation and from the aberrations of the instrument, is necessarily variable, since it depends upon variable elements.—Translated from the Memoirs of the Brussels Academy; *Philosophical Magazine.*

MICA FOR POLARIZING LIGHT.

Prof. Forbes has read to the British Association a paper "On the use of Mica in Polarizing Light;" and, after stating his method of employing it, expressed his opinion that it might be made very important towards explaining the nature of metallic reflection. Prof. Lloyd agreed that this process might throw light upon the internal structure of metallic substances, and in other respects considered it to be highly important. By polarizing light, either elliptically or circularly, in a superior way to any hitherto practised, it might lead to a termination of some of the difficulties which beset the inquiry. As Prof. Forbes had found, on exposing mica to heat, that exceedingly thin laminæ were separated, the surfaces of which furnished these reflections, he supposed it likely that all metallic plates presented similar laminæ, on which depended their polarizing influence. Prof. Powell confirmed these opinions, and adduced instances of such laminæ polarizing light, without affecting the image thrown upon the metal.—*Literary Gazette.*

NEW SOURCE OF LIGHT.

M. SEGUIN has communicated to the Academy of Sciences at Paris, a memoir on the distillation of animal substances, in which he states that he has reduced the process to such a degree of simplicity as to render it profitable for the sake of the products of the distillation. Thus, from the carcass of a horse he obtained, by destructive distillation, 700 cubic feet of gas, suitable for purposes of illumination, 24 lb. of sal ammoniac, and 33 lb. of animal black. The gas obtained was found to be composed of one part of olefiant gas and four of carburetted hydrogen; and might be preserved for months in contact with water without being in any way injured, or its brilliancy, as a combustible, impaired. M. Seguin found that 3,234 cubic inches of this gas, when burnt for one hour, gave twice and a half as much light as a Carcel lamp.—*Athenæum.*

NEW EXPERIMENTS ON LIGHT.

M. ARAGO has read to the French Academy of Sciences a note from M. Soliel, jun., claiming priority in all that relates to the construction of chromatic apparatus, for exhibiting, on a large scale, the experiments of polarized rays of light through crystalline laminæ. By aid of this apparatus, which is simple, and reasonable in price, 1500 persons have been enabled to see and admire the system of admirable fringes produced by the interference of rays. M. Soliel's instruments are equally adapted to the demonstration of the laws of double refraction, and their identity with luminous polarization. By interposing crystallized laminæ in the path of rays, polarized by tourmaline, it exhibits, on a large scale, the brilliant complementary colours discovered, in this case, by M. Arago. This same crowd of spectators witnessed the surprising results shown by crystals with one or two axes, which threw on a white screen, systems either circular, hyperbolic, or lemnisbates, curves tinted with 1000 colours, and of a brilliancy which no other process can even approach. Thus, all the magnificent images, obtained in France by Fresnel, Biot, and Arago; and in England by Brewster, Herschel, and Airy; are no longer cabinet phenomena, but may be exhibited to large assemblages.

M. Soliel has also discovered a very curious effect produced by the dilatation of carbonate of lead on its optical axes and their superb curves.

ROCK SALT THREAD.

M. GAUDIN has submitted to the French Academy of Sciences, a process for converting Rock Salt into ductile thread, to be employed as a substitute for the uncertain metallic wire now used in the Torsion balance, and which is rendered uncertain by its variable elasticity in supporting the electric or magnetic needle. Other applications of this new thread are proposed, as in giving mathematical certainty to the oscillations of the pendulum. The importance of M. Gaudin's discovery, as relates to the chemical sciences, is also great. There is a class of phenomena—those that have reference to the internal organization or molecular juxta-position of bodies, with the nature of which we are but imperfectly acquainted. The modern discoveries respecting light and colours show that certain effects result from the properties of the molecules themselves, others from their agglomeration : and, it is more than probable that this discovery will throw farther light on the subject.—*Athenæum.*

HEAT IN LIQUIDS.

THE experiments of M. Despretz concerning the propagation of heat in liquids have been attended with the most satisfactory results. From these it appears, that a liquid column being heated at the upper part, the heat is propagated according to the same laws as those belonging to solid bodies: that the temperature decreases from the axis to the surface, and from the surface to the wall of the side. The depth to which solar heat penetrates in a given time may be easily calculated in these data, in large seas, and isolated seas.—*Athenæum.*

TEMPERATURE OF FLOWERS.

MM. VAN BEEK and BERGSMA have been making some curious observations on the temperature of the flowers of the *Colocasia odora,* with the thermometric needles of MM. Becquerel and Breschet. On the 5th of September, 1838, the spadix had acquired the extraordinary temperature of 108° Fahr., while the atmosphere stood at 70°, thereby forming a difference of 38°.—*Athenæum.*

THE DAGUERREOTYPE—PHOTOGENIC DRAWING.

THIS invention was noticed in the *Year-Book of Facts,* 1839, (p. 29). During the past year, every scientific journal in England has teemed with contributions to the history of "the New Art," and to its practical details; which have likewise been made the staple of many lectures, and papers read before public societies. Valuable as may be these illustrations, it becomes impossible to adapt them to our pages; since they would occupy a considerable space, altogether disproportionate to their intrinsic value; especially when it is added that at the moment we are writing, (January, 1840,) the processes of the art are in course of successive improvement, and new adaptations of the discovery are constantly being discovered.

The most important fact, however, is the publication of M. Daguerre's own Account of his Discovery, in Paris; and, within a few days, the appearance of a translation of the same in England, entitled

"History and Practice of Photogenic Drawing on the true Principles of the Daguerréotype." (Smith, Elder, and Co., Cornhill.) The contents are classified in four chapters. The first comprises the Bill for rewarding the inventors, granting to M. Daguerre an annual pension for life of 6,000 francs, (250*l.* sterling); to M. Niepce, jun., a similar pension of 4,000 francs, (166*l.* 13*s.* 4*d.*); these pensions being one-half in reversion to the widows of the inventors. Next is the bill presented by the Minister of the Interior, M. Duchâtel, to the Chamber of Deputies, in June last, proposing the above grants, explaining the partnership of MM. Daguerre and Niepce, and bearing M. Arago's guarantee for accuracy. The commission appointed to examine the discovery were the following members of the Chamber, MM. Arago, Etienne, Carl, Vatout, de Beaumont, Tournouer, Delessert (François), Combarel de Leyval, and Vitet, all names distinguished in science. The special commission of Peers was composed of Barons Athalin, Besson, Gay-Lussac, the Marquis de Laplace, Vicomte Siméon, Baron Thénard, and the Comte de Noé. The next documents are Arago's very minute and interesting Report to the Deputies, and a similar Report from the Special Commission to the Peers.

Chapter II. includes the practical and historical details of the invention; and relates the partnership of Niepce and Daguerre; with experiments and improvements. A proof of Daguerre's priority to Niepce is the application of iodine, which constitutes the great distinction between the processes of the two experimenters; "in a word," says the translator, "between the approximation and the real principle."

Chapter III. is, however, still more practical in its details, as a quotation will show:—The designs are executed upon thin plates of silver, plated on copper. Although the copper serves principally to support the silver foil, the combination of the two metals tends to the perfection of the effect. The silver must be the purest that can be procured. As to the copper, its thickness ought to be sufficient to maintain the perfect smoothness and flatness of the plate, so that the images may not be distorted by the warping of the tablet; but unnecessary thickness beyond this is to be avoided, on account of the weight. The thickness of the two metals united ought not to exceed that of a stout card.

The process is divided into five operations:

1. The first consists in polishing and cleaning the plate, in order to prepare it for receiving the sensitive coating, upon which the light traces the design.

2. The second is to apply this coating.

3. The third is the placing the prepared plate properly in the camera obscura to the action of light, for the purpose of receiving the image of nature.

4. The fourth brings out this image, which at first is not visible on the plate being withdrawn from the camera obscura.

5. The fifth and last operation has for its object, to remove the sensitive coating on which the design is first impressed, because this coating would continue to be affected by the rays of light, a property which would necessarily and quickly destroy the picture.

The operations are then minutely described; under the third of which is illustrated the time necessary for producing a design, which depends entirely on the intensity of light on the objects, the imagery of which is to be re-produced. At Paris, for example, this varies from three to thirty minutes.

The several instructions are illustrated by six pages of outline diagrams of the requisite apparatus. Thus, Plate I. shows the wire frame for supporting the plate while heating; the "plate of plated silver" on which the design is made; the board upon which the plate is laid; the spirit-lamp, and the muslin bag, with pumice powder for polishing. Plate II. shows the box for iodine, used in the second operation; and a grooved case for preserving the plates from injury. Plate III. "represents four different positions of the frame into which the plate with its wooden tablet is put, on removal from the iodine process;" the objects being to adapt the plate to the camera obscura, and to protect the iodine coating from the action of light till the moment in which it receives the focal image. Plate IV. shows the camera obscura, as adapted to photogenic delineation. Plate V. represents three views of the apparatus for submitting the plate to the vapour of mercury; a kind of case, provided with a spirit-lamp, and a thermometer on one side to denote the rate of the process. Plate VI. shows various apparatus for the last operation of washing the plate: as three troughs, with the plate placed therein; the funnel for filtering the saline wash; a little hook for shaking the plate while in the wash; and a wide-mouthed vessel for warming the distilled water. The fourth and last chapter elucidates another new art, invented by Daguerre—the principle and practical details of Dioramic Painting.

The *recipe* is briefly as follows:—A copper-plate plated with silver, its surface well cleansed with diluted nitric acid, is exposed to the action of the vapour of iodine: this forms the first coating, or ground; which is inconceivably thin, and requires to be perfectly even. The plate thus prepared is placed on the table of the camera obscura; and after remaining eight or ten minutes, according as the subject or the degree of light may require, is withdrawn. At this stage of the process, however, the most practised eye will not discern the slightest trace of the action of the light on the prepared surface. The plate is then exposed, in a proper apparatus, to the vapour of mercury; and, when heated to sixty degrees, the picture appears as if by magic. A singular, and hitherto inexplicable circumstance, requires to be noticed in reference to this part of the process; viz., that the plate must be in an inclined position; and that, if it be placed directly opposite the aperture whence the vapour of the mercury escapes, the result will not be satisfactory. Lastly, the plate must be dipped in hyposulphate of soda, and afterwards well sluiced with distilled water: the operation is then complete.

The first experiment made in this country with the Daguerréotype, was exhibited by M. St. Croix in London, on the 13th of September. The place of exhibition was No. 7, Piccadilly, nearly opposite the southern crescent of Regent-street; and the picture produced was a beautiful miniature representation of the houses, pathway, sky, &c., resembling an exquisite mezzotint.

In the *Philosophical Magazine*, November, Mr. Towson explains an important fact which had hitherto escaped observation. It appears that M. Daguerre does not use an achromatic lens; and that the focus he employs is obtained by advancing or withdrawing the frame of the obscured glass, until he obtains the outlines of the subjects with the greatest neatness. This method would be most correct, if the chemical rays were identical with the luminous rays. If such were the case, the effect produced on this plate would be precisely that which had appeared on his obscured glass. But, it is a well known fact, that the chemical rays are more susceptible of refraction than the luminous rays; wherefore, in order to obtain the neatest effect, it is indispensable that the camera should be adjusted to the focus of the chemical rays. The author then details his improved method, by which the time of exposing the plates in the camera may be reduced from three or five minutes to ten or twelve seconds. The use of larger lenses, which the correction of the focus enables him to adopt, would, he considers, render the Daguerréotype applicable to taking portraits from life; allowed to be a great desideratum, in the discussion at the Institute, after M. Arago had announced Daguerre's process.

Dr. Donné is stated to have applied, with success, the ordinary process of engraving directly to the proofs taken with the Daguerréotype; a discovery almost as important as the invention of the apparatus itself.

For preserving the impressions of the Daguerréotype, M. Dumas has found to succeed, better than any other composition, a varnish (one part dextrine to five of water) applied hot to the plate. The photogenic impression will be thus free from any danger of friction, and may be copied by means of transparent paper, &c. M. Sylvester has employed a composition of dextrine, two parts; water, six parts; and alcohol, one part: but, both compositions interfere with the brilliancy of the plate.

M. Arago has since stated that instead of placing the iodine in the box with the plate, the latter is first impregnated with the vapour, and this is placed in a flat box, within half an inch of the plate on which the drawing is to be taken. The box is then to be shut, and in two minutes the silver plate will have acquired the proper tint.

Dr. Schafhentl, of Munich, has exhibited a new process of producing photogenic drawings combining Daguerre's minuteness with the light and shadow of an original drawing, by means of Indian ink. The preparation of these new photogenic plates is, however, as yet too complicated for popular practice.

⁂ Among the more popular as well as important papers on the New Art, are letters from Mr. Bauer, F.R.S., and M. Niepce, on the originality of the invention; reported in the *Literary Gazette*, 1839, p. 137: also, an elaborate paper by Mr. A. Smee, at p. 314 of the same journal. A cheap and simple method of preparing papers for photogenic drawings, in which the use of any salt of silver is dispensed with, has been communicated to the Society of Arts for Scotland, by Mr. Mungo Ponton, F.R.S.L., and is reprinted in the *Literary World*, Vol. i. p. 254. In the Reports of the Proceedings of the British Association will be found many important observations by Mr. H. F. Talbot, and Sir John Herschel; the latter describing a remarkable property in the extreme red rays of the prismatic spectrum, which had occurred to him in his experiments on Mr. Talbot's photogenic paper. The *Athenæum* for the year contains, in addition to the reports of papers read to Societies, experiments and new preparations, by correspondents; too numerous to particularize. In *Jameson's Journal*, Vol. xxvii., will be found notes by Sir John Robinson, p. 155; and observations by Dr. Fyfe, p. 144. In the *Philosophical Magazine*, Vol. xiv., p. 196, will be found Mr. H. Talbot's "Account," extending through fifteen pages; at p. 365, of the same volume is a "Note" communicated to the Royal Society by Sir John Herschel: and at p. 475, are extracts from the proceedings of the Royal Irish Academy, proposing the light of incandescent coke to blacken photogenic paper, as a substitute for solar light, or that from the oxy-hydrogen blow-pipe with lime. And in the *Philosophical Magazine*, Vol. xv., p. 381, will be found Mr. Towson's paper on the proper focus for the Daguerreotype, above quoted. An article on the application of the art to botanical purposes by Dr. Golding Bird, appears in the *Magazine of Natural History*, April, 1839. In the *Mechanics' Magazine*, Vol. xxx., p. 329—455, are letters from Sir Anthony Carlisle, and Mr. T. Oxley; at p. 428, a letter from Mr. Egerton Smith; and an Abstract of Daguerre's Treatise, p. 465—471, &c.

OPERATION OF POISONS.

On June 13, was read before the Royal Society, "a Series of Experimental Researches on the mode of operation of Poisons," by J. Blake, Esq. In this paper, the author examines more particularly the action of those poisons which appear to produce death by affecting the nervous system. After reviewing the evidence adduced in support of the opinion that the effects of some poisons are owing to an impression made on the nerves of the part to which they are directly applied, he proceeds to relate a series of experiments undertaken in order to show with what rapidity the blood is circulated through the body, and tending to prove that a substance may be generally diffused through the system in nine seconds after its introduction into the veins. Experiments are then related in which the more rapidly fatal poisons had been used, and in which it was found that an interval of more than nine seconds always elapsed between the administration of a poison and the appearance of the first systems of its action. The mere contact of a poison with a large surface of the body appears to be insufficient to give rise to general effects, as long as it is prevented from entering into the general circulation. Various causes of fallacy in experiments of a similar kind, which have been adduced in support of an opposite opinion, are pointed out. The following is a summary of the conclusions arrived at by the author:

1. The time required for a substance to penetrate the capillary vessels may be considered as inappreciable.
2. The interval elapsing between the absorption of a substance by the capillaries, and its general diffusion through the body, may not exceed nine seconds.
3. An interval of more than nine seconds always elapses between the introduction of a poison into the capillaries, or veins, and the appearance of its first effects.
4. If a poison be introduced into a part of the vascular system nearer the nervous centres, its effects are produced more rapidly.
5. The contact of a poison with a large surface of the body is not sufficient to give rise to general symptoms, as long as its diffusion through the body is prevented.—*Athenæum.*

Electrical Science.

VOLTAIC COMBINATIONS.

On May 30, was read to the Royal Society, the "Fifth letter on Voltaic Combinations; with some account of the effects of a large Constant Battery;" by J. F. Daniell, Esq. The author, pursuing the train of reasoning detailed in his preceding letters, enters into the farther investigation of the variable conditions in a voltaic combination on which its efficiency depends: and the determination of the proper proportions of its elements for the economical application of its power to useful purposes. He finds that the action of the battery is by no means proportioned to the surfaces of the conducting hemispheres, but approximates to the simple ratio of their diameters; and hence concludes that the circulating force of both simple and compound voltaic circuits increases with the surface of the conducting plates surrounding the active centres. On these principles, he constructed a constant battery consisting of seventy cells in a single series, which gave, between charcoal points, separated to a distance of three-quarters of an inch, a flame of considerable volume, forming a continuous arch, and emitting radiant heat and light of the greatest intensity. The latter, indeed, proved highly injurious to the eyes of the spectators, in which, although they were protected by grey glasses of double thickness, a state of very active inflammation was induced. The whole of the face of the author became scorched and inflamed, as if it had been exposed for many hours to a bright midsummer sun. The rays, when reflected from an imperfect parabolic metallic mirror in a lantern, and collected into a focus by a glass lens, readily burned a hole in a paper at a distance of many feet from their source. The heat was quite intolerable to the hand held near the lantern. Paper steeped in nitrate of silver, and afterwards dried, was speedily turned brown by this light: and, when a piece of *fine* wire-gauze was held before it, the pattern of the latter appeared in white lines, corresponding to the parts which it protected. The phenomenon of the transfer of the charcoal from one electrode to the other first observed by Dr. Hare, was abundantly apparent; taking place from the *zincode* (or positive pole) to the *platinode* (or negative pole). The arch of flame between the electrodes was attracted or repelled by the poles of a magnet, according as the one or the other pole was held above or below it; and the repulsion was at times so great as to extinguish the flame. When the flame was drawn from the pole of the magnet itself, included in the circuit, it rotated in a beautiful manner. The heating power of this battery was so great as to fuse, with the utmost readiness, a bar of platinum, one-eighth of an inch square: and the most infusible metals, such as pure rhodium, iridium, titanium, the native alloy of iridium and osmium, and the native ore of platinum, placed in a cavity scooped out of hard carbon, freely melted in considerable quantities. In conclusion, the author briefly describes the results of some experiments on the evolution of the mixed gases from water in a confined space, and consequently under high pressure; with a view to ascertain, first, in what manner conduction would be carried on, supposing that the tube in which the electrodes were introduced was quite filled with the electrolyte, and there were no space for the accumulation of the gases; secondly, whether decomposition having been effected, recombination would take place at any given pressure; and, lastly, whether any re-action on the current-force of the battery would arise from the additional mechanical force which it would have to overcome.—*Proc. Royal Soc.; Athenæum.*

ELECTRO-CHEMICAL PROTECTION OF METALS FROM CORROSION.

PROF. SCHŒNBEIN has read to the British Association a paper "On the Theory of the Electro-chemical Protection of Metals from Corrosion." After glancing at the theories of electro-chemistry proposed by Berzelius and Sir H. Davy, the Professor adduced a numerous series of experiments, bearing upon the fact of the apparent necessary existence of a current to prevent the corrosion of the lesser oxydizable metal; for when pieces of copper and zinc were placed in metallic connection, and immersed in vessels of brine not communicating with each other, the copper was invariably more or less attacked, unless a portion of *both* metals was immersed in each vessel: when, however, the copper and zinc legs of a compound arc were immersed in communicating vessels, the zinc was attacked, and the copper perfectly protected from corrosion. This, and other facts bearing upon it, being proved, the Professor considered the possibility of any development of polar properties influencing the behaviour of metals towards corroding liquids: this idea he experimentally proved to be untenable, as he had in vain sought for any case in which the chemical affinities of bodies were materially affected by their elastic conditions. Prof. Schœnbein stated that he did not believe the protection of copper from the action of sea-water, by means of zinc or iron, depended upon the existence of a continuous current of feeble intensity, continually circulating through the metals, but rather on the induction of a kind of passive state in the copper, in the following manner:—A piece of copper being placed in metallic communication with a piece of zinc, and immersed in sea-water, slowly develops a current, which, although of feeble intensity, is, nevertheless, of sufficient electrolytic effect to decompose a portion of water, resolving it into its constituent elements; the oxygen combines with the zinc; the hydrogen being evolved, and becoming diffused over the surface of the copper, to which it adheres, this gaseous layer affords a mechanical protection from the corroding influence of the sea-water. The Professor drew some analogy between this state and the extraordinary passive state of iron, under certain conditions, towards nitric acid, previously described by him. In this case, Prof. Schœnbein thought it not improbable that peroxide of hydrogen became developed, and formed the protecting agent to the iron wire.—*Literary Gazette.*

ELECTRICAL CURRENTS IN METALLIFEROUS VEINS.

PROF. REICH has reported to the British Association some experiments similar to those of Mr. Fox and Mr. Henwood in Cornwall; and the electrical current in mines of copper or lead as subjected to tests in the mine of Himmelsfurst near Freyburg. The two points to be connected were wrought to a fresh surface, and a copper disc, three inches wide, pressed tightly upon them, by means of a wooden block. Upon this plate, the bare end of a copper wire, the rest being spun over with silk, was held by a clamp; and a long wire was reeled off, till it came into contact with the second point, where Schweigger's delicate multiplier ascertained the effect of the current as follows:—"Two ore points, separated by a non-metalliferous mass, a cross vein, or a vein wrought out, give rise to an electric current in the metallic wire connecting them. Two ore points, in uninterrupted metallic connection with each other, induce no electrical current through a wire connecting them. If only one disc be connected with an ore point, and the other with the timbering, or be held in the hand, there is no effect produced on the multiplier. If an ore point be connected with masses of ore already won, a current sometimes manifests itself, and sometimes there is none. When an ore point is connected with non-metalliferous rock, frequently no current takes place; frequently, however, a feeble current occurs in the connecting wire. (This result does not agree with that of Fox and Henwood, who never detected a current; but Prof. Reich performed the experiment eighteen times, and always obtained the same result.) With respect to the cause of the electrical currents, observed in metalliferous veins, three different opinions have been broached. They have been ascribed, 1, to general electric currents at the earth's surface, produced either entirely, or in part, by the earth's magnetism; 2, to hydro-electric, and, 3, to thermo-electric, actions of the various metallic components of the vein. The first hypothesis, according to Reich, is refuted by the independence of the direction of the currents on their position relatively to the earth's axis. Thermo-magnetism he holds to be incapable of producing such strong currents: as the strongest currents are observed exactly where the two points were separated by a non-metallic conductor; and he concludes that there remains only the hydro-electric action of the metallic components of the vein to account for the phenomena."—*Literary Gazette.*

ELECTRIC GIRLS.

Two Girls, natives of Smyrna, have exhibited themselves in Marseilles, and other towns of France. The girls stationed themselves facing each other, at the end of a large table, keeping at a distance of one or two feet from it, according to their electric dispositions. When a few minutes had elapsed, a crackling, resembling electricity upon a sheet of gold paper, was heard; when the table received a strong shake, which always caused it to advance from the elder to the younger sister. A key, nail, or any other piece of iron, placed on the table, instantaneously stopped the phenomenon. When the iron was adapted to the under part of the table, it produced no effect upon the experiment. Saving this singular circumstance, the facts observed constantly followed the known laws of electricity, whether glass insulators were used, or whether one of the girls wore silk garments. In the latter case, the electric properties of both were neutralized. Such were the results for some days after the arrival of the young Greeks; but the temperature having become lower, and the atmosphere charged with humidity, all perceptible electricity left them.—*Marseilles Letter, Sept.* 1839.

SHIP STRUCK BY LIGHTNING.

AFTER the mainmast of H. M. S. *Rodney* was struck by lightning in her passage from Athens to Malta, the broken hoops surrounding it were all found to be magnetized, in the same uniformity of direction as if they had been operated on in one direction also by the galvanic helix. Thus, in a hoop broken in two athwart-ships (speaking with reference to the ship's head), the larboard end of the foremost portion was a south and its starboard end a north pole; the end of the aftermost portion in contact with the south pole of the foremost portion being, consequently, a north pole, and the other end thereof a south; and so uniformly with all the other hoops, at whatever part they were broken; similar poles in each hoop always pointing in similar directions in the circumference of the respective circles.—*Athenæum.*

PROTECTING EFFECTS OF GREAT FIRES IN THUNDER-STORMS.

WHEN, seven years ago, M. Arago alluded, in *éloge* of Volta, to the ideas of that illustrious philosopher concerning the advantages which might be derived from great flames during thunder-storms,—he "conceived that we might obtain on this point some information by comparing the meteorological observations made in those counties of England, where so many high furnaces and workshops transformed night and day into great oceans of fire, with the agricultural ones which surrounded them." This comparison has been made, and it has been found that the agricultural districts distinctly afford evidence of a larger number of such storms than the mineral ones; and yet M. Arago does not consider the question settled. The great furnaces in England abound especially where there are many metallic mines; and the rare occurrence of thunder-storms in these localities may, probably, with more truth be ascribed to the *nature of the soil* than to the action of those enormous fires which the working of the minerals requires.—From an elaborate paper on Thunder and Lightning, in the *Annuaire du Bureau des Longitudes,* 1838; translated in *Jameson's Journal,* Nos. 51 and 52.

THE GYMNOTUS AND TORPEDO.

ON Jan. 18, at the Royal Institution, Prof. Faraday lectured "On the Gymnotus and Torpedo." The gymnotus, or electric eel, has a large bull head, with a long tapering body, underneath which are most beautiful fringes or abdominal fins. By these are obviated the tortuous progress of the common eel, and the movements of the gymnotus are rendered exceedingly graceful. The essential organs of this creature, those requisite for vital functions, are situated immediately behind the head (within a short space of which the alimentary canal begins and ends), and are very small in proportion to the electrical organs. The latter occupy the whole remaining length. They consist of four separate organs; a large one on each upper side, and smaller corresponding ones underneath. Their structure is cellular, intersected with horizontal laminæ, and composed of matter different from any other in the body of the fish. The nerves that lead from the brain and spinal marrow to the electrical organs, are enormous in comparison with those that supply the fluid, or force, or influence necessary for vitality. They may be cut through, and thus the electrical organs separated from the vital parts; the creature still lives and flourishes, and becomes even more lively than when in its natural state, and when it may be said to be inconvenienced by the large demand of the organs for the nervous fluid, or the something which is doubtless used, worked up in them to produce electrical effects. In proportion to the using up of this nervous influence is the exhaustion of the eel after every electrical discharge. Now, in all common electrical phenomena, the converse can be obtained. For instance, heat from electricity, electricity from heat, chemical action from electricity, and *vice versâ*; light from electricity; but the opposite, it is to be regretted, is as yet more an assertion than an established experimental fact. However, enough is known to support the axiom above stated. Neither can be named as the original cause: each is cause and effect; the thing produced can be converted into the thing producing; they all may be the effects of one common cause, but, to the knowledge of this, experimental philosophy has not yet come. Upon this principle of reconversion, then, will future investigations be conducted; and the expectations are, that, after having completely exhausted the fish, by passing a current of fluid through its electrical organs, and in a contrary direction to that in which it flows in the fish, experimenters will be able to bring back this nervous influence, to reconvert electrical fluid into nervous fluid, or natural substance, and thus arrive at a knowledge of the nature of that wonderful agent.—*Literary Gazette.*

THE GYMNOTUS.

On Oct. 1, was read to the Electrical Society, a communication from Mr. Gassiot, "On the Attractive Force manifested by the Electricity of the Gymnotus Electricus." In a former paper, Mr. Gassiot had submitted to the Society experiments of which, with the aid of Mr. Snow Harris's beautiful apparatus, the thermo-electrometer, heating effects had been obtained from the electric action of the gymnotus. The present one describes experiments by which attractions of gold leaves were manifested; it being, Mr. Gassiot believed, the first instance on record of such an effect from animal electricity, and consequently enabling the electrician to fill up that blank in Faraday's celebrated table of identity in the different descriptions of electrical action. Attraction was evidenced with Bohnenberger's single leaf electroscope; and also, but much more satisfactorily, with an electroscope made of a common glass tumbler, placed in an inverted position, having two small holes drilled in the opposite sides, through which passed two wires with brass balls attached: to each ball a gold leaf was fixed, about one inch long and one-eighth of an inch wide, the leaves being placed parallel to each other. The leaves were approximated as close as the eye could observe, without contact, or about 1-30th to 1-40th of an inch apart; and, on contact with the eel being made, they were not only attracted with considerable force, but were actually fused, scintillating in the most beautiful manner.

Extract of a letter from J. Samo, Esq., dated Surinam, 31st Dec. 1838. "The Gymnotus Electricus is found in the rivers of Surinam. It is very difficult to select the true from the fabulous accounts of this creature. It is generally found in shallow rivulets of fresh water, having a rocky, uneven bottom, and principally in those parts which are shaded by high trees. One of the three which Mr. Samo, by the first opportunity, intended forwarding to the Electrical Society, had been attacked and destroyed by water-rats; which animals, Mr. Samo concludes, must be insensible to the electric shocks of the gymnotus, although horses can be paralyzed by them." Mr. Clarke suggested a trial with the one in the Adelaide Gallery, at the same time observing, that fish frequently recover from the shock of the gymnotus; that in all probability the first rats that attacked the eel had been paralyzed, and had afterwards recovered; but that the gymnotus, becoming exhausted by the repeated electrical discharges to repel the first assailants, had fallen an easy prey to the others.

CONSTANT BATTERY OF THIRTY ELEMENTS.

On May 7, was read to the Electrical Society, a paper detailing experiments made by Mr. Mason, and repeated before the Society, with his constant battery, consisting of thirty elements. First, the vegetation of carbon, on the positive pole with the water battery, was reversed with the constant, and produced on the negative. The deflagration of mercury with both the electrodes followed: the positive producing a red light and a chattering noise; the negative, a bluish light, throwing off dense fumes of the oxide. One-eighth of an inch of flame was gained on making contact with the charcoal electrodes, after having been placed for a second or two in bichromate of potash; and it obeyed the same law of rotation on the poles of a magnet, as under the ordinary circumstance of simple charcoal points. Mr. Mason considered that thus the flame may be assisted for experiment, without in any way diminishing the accuracy of the results. In an attempt to procure an amalgamation of chrome and potassium, or a triple compound of chromium, potassium, and mercury, the two latter combined; but the former appeared to be at the positive electrode, and scintillated as iron. An amalgam of mercury and potassium was next procured. After this experiment, on taking the negative electrode out of the mercury, and throwing that which adhered to it into water, it instantly took fire, proving, according to Mr. Mason, that a portion of the potassium must have passed through the mercury without combining with it, and attached itself to the negative electrode. In carrying out experiments with the constant battery, Mr. Mason recommends that one solution should be exhausted before another is supplied. —*Literary Gazette.*

THEORY OF THE VOLTAIC CIRCLE.

PROF. GRAHAM has explained to the British Association the views now general as to the propagation of electrical induction through the fluid and solid elements of the voltaic circle, by no circulation of the electricities, but by their displacement in the polar molecule. In a molecule of hydrochloric acid, for instance, the electricities are displaced, the positive goes to the atom of chlorine, and the negative to that of the hydrogen, and each retains the chemical affinity of the chlorine and hydrogen respectively. Prof. Graham wished to refer the phenomena directly to the chemical affinities, and to omit entirely the idea of electricities being possessed by the molecules. In short, he observed the term electricity, or chemical affinity, must give way. He then took hydrochloric acid as the type of exciting fluids; he gave to each molecule a pole, the one with an affinity for chlorine, which he termed *chlorous* affinity, instead of negative electricity; and the other with an affinity like that of zinc or hydrogen, which he termed *zincous* affinity instead of positive electricity. Extensive diagrams illustrated his views; but the purport of his paper was to show that an advantage would be gained by change of terms, and that the condition of the terminal affinities would convey more readily the states of the elementary bodies. The change in the nomenclature proposed was as follows: —

Synonymes.

Zincous	Positive.
Chlorous	Negative.
Zincoid	Positive pole. Anode, zincode.
Chloroid	Negative pole. Cathode, platinode.

The term zincoid, Prof. Graham observed, means what possesses the properties of zinc, quasi zinc; and chloroid, what possesses the properties of chlorine, quasi chlorine. Zinc was the head of its class; and of all salified bodies, chlorine was the head. They were types of their families. The same views were applicable to electrolytes, the elements of which he also wished to be called zincous and chlorous. Thus the concluded induction is an assumption, and we get rid of positive and negative electricity.—*Ibid.*

SMALL BUT VERY POWERFUL VOLTAIC BATTERY.

MR. W. R. GROVE, M.A., has constructed a small battery consisting of seven liqueur-glasses, containing the bowls of common tobacco-pipes; the metals, zinc and platinum; and the electrolytes, concentrated nitric and dilute muriatic acids. This little apparatus has produced effects of decomposition, equal to the most powerful batteries of the old construction. (See *Philosophical Magazine*, May, 1839, p. 388.) Mr. Grove has since tried various combinations upon the same principles, and though some of the rarer substances, such, for instance, as chloric acid, have produced powerful combinations, he has found none superior, and few equal, to the above. Mr. Grove has, therefore, economized the materials: thus, on the side of the zinc, salt and water has been found little inferior to dilute muriatic acid; and it dispenses with the amalgamation of the zinc. By using flattened parallelopiped shaped vessels, instead of cylindrical, the concentrated acid is much economized, the space diminished, and the metals approximated. (According to Prof. Ritchie, the power is inversely as the square root of the distance between the metals.)

A hastily-constructed battery, upon this principle, has been presented to the British Association. It consists of an outer case of wood, (glazed earthenware is better), 7½ in. by 5 and 3, separated into four compartments by glass divisions; into which are placed four flat porous vessels, measuring, in the interior, 7, 2½, and 3-10ths of an inch: they contain each 3 ounces, by measure; the metals, four pair, expose each a surface of 16 square inches; and the battery gives, by decomposition of acidulated water, 3 cubic inches of mixed gases per minute; charcoal points burn brilliantly; and it heats 6 inches of platinum wire 1-56th of an inch diameter; its effect upon the magnet, when arranged as a single pair, is proportionately energetic; it is constant for about an hour, without any fresh supply of acids. The porous vessels are identical in their constitution with the common tobacco-pipe. Its power, with reference to the common constant battery, is, *cæteris paribus*, as 6 to 1, but the proportions vary with the series. The cost of the whole apparatus is about 2*l*. 2*s*. During the operation of this battery, the nitric acid, by losing successive portions of oxygen, assumes, first, a yellow, then a green, then a blue colour, and, lastly, becomes perfectly aqueous; hydrogen is now evolved from the platina; the energy lowers, and the action becomes inconstant. This valuable instrument of chemical research is here made portable; and, by increased power in diminished space, its adaptation to mechanical, and especially to locomotive purposes, becomes more feasible.—*Ibid.*

FLAME OF A CANDLE AND VOLTAIC BATTERY.

On April 16, was read to the Electrical Society, a paper "On the action produced in the flame of a Candle by the Voltaic Battery," by Mr. Gassiot. Dr. Paris, in his *Life of Sir Humphry Davy*, states that in the "Laboratory Register of the Royal Institution," October 6, 1807, there is described a beautiful experiment of Sir Humphry's—"that of producing the vegetation of the carbon of the Wick of a Candle by placing it between the wires of a voltaic battery." This experiment has been repeated by Mr. Gassiot, who says: "it is difficult to describe the beautiful appearance displayed while the carbon is depositing on the electrodes." The result of Mr. Gassiot's experiments is, that when the ends of the terminal wires or electrodes of a voltaic battery are introduced into the flame of a candle, certain constant and distinct effects are produced on each electrode, although no evidence of the completion of the circuit can be detected, either by a delicate galvanometer, or by the evolution of iodine from hyd. potass. On the other hand, while with single plates, electro-magnetical and electro-chemical actions are distinctly shown, and are so easily developed, no effect can be produced on the carbon of a candle. The battery used by Mr. Gassiot consisted of one hundred cells, each containing one quart of rain-water: it was composed of the usual elements, copper and zinc rolled one within the other; metallic contact being prevented by linen interposed between the plates.—*Literary Gazette.*

LENZ'S GALVANIC APPARATUS.

It has appeared to M. Lenz to be especially necessary to be enabled accurately to ascertain the force of the galvanic stream at each instant, and on each experiment; the practicability of which M. Lenz has established to his satisfaction, by means of the multiplier of M. Nervander Von Helsingfors. His first endeavour was, therefore, to procure a similar instrument, and the most perfect of its kind.

M. Lenz has discovered, by experiments, that all kinds of native copper act upon the magnets of other multipliers; although by chemical test, he could discover no trace of the presence of iron. Hence he is convinced that copper is slightly magnetic, or, to follow the expressed opinion of Faraday, that the ordinary temperature of about 65° Fahr. is not sufficiently high to deprive copper of its magnetism. M. Lenz has found brass to exhibit, in the least degree, this counteracting power; and his apparatus constructed of this material bids fair to answer the desired end, as far as may be judged by the effect upon a single needle. On account of its great flexibility, silver has been chosen for the winding wire. Appended to the paper whence these details have been obtained are several interesting remarks on certain points in the doctrine of galvanism; for which see the translation from Poggendorff's *Annalen der Physik; Polytechnic Journal,* No. 4, pp. 302—306.

PLACING MAGNETS.

Prof. Lloyd has observed to the British Association, that the phenomena of terrestrial magnetism could not be determined by one magnet. His researches had, therefore, been directed to ascertain the best position of three magnets, so as in the least degree to affect each other, and work out this problem. The learned Professor then went into a demonstration of the fundamental theorem included in this proposition, and gave a formula, which he illustrated by diagrams on the board beside him, and of which we give the substance. One magnet should be placed in the line of the magnetic meridian, and is termed the declination magnet; the second, perpendicular to it, representing the horizontal force; and the third, representing the vertical force, at the angle opposite the base of the other two. The limits of distance can, at present, be determined by experiment only. By the positions of these three magnets, we have four indeterminate arbitrary angles, by which we are enabled to fulfil four equations of conditions; and thus the relative action is rendered nothing, and the mutual action of the three magnets destroyed. A gallery of about forty feet in length is most practically convenient for the placing of the three magnets. The result arrived at by Prof. Lloyd is exceedingly interesting, in relation to the observatories about to be erected in the British colonies, and to the solution of the difficult problem of terrestrial magnetism.

ELECTRIC POLARIZATION OF SOLID AND LIQUID CONDUCTORS.

It appears, from numerous experiments, that the polarization of polar wires depends on two circumstances: first, the current of the pile; secondly, the nature of the liquid in which the wires are plunged. —*Bibliothèque Universelle de Genève; Literary Gazette.*

ELECTRIC SPARK.

M. Edmund Becquerel has been making experiments on the calorific radiation of the electric spark, and comes to the following conclusions: whether this spark proceeds from a battery or not, there is no elevation of temperature, let the distance be what it will; but, as the electric spark excites or revives the phosphorescence of a body gifted with this quality, it is reasonable to suppose that it affects it by some peculiar radiation, differing from that which produces the sensation of heat.—*Athenæum.*

PHOSPHORESCENCE.

M. Becquerel is of opinion that electric light renders certain bodies phosphorescent, when they have for some time been exposed to its action. The violet rays possess the greatest degree of this power, while the red rays are entirely destitute of it. Those substances which suffer almost all white rays to pass through them, reduce their phosphorescent property to nearly one half.

FIRE DAMP.

A correspondent of the *Athenæum* observes:—It is well known that fire damp explodes on ignition by an electric spark; and on this principle it is proposed, that an experiment be made with an apparatus consisting of Professor Daniell's voltaic battery and electrical wires, for the purpose of firing the gallery of a mine charged with an explosive mixture. By means of this ingenious and scientific contrivance, an explosion of fire damp can at all times be effected with perfect safety, whenever the gas is evolved in sufficient quantities to generate this destructive element; and as the wires can be conveyed to the remotest chamber of the mine at a trifling cost, the experiment may be repeated with great facility by the aid of a powerful battery, in any place where it indicates its appearance, or creates a suspicion of danger. The actual presence, quantity, and position of the fire damp, can always be ascertained with precision and certainty, through the agency of the safety lamp; but care must be taken that the carbonic acid and azote remaining in the mine after the inflammation, be got rid of, either by decomposition, absorption, or ventilation, before the workmen resume their labours. This formidable and treacherous enemy will thus be effectually and instantaneously annihilated, that otherwise could be but slowly, partially, and progressively consumed; combined with the advantage, that the health of the miners will cease to be impaired from respiration in a foul atmosphere.

ELECTRIC EFFECTS OF CONTACT.

M. Becquerel has read the first part of a memoir before the French Academy of Sciences, containing an account of his new experiments on the Electric Effects of Contact. These lead him to conclusions concerning the causes of certain effects which differ from those of Davy, who ascribed them to mere contact. M. Becquerel refers them to friction, and hopes to prove it by the following experiments. First, two condensing plates made of platina were adapted to an excellent electroscope; one of them was touched with a very dry piece of chalk, and the other with the finger: on separating the plates, it was found, that no electric effect had been produced. Secondly, a layer of calcined lime, highly dried, was spread on a piece of wood, equally dry; upon this was carefully placed, so as not to allow of friction, a plate of copper, fixed to a handle; it was then put in contact with one of the plates of the condensator, while the other was touched with the finger. After repeating this several times, no electric discharge was obtained; but, if the copper disc were placed on the lime with friction, the condensator was charged after touching a few times, and the greater the friction the stronger the charge. The lime assumed positive, and the metal negative, electricity. Thirdly, if oxalic, succinic, benzoic, or boracic acid, were substituted for the lime, in the form of a very dry powder, electricity was also obtained by friction, but not by mere contact; and, in this case, the metal took the positive, and the acid the negative electricity. Lastly, the same result was obtained from crystals of oxalic acid and lime. Therefore, says M. Becquerel, the results obtained by Davy are due to friction, and not to the electro-motive action of Volta.—*Athenæum.*

EXPERIMENTS ON IRON-BUILT SHIPS.

On April 25, was read to the Royal Society an "Account of Experiments on Iron-built Ships, instituted for the purpose of discovering a Correction for the Deviation of the Compass produced by the Iron of the Ships;" by G. Biddell Airy, Esq., A. M. In this paper, the problem of the deviation of a ship's compass, arising from the influence of the iron in the ship, more particularly in iron-built ships, is fully investigated; and the principles on which the correction for this deviation depends having been determined, practical methods for neutralizing the deviating forces are deduced and illustrated by experimental application. The author states that, for the purpose of ascertaining the laws of the deviation of the compass in the iron-built steam-ship, *the Rainbow*, four stations were selected in that vessel, about four feet above the deck; and at these the deviations of the horizontal compasses were determined in the various positions of the ship's head. All these stations were in the vertical plane, passing through the ship's keel, three being in the after part of the ship and one near the bow. Observations were also made for determining the horizontal intensity at each of these stations. The deviations of dipping needles at three of these stations were also determined, when the plane of vibration coincided with that of the ship's keel, and also when at right angles to it. After describing the particular method of observing rendered necessary by the nature of the vessel and the circumstances of her position, the author gives the disturbance of the horizontal compass at the four stations deduced from the observations. The most striking features in these results are, the very great apparent change in the direction of the ship's head, as indicated by the compass nearest the stern, corresponding to a small real change in one particular position, the former change being 97° ; whereas the latter was only 23° ; and the small amount of disturbance indicated by the compass near the bow. After giving the observations for the determination of the influence of the ship on the horizontal intensity of a needle suspended at each of the stations, in four different positions of the ship's head, and the disturbances of the dipping needle at three of these stations, the author enters upon the theoretical investigation. The fundamental supposition of the theory of induced magnetism, on which Mr. Airy states his calculation to rest, is, that, by the action of terrestrial magnetism, every particle of iron is converted into a magnet, whose direction is parallel to that of the dipping needle, and whose intensity is proportional to that of terrestrial magnetism; the upper end having the property of attracting the north end of the needle, and the lower end that of repelling it. The attractive and repulsive forces of a particle on the north end of the needle, in the directions of rectangular axes towards north, towards east, and vertically downwards, and of which the compass is taken as the origin, are first determined on this supposition in terms of the co-ordinates ; and thence the true disturbing forces of the particle in these directions. The disturbing forces produced by the whole of the iron of the ship are the sums of the expressions for every particle.

These statements suggest the following as rules which it is desirable to observe in the present infancy of iron-ship building. It appears desirable that,—1, Every iron sea-going ship should be examined by a competent person for the accurate determination of the four constants above mentioned for each of the compasses of the ship; and a careful record of these determinations should be preserved as a magnetic register of the ship. 2, The same person should be employed to examine the vessel at different times, with the view of ascertaining whether either of the constants changes in the course of time. 3, In the case of vessels going to different magnetic latitudes, the same person should make arrangements for the examination of the compasses in other places with a view to the determination of the constant N. 4, The same person should examine and register the general construction of the ship, the position and circumstances of her building, &c., with a view to ascertain how far the values of the magnetic constants depend on these circumstances, and in particular to ascertain their connection with the value of the prejudicial constant M. 5, The same person should see to the proper application of the corrections and the proper measures for preserving the permanency of their magnetism. The most remarkable result in a scientific view from the experiments detailed in the present paper is the great intensity of the permanent magnetism of the malleable iron of which the ship is composed.—*Abridged from the Athenæum,* No. 605; *which see for the formulæ.*

FARADAY'S EXPERIMENTAL RESEARCHES IN ELECTRICITY.

On Dec. 6th and 13th, 1838, was read to the Royal Society the *Fifteenth Series*,* of "Experimental Researches in Electricity," entitled, "Note of the character and Direction of the Electric Force of the Gymnotus;" by Prof. Faraday.

The author briefly refers to what has been done by others in establishing the identity of the peculiar power of the Gymnotus and Torpedo with ordinary electricity. The Professor's researches were made upon a living Gymnotus, now at the Gallery of Science, in Adelaide-street; when he proceeded, with suitable apparatus, to compare its power with ordinary and voltaic electricity, and to obtain the direction of the force. Without removing it from the water, he was able to obtain, not only the results procured by others, but the other electrical phenomena required, so as to leave no gap or deficiency in the evidence of identity. The shock, in very varied circumstances of position, was procured: the galvanometer affected; magnets were made; a wire was heated; polar decomposition was effected; and the spark obtained. By comparative experiments made with the animal and a powerful Leyden battery, it was concluded that the quantity of force in each shock of the former was very great. It was also ascertained by all available tests, that the current of electricity was, in every case, from the anterior parts of the animal through the water or surrounding conductors to the posterior parts. The author then expresses his hope that by means of these organs and similar parts of the Torpedo, a relation as to *action and re-action* of the electric and nervous powers, may be established experimentally.—*Philosophical Magazine.* For the details of the experiments, see *Philos. Mag.* No. 97, p. 358.

EVOLUTION OF HEAT BY THERMO-ELECTRICITY.

Mr. F. Watkins has observed this phenomenon with a delicate air electro-thermometer of the construction suggested by Mr. W. S. Harris, and also with a Breguet's metallic thermometer, arranged with M. de la Rivé's contrivance for passing a current through its helix. Mr. Watkins employed a massive thermo-electric battery, consisting of 18 pairs of bismuth and antimony prismatic rods, four inches long, united consecutively, similar to the arrangement of M. Van der Voort, of Amsterdam. When the electricity was exerted by applying a hot iron heater near one extremity of the battery, and ice (placed around the wires) at the opposite extremity, with the circuit complete through the air electro-thermometer, the heating effect on the fine platinum wire in the spherical reservoir was immediately visible by the ascent of the coloured liquid up the fine glass tube communicating with it. The elevation of the liquid column was about 20°, equal to one inch on the scale. When the Breguet's thermometer was placed in the circuit, the index attached to the bottom of the compound metallic helix moved round 10° in the direction of the coils; the helix expanding by the elevation of temperature conferred by the passage of the thermo-electric current. The elevation of temperature of the metals

* For the Eleventh, Twelfth, Thirteenth, and Fourteenth, Series, see Year-Book of Facts, 1839, p. 104—106.

forming the electric circuit in both instruments was always manifested when the circuit was completed, and remained constant; but, on breaking the circuit, the loss of heat was very apparent.—*Philos. Magazine.*

COMPOSITION AND DECOMPOSITION OF WATER.

Mr. Grove has succeeded, by a very simple experiment, in decomposing water by platina wires, in communication with a voltaic pile; which, it is believed, has never before been done. He takes two tubes of glass, closed at one end, and about half filled, the one with oxygen and the other with hydrogen; and having passed a platina wire into each, plunges the tube into water, slightly acidulated. He then brings the wire in the oxygen tube into contact with the zinc plate of a voltaic pair, and the other with the copper plate. Shortly, the water rises in both tubes, but twice as high in the hydrogen as in the oxygen tube. Here, (says Mr. Grove,) are both the formation and decomposition of water, two actions depending on each other.—*Railway Magazine.*

ELECTRICAL DISCHARGES THROUGH WIRES OF SMALL DIAMETER.

Nairne remarked, that powerful discharges of electric batteries through small iron or silver wires, when they were long enough to become red, only contracted the length of the wires without diminishing their weight. M. Becquerel has repeated the experiment on a platina wire of 0·72 millemetres diameter, and observed the same effect. He has also determined the law of the contractions, and found that the diminutions per cent. of length are reciprocally proportional to the cubes of the diameters of the wires; the electric discharges being, of course, equal.—*Railway Magazine.*

VAST ELECTRICAL APPARATUS.

An immense electrical apparatus, stated to be "the largest in the world," has been constructed by Clarke, of the Strand, for an exhibition entitled, "The Gallery of Natural Magic," at the Colosseum, Regent's Park. Its plate measures 7 feet in diameter, and consequently exposes an electric surface of upwards of 80 feet. The instrument is mounted in a novel and scientific manner. Its conductors are of varnished copper, and give a striking distance, or length of spark hitherto deemed unattainable. The terminating balls of the conductors are strongly gilt, so as to prevent dissipation. Its single pair of rubbers deserves especial attention, from the superior and simple manner by which they are supported, and the firmness and perfect control of the instrument. With the battery are produced almost terrific effects, of intense heat, igniting and fusing metals, &c.; and the charge may be sent through five miles of copper wire.

NOVEL ELECTRICAL MACHINE.

This is a leather strap, connecting two drums in a worsted-mill, in the town of Keighley, Yorkshire. The strap is 24 feet in length, 6 inches in breadth, and $\frac{1}{8}$ inch in thickness; making 100 revolutions in a minute. The drums, over which it passes at both ends, are 2 feet in diameter, made of wood, fastened to iron hoops, and turning on iron axles; these drums are 10 feet distant from each other, and the strap crosses in the middle between them, where there is some friction; the strap forming a figure of 8. There is no metal in connection with the strap, but it is oiled. On presenting your knuckle to the strap, above the point of crossing, flashes of electrical light are given off in abundance; and when the points of a prime conductor are held near the strap, most pungent sparks are given off to a knuckle at about two inches: a Leyden jar of considerable size has also been charged in a few seconds, by presenting it to the prime conductor. An electrical battery has been frequently charged from this strap, in a very short time; and it is always the same, generating electricity from morning to night.—*Rev. T. Drury to Prof. Faraday; Philos. Mag.; abd.*

ENGRAVING BY ELECTRICITY.

M. Jacobi, in a letter to Prof. Faraday, (communicated to the *Philosophical Magazine*), states, that, by a fortunate accident, he has discovered the means of making copies in relief of an engraved copper-plate, by means of voltaic action; and of obtaining a new inverted copy of those in relief by the same process, so as to insure the power of multiplying the copper copies to any extent. By this voltaic process, the most delicate and even microscopic lines are reproduced. The apparatus is simply a voltaic pair, *à cloison;* where the engraved plate is used in the place of the ordinary copper-plate, being plunged in the solution of sulphate of copper. A galvanometer with short wires should always make part of the circuit, so that one may judge of the force of the current, and direct the action; the latter being effected by separating the electromotive plates more or less from each other, or modifying the length of the conjunctive wire; or, finally, diminishing more or less the conducting power of the liquid on the zinc side; but for the success of the operation, the solution of copper should be always perfectly saturated. The action should not be too rapid; from 50 to 60 grains of copper being reduced on each square inch in 24 hours.

We may reduce the sulphate of copper by making the current of a single voltaic pair pass through the solution by copper electrodes; as the anode is oxidized, the cathode becomes covered with reduced copper, and the supply of concentrated solution may then be dispensed with. A thoroughly concentrated solution of sulphate of copper is not decomposable by electrodes of the same metal, even on employing a battery of three or four pairs of plates. The needle is certainly strongly affected as soon as the circuit is completed; but the deviation visibly diminishes, and very soon returns almost to zero. If the solution be diluted with water to which a few drops of sulphuric acid have been added, the current becomes very strong and constant; the decomposition goes on very regularly; and the engraved cathode becomes covered with copper of a fine pink red colour. If we replace the solution of a sulphate of copper by pure water acidulated with sulphuric acid, there is a strong decomposition of water, even on employing a single voltaic couple. The anode is oxadized, and hydrogen is disengaged at the cathode. At the commencement, the reduction of copper does not take place; it begins as soon as the liquid acquires a blue colour, but its state of aggregation is always incoherent. M. Jacobi has continued this experiment for three days, until the anode was nearly dissolved: the colour of the liquid became continually deeper, but the disengagement of hydrogen, though it diminished in quantity, did not cease. M. Jacobi considers we may conclude from this experiment that in secondary voltaic actions there is neither that simultaneity of effect, nor that necessity of entering into combination or of being disengaged from it, which has place in primary electrolytic actions. With respect to the technical importance of these voltaic copies, M. Jacobi observes, that we may use the engraved cathode, not only of metals more negative than copper, but also of positive metals and their alloys, (excepting brass); notwithstanding that these metals, &c., decompose the salts of copper with too much energy when alone. Thus, one may make, for example, stereotypes in copper which may be multiplied as often as we please.

Immediately after the publication of the above details, Mr. Thomas Spencer, of Liverpool, stated in the *Athenæum*, that he had not only succeeded in doing all that M. Jacobi had done, but had successfully overcome those difficulties which arrested the progress of the latter. The objects which Mr. Spencer says he proposed to effect were the following: — "To engrave in relief upon a plate of copper—to deposit a voltaic copperplate, having the lines in relief—to obtain a fac-simile of a medal, reverse or obverse, or of a bronze cast—to obtain a voltaic impression from plaster or clay—and to multiply the number of already engraved copperplates." The results which he has obtained are very beautiful; and some copies of medals are remarkably sharp and distinct, particularly the letters, which have all the appearance of having been struck by a die.—For the details of Mr. Spencer's process, see the *Athenæum*, No. 625, p. 811.

PRINTING BY ELECTRICITY.

The production of drawings by electricity seems to have engaged more attention abroad than in this country. In Russia, have long been engraved what are called "Russian snuff-boxes," which are formed of a kind of imitation platinum, and have drawings made upon them by an application to their conducting powers. Recently, M. Jacobi has copied copper-plates by galvanism, as above stated.

The sympathy and antipathy of electricity to particular colours seem, however, to point out means of more easily effecting the process of copying. It has long been known that electricity is repelled by a black surface, and attracted by white; and some interesting illustrations of the effects of a thunder-storm upon black cattle are related in the *Philosophical Transactions*. This effect has been farther confirmed by the operation of lightning on the blackened masts of men-of-war. (See page 33 of the present volume.)

This property of colour might be so applied as, by electrical power, to produce engraved plates from prints, impressions of prints from plates, or even from other prints; and an operation introduced which might, in some cases, compete with photography, and, in others, supersede the printing-press.—*Mr. Hyde Clarke; Railway Magazine.*

EXPLOSION OF GUNPOWDER BY THE VOLTAIC BATTERY.

COL. PASLEY, R. E., has made several successful subaqueous and subterraneous experiments at Chatham, in firing gunpowder by means of the voltaic battery, at the distance of 170 yards. The conducting wires were nearly the whole distance either under ground or water. Large stones, prepared for the purpose, and let down 14 feet under water, were blasted in the Medway by this apparatus, some charges having been under water for a few days previous to the explosion. The conducting wires were secured to a tarred rope with hemp yarn sewed round them. It was found difficult to attach the ropes in such a manner as to avoid breaking the small platinum wire enclosed in the charge. It was also found that the force of the battery considerably depended on the distances of the objects, and the thickness of the wires used in the experiments. Conducting wires of small diameter, used for a given distance, were found to require a much more powerful battery to ignite the charge than those of greater diameter.—*Railway Magazine.*

The details of Col. Pasley's operations against the wreck of the *Royal George*, at Spithead, with the voltaic battery and gunpowder, have been very successful. On Sept. 23rd, a cylinder containing 2,320 lb. of powder was lowered to the bottom, and placed alongside the most compact portion of the wreck by divers attaching to it hauling-lines rove through blocks. The vessel in which the battery was placed, was then drawn off 500 feet, which was the length of the connecting wires; and instantaneously, on the circuit being completed, the explosion took place. The most remarkable effect on the water was its uprising to 28 or 30 feet in a compact mass, from a depth of 90 feet. Neither the shock nor the sound was so great as in an explosion with 45 lb. of powder. Altogether, the Colonel has recovered from this wreck 12 guns, 5 gun-carriages, 100 beams or riders, or fragments of them, exclusive of other timbers, planks, and coppers; besides the working places and boilers complete, the stern and great part of the bows, the two capstans, part of the main-mast, and all that remained of the fore-mast, of the *Royal George.*

The above is but one of the many adaptations of electricity to useful purposes, during the past year. A correspondent of the *Mechanics' Magazine*, we perceive, proposes to light street lamps by electricity.

ELECTRO-MAGNETISM AS A MOTIVE POWER.

THE emperor of Russia lately appointed a commission to inquire into the applicability of electro-magnetism as a moving power; and from an official report by this commission, the substance of which is given in the last number of *The United Service Journal*, it appears that Prof. Jacobi has actually succeeded in impelling a vessel by electro-magnetic power.

The vessel was of that species of galley which is well known in the

Russian navy; its measurement, 26 feet in length and $8\frac{1}{2}$ in width. On smooth water, it was impelled at the rate of more that 3 feet per second of time, or somewhat above 2 miles per hour; and the average of a number of experiments was from 2 to 3 feet per second. It performed a distance of rather less than 5 miles along the Neva and the town canals in about $2\frac{1}{2}$ hours. The space occupied by the machinery was 1 foot 2 inches in breadth and 2 feet 1 inch in length. The galvanic batteries, consisting of 320 pairs of plates, were arranged along the sides of the vessel, within which twelve persons were seated at their ease. These batteries were used several days consecutively, and did not exhibit any diminution of power. The experiments which have been made have opened out much that was unknown on the subject both of electricity and magnetism, with regard to their practical bearings, and suggested the introduction of very considerable improvements in the construction of the machinery upon a larger scale. From two to three months were consumed in the various trials hitherto made, but as yet they have not been sufficient to determine what quantity of zinc a machine will require per day to every horse-power, or how much of it will be converted into vitriol; it was ascertained, however, that the plates, whose whole weight was originally 400 lb. and superficies 96 feet square, had not lost more than 24 at the termination of the experiments.

An American gentleman, (Capt. Taylor,) states that he has been equally successful in applying electro-magnetism, as a driving power to machinery on shore.

ELEMENTARY LAWS OF ELECTRICITY.

Mr. W. Harris, F. R. S., on June 20, presented to the Royal Society the *Third Series* of his "Inquiries concerning the Elementary Laws of Electricity;" in which he proposes to perfect the methods of electrical measurement, whether relating to the quantity of electricity, intensity, inductive power, or any other element requiring an exact numerical value; and, by operating with large statical forces both attractive and repulsive, to avoid many sources of error inseparable from the employment of extremely small quantities of electricity, such as those affecting the delicate balance used by Coulomb. The author then describes some improvements in his hydrostatic electrometer; the indications of which depend on the force between two opposed planes operating on each other under given conditions, are reducible to simple laws, and are hence invariable and certain; the attractive force between the two discs is not subject to any oblique action, is referable to any given distance, and may be estimated in terms of a known standard of weight. The author next proceeds to the elementary laws of electrical action; and proves, by experiments, that induction invariably precedes, or at least, accompanies, attraction and repulsion. Experiments also show this inductive influence to be, probably, in some way dependent on the presence of an exquisitely subtle form of matter, which may become disturbed in bodies, and assume new states or conditions of distribution. Very numerous experiments are detailed, showing the influence of changes of different intensity, of changes in the dimensions and distances of the opposed discs, of interposed bodies of different forms, &c., on the phenomena of induction. The paper concludes with formulæ as the results of the author's investigations.—*Philos. Mag.*

PECLET'S NEW ELECTRIC CONDENSER.

This new condenser is composed of three plates of glass, roughened by rubbing the surfaces carefully one upon another. They are entirely covered with gold leaf pasted on with alumine. One of these plates, A, is fixed to a common gold leaf electrometer, its upper surface being varnished. The second, B, is placed on the first; it is varnished on both sides; a small, gilt, unvarnished, copper stem is fixed horizontally at a point in the circumference; it carries in its centre, like the moveable plate of common condensers, a glass stem, which serves as a handle. Finally, on this last plate is a third plate, C, with a hole in its centre, through which passes the stem of the plate B. The plate C is varnished on the under side only, and its central orifice is furnished with a glass tube which encloses the stem of the plate B, but of a less height.

This apparatus is used as follows: we touch the upper plate with the metal whose action upon gold we wish to determine, and put the plate B in connection with the ground:—this connection is then broken, the plate C is raised, and we touch the plate A. This manœuvre is repeated a certain number of times. Lastly, by means of the stem of plate B, we raise at once the plates B and C; when the gold leaves of the electrometer diverge to a distance dependent on the number of contacts.

The cage which encloses the gold leaves is formed of parallel glass plates, and is placed on a screw tripod, furnished, on one side, with a vertical plate, pierced with a small hole, and on the other with a portion of a divided vertical circle, whose centre is at the same height as the hole of the plate and the upper extremity of the gold leaves: in looking through the hole of the plate we observe the deviation.

The following two series of experiments will give an idea of the power of this apparatus. By touching the upper plate with an iron wire, after 1, 2, 3, 4, 5, and 10 contacts, the leaves separated 9 and 2-5ths, 20, 25, 31, 41, and 88°. By touching the upper plate with a platina wire, a single contact produced but a feeble deviation, which increased to 15° after three contacts, and to 53° after 20. The experiments with platina were made by using a platina wire, which had just been reddened in the flame of alcohol, after washing the hands in distilled water. The experimenter had previously assured himself, by a great number of successive contacts, in which he touched the upper plate with a finger, that the plates did not contain any electricity.

The new fact of the development of electricity by the contact of gold and of platina, was also directly proved by means of a simple condenser of extreme sensibility obtained by giving to the coats of varnish a suitable thickness, and rendering their surfaces perfectly plane.

It is evident, from the disposition of the apparatus, that the quantity of electricity set at liberty, which causes the divergence of the leaves, is proportionate to the number of contacts: now, it results from numerous experiments, that as far as about 20° the deviation is proportionate to the number of contacts, hence to this extent the deviation is proportionate to the quantity of electricity. It would be easy to make a table which would give the quantity of electricity corresponding to the deviations which exceed 20°, since these quantities are proportionate to the number of contacts.—*Ann. de Chimie; Franklin Journal.*

LIGHTNING CONDUCTORS.

Mr. W. S. Harris, F. R. S., has communicated to the *Philosophical Magazine*, No. 98, an elaborate memoir "on Lightning Conductors, and on certain Principles in Electrical Science," with experiments and details, too numerous to quote.

In conclusion, Mr. Harris states as his confirmed opinion, after a long and severe examination of the laws of electrical action, and of cases of ships and buildings struck by lightning;—that a lightning rod is purely passive; that it operates simply in carrying off the lightning which falls on it, without any lateral explosion *whatever*. Mr. Harris does not, however, deny the general inductive effect mentioned by Lord Stanhope on bodies opposed to the influence of the thunder-cloud, and that the displaced electricity will again find its equilibrium of distribution, and return to those bodies, which effect would necessarily take place, whether we had a lightning conductor or not; an additional reason for linking the detached conductors in a ship's hull into one great mass, so as to have as few interrupted circuits as possible in any direction: this opinion Mr. Harris is prepared to substantiate by striking cases in which ships have been struck by lightning.

GAUSS ON THE THEORY OF MAGNETISM.

In our theory it is assumed that every determinate magnetized particle of the earth contains precisely equal quantities of positive and negative fluid. Supposing the magnetic fluids to have no reality, but to be merely a fictitious substitute for galvanic currents in the smallest particles of the earth, this quality is necessarily part of the substitution; but, if we attribute to the magnetic fluids an actual existence, there might, without absurdity, be a doubt as to the perfect equality of the quantities of the two fluids.

In regard to detached magnetic bodies, (natural or artificial magnets,) the question as to whether they do or do not contain a sensible excess of either magnetic fluid might easily be decided by very exact and delicate experiments.

In case of the existence of any such excess in a body of this nature, a plumb-line to which it should be attached should deviate from the true vertical position in the direction of the magnetic meridian.

If experiments of this kind, made with a great number of artificial magnets, and in a locality sufficiently distant from iron, never showed the slightest deviation, (which we should rather expect,) the equality of the two fluids might, with the highest degree of probability, be inferred for the whole earth; though, without wholly excluding the possibility of some inequality.—*Taylor's Scientific Memoirs*, vol. ii. (part vi.) p. 228.

Chemical Science.

CHEMICAL COMBINATION.

On May 7 and 27, Prof. Daubeny explained to the Ashmolean Society certain views with respect to the Constitution of Matter and the laws of Chemical Combination, which had been brought forward within the last few years. He showed, that although matter may consist of atoms, yet chemical union must be supposed to take place between groups or assemblages of such atoms, and not between the individual atoms themselves. He then showed, that the bodies which are to be regarded in the light of elementary substances, in the case of vegetable or animal matter, are themselves compounds, consisting of carbon, nitrogen, hydrogen, and oxygen; and, on the other hand, that the chemical properties of those bodies, which, as far as we know, are simple, admit of being modified very considerably by electrical and other agencies. He likewise pointed out that several distinct chemical compounds may exist made up of the same elements, united in the same proportions; and, on the other hand, that simple as well as compound substances assume various distinct crystalline forms, according to the circumstances under which they happen to become solid.

Dr. Daubeny next explained the doctrine of Catalysis, or the influence which certain bodies exert in producing chemical changes, without entering into combination with either of the constituents of the compound substances which they decompose. This doctrine was illustrated by the action of spongy platinum in kindling hydrogen gas, by the conversion of starch into gum and sugar under the influence of diastase or of sulphuric acid, and probably in the process of fermentation produced in saccharine matter by the presence of a minute proportion of gluten. He next pointed out a species of attraction which seems to operate amongst homogeneous masses of matter, as cohesive attraction does amongst their particles; and which he proposed to distinguish by the term *adhesive attraction*. It is by virtue of this force that portions of flinty matter disseminated through a mass of clay, become gradually collected into lumps or nuclei. He then noticed the new views of Prof. Graham of London, who has shown that water acts an important part in the constitution of acids, salts, &c., and influences their chemical properties no less than their crystallization. He concluded by observing that it was probable that the various products of animal and vegetable life result from the operation of these and other similar natural laws, which pervade the organic as well as the inorganic kingdoms; and that many of the compounds hitherto obtained exclusively from the processes of living matter may be created by artificial means. Two or more of these compounds have been already produced by chemists of the present day.

Dr. Daubeny having, in the course of his paper, alluded to the new method of distinguishing certain organic compounds by observing their respective effects upon polarized light, and having exhibited an apparatus constructed for the illustration of this new method of analysis, Prof. Powell rose, and explained briefly the principles upon which this investigation proceeded, by stating the laws of plane and circular polarization. It appears that, whilst sugar derived from the cane deflects the polarized ray to the right, that obtained from honey or pure starch causes a deviation to the left; and the amount of deflection in a given sample will vary with the relative proportion of the different kinds of sugar.—*Athenæum.*

CHEMICAL VIEWS REGARDING THE FORMATION OF ROCKS.

Prof. Nepomuk Fuchs, in some elaborate "Chemical views regarding the Formation of Rocks, which seem to afford new arguments in favour of Neptunism," having briefly investigated the siliceous, calcareous, and carbonaceous series, considers the accompanying collateral and intermediate geological events. During the crystallization, the pasty or semi-solid masses must have been contracted into smaller space. The consequence of this must have been the formation of rents and fissures into which the existing amorphous mass penetrated, and in which it would freely crystallize; and, in this way, veins were formed, as, for example, those of granite. In a similar manner, great caverns and empty spaces were produced; and this caused sinkings and fallings-in, by which strata were dislocated, had their original position altered, and acquired the appearance of having been elevated. Sinkings of masses of rock also produced valleys and ravines, besides circular hollows in which water was collected, some of which still remain, and others were broken through. By means of earthquakes, the falling-in of caverns, and the bursting of lakes, huge masses of debris were formed, which became the sport of the waters, whose power was often combined with that of hurricanes and violent rains. Great devastation was produced by such agencies, assisted by volcanoes.

During such prodigious processes as those by which the earth and atmosphere were formed, the imponderables must have been in very active operation; sometimes assisting in forming, and sometimes in destroying what had been already formed. The electrical meteors, more especially, must have acted with a vigour and power, of which no conception can be obtained from their present displays. Since, even at present, lightning sometimes shatters rocks, and melts quartzose sand, we may believe that, at such a time, the electrical masses of fire descending from the heavens, may have vitrified rocks, and caused actions in the depths of the earth, of such a nature as to lead to the belief that they were produced by subterranean fire. Since, even at the present day, water-spouts uproot the strongest trees, and remove them far from their original situations; we may conceive water-spouts, of the period we have been discussing, powerful enough to remove large fragments of rock from their beds, and to convey them to remote regions, where we are now surprised to find such strangers. Lastly, we ought not to judge of the scale on which such operations proceeded at these early periods of the world's history, from what we observe now taking place on the surface of our globe.—Compiled and curtailed from Leonhard's *Jahrbuch*, and the Munich *Galehrte Anzeigen*; in *Jameson's Journal*; No. 51.

CLASSIFICATION OF SOILS.

M. De Gasparin has presented to the French Academy of Sciences a memoir on this important subject; which is the first part of a work upon *Agronomy*, or that branch of the science of Agriculture which has for its object the study of soils; their susceptibility and relative value being reserved for another publication. We must, however, confine ourselves to the following facts, which result from M. De Gasparin's investigation:

1. A small quantity of carbonate of lime is sufficient to change the character of soils. The 5 or 6 per cent. of this substance which is supplied by marling, produces very remarkable effects; whilst the 100th part, which is contained in the soil of Lille, as analyzed by M. Berthier, likewise sensibly affects its nature and vegetative power. Lime gradually disappears from soils, being changed into a bi-carbonate. The enclosure of *La Grande Chartreuse*, which is formed of the debris of rocks containing lime, does not now furnish a single particle of this earth.

2. The carbonate of magnesia modifies soils in the same way as carbonate of lime.

3. It has often been attempted to ascertain the characters which distinguish those soils in which gypsum produces an effect upon vegetables, and those in which it has none; but hitherto without success. The author has ascertained that gypsum has no action upon recent alluvial soils, and that it is beneficial for all more ancient soils, beginning with the *diluvian*.

4. He has found sal-ammoniac in all the clays belonging to the vegetable stratum of soils; thus showing the importance of clay as a magazine of the materials which favour vegetation.

5. If by washing we separate into several portions the coarser and finer parts of earth, we find that the tenacity of such soil is in proportion to the quantity of the latter kind, except in a small number of cases.

6. Upon examination with the microscope, it is ascertained that these exceptions are owing to a coating of ferruginous clay which adheres to the surface of the mineral particles, that washing separates it with difficulty, and that it serves as a cement, forcibly agglutinating, and increasing the tenacity of the whole.—*Jameson's Journal.*

CHEMICAL THEORY OF VOLCANOES.

Prof. Daubeny, in reply to Prof. Bischof's objections to his Chemical Theory of Volcanoes, sums up with the following enumeration of phenomena of constant occurrence, which follow directly from the principles of his views:

1. The evolution of sulphuretted hydrogen, in quantities far exceeding what are to be explained by the re-action of carbonaceous

matter upon sulphates, or any of those other processes which sometimes produce it on the surface of the earth.

2. The disengagement of sal-ammoniac; for, although one of the constituents of this compound, the muriatic acid, might arise from the decomposition of sea-salt by aqueous vapour, the other one, the ammonia, implies the presence of free hydrogen as well as of nitrogen gas, near the focus of the volcanic action.

3. The circumstance which Dr. Daubeny has substantiated in so many cases that he begins to believe it almost universally true, is that the atmospheric air exhaled from volcanoes, and, indeed, generally from the interior of the earth, is deprived in a greater or less degree of its proper proportion of oxygen. That processes, therefore, by which this principle is abstracted, are going on extensively within the globe cannot be denied; and hence Dr. Daubeny conceives that any theory which attempts to account for volcanic action without taking notice of so essential a phenomenon, ought to be regarded as imperfect and unsatisfactory.—*Jameson's Journal.*

Dr. Daubeny, in a note to the preceding Memoir, adds: "that with regard to the precise nature of the chemical processes which give birth to volcanoes, he has always studiously distinguished between the degree of probability belonging to that part of the hypothesis in question which asserts the fact of the general absorption of oxygen gas throughout the interior of the globe, and that which undertakes to explain the latter phenomena, by supposing certain bodies, capable at once of decomposing water, and of combining with its oxygen without eliminating a volatile product, to exist in the spots at which the action originates. Whatever difficulty there may be in imagining the earthy and alkaline metalloids to exist among the number of these bodies, that difficulty will, at least, not be enhanced by supposing them to be as generally distributed as the phenomena themselves appear to warrant us in supposing. If it be once granted, that potassium and soda contribute, by their combustion, to feed the fires that glow between Vesuvius and Etna, there seems no reason why we should refuse to believe that the same substances may be present, wherever the most waters or earthquakes indicate a similar train of phenomena.—*Jameson's Journal,* No. 53; *abridged.*

CHEMICAL POWERS OF LIGHT.

M. EDMUND BECQUEREL has communicated to the French Academy of Sciences some important investigations on the Chemical Powers of Solar Light, which will probably lead to new and valuable results. It has been long known that light has the power of variously affecting certain chemical compounds; sometimes causing combination between two elements, and in other cases effecting the decomposition of compound substances; and it has been found that, when a pencil of light is decomposed by passing through a prism of glass, those rays which possess this power are differently refracted from the coloured rays; and hence the existence of peculiar rays, to which the name of chemical rays is given, has been deduced. The chief difficulty in experiments on these rays has been the slow nature of the actions caused, and the difficulty of appreciating them. M. Becquerel has overcome these sources of uncertainty, and is enabled to study the chemical power of light with ease, and measure the effects produced with considerable accuracy. The manner in which this is done is very simple. Two liquids of different densities, but both conductors of electricity, and of such a nature as to act chemically upon each other when exposed to the influence of solar light, are selected; and a portion of both is put into a cylindrical vessel blackened on the exterior. A plate of platinum is placed in the denser of the two fluids, and another similar plate is also immersed in the lighter liquid; these plates being then connected by means of platinum wires with the two terminations of a very delicate galvanometer, the apparatus is complete. If when thus arranged, a ray of light is suffered to pass through the mass of fluid, it causes chemical action to take place at the surface of contact between the two liquids, and a current of electricity which this sets in circulation is immediately rendered evident by the galvanometer. As the angle of deflection of the galvanometer indicates the power of the electric current, and as that is in exact proportion to the chemical action which originates it, it is evident that this arrangement gives an accurate measure of the power of the chemical rays of light, at different times, from different sources, and under various circumstances. M. Becquerel details some experiments on the quantity of these chemical rays, which is intercepted when a ray of light is made to pass through screens of different substances, such as rock crystal, mica, and variously coloured glasses; and states that he is still engaged in experimenting on the subject.—*Athenæum.*

COLOURED FILMS PRODUCED BY ELECTRO-CHEMICAL AGENCY AND BY HEAT.

MR. WARINGTON has communicated to the *Philosophical Magazine*, No. 100, a paper upon this interesting inquiry, which is intended as a reply to a memoir by the late Prof. Nobili on the subject; the translation of which appeared in the *Scientific Memoirs*, vol. i., p. 94, under the title of a "Memoir on Colours in general, and particularly on a new Chromatic Scale deduced from Metallochromy, for scientific and practical purposes."

We have not space for the details, but quote the summary of this valuable contribution. The author infers: 1st, that the appearances called electro-chemical are not films of oxygen and acid, but lead in a high state of oxidation, thrown down on the surface of the metal by means of a voltaic combination acting through a medium formed by a solution of acetate of lead; 2ndly, that these colours owe their varied tints to the varying thickness of the precipitated film, and that the light is reflected through them from the polished surface of the metallic below; 3rdly, that the colours produced on the surface of metals by the application of heat are owing to the formation of thin films of oxide on the metal, in consequence of exposure to the air during the process; that this does not involve the necessity of any one oxide being always formed, as this must vary according to the affinity of the metal used for oxygen, under the influence of a raised temperature;

4thly, that the opacity of the metal is not in the slightest degree an argument against the transparency of the oxide; and we have, both in nature and art, numerous cases which place this question beyond a doubt; 5thly, that we can produce analogous appearances by substituting other elements for oxygen, such as iodine, chlorine, bromine, sulphur, phosphorus, carbon, &c.

COLOUR OF BLOOD DURING COAGULATION.

DR. P. K. NEWBIGGING, in a paper read before the Royal Society of Edinburgh, on April 15, describes certain anomalous appearances presented by venous blood when left in contact with coloured porcelain. When blood drawn from a vein is either allowed to coagulate in a porcelain cup, or after coagulation is left in it for some hours, the dark purple tint, characteristic of venous blood, is found to be altered to the bright arterial hue, wherever it was in contact with any elevated degree of the green colour, which is communicated by means of the protoxide of chrome; and in one instance only, the same effect was produced by a device of a crimson tint. The author cannot refer this effect to the porcelain being of raised pattern, adhesion of oxygen to its surface, or to any peculiar change of the molecules of the blood: he, therefore, merely records the fact, adding that "the change of colour which takes place seems identical with the florid hue occasioned by arterialization of the blood from ordinary causes."—*Jameson's Journal.*

VINOUS FERMENTATION.

ON June 6 was read to the Royal Society a series of "Experiments on the Chemical Constitution of several bodies which undergo the Vinous Fermentation, and on certain results of the Chemical Action:" by R. Rigg, Esq. The special object of this paper is to show, 1st, that sugar is not constituted of carbon and water only; 2dly, that during the vinous fermentation, water is decomposed; 3dly, that neither pure carbonic acid nor alcohol, in the common acceptation of the term, is the product of this chemical action; and, 4thly, that fermented liquors owe some of their valuable qualities to peculiar products formed during fermentation. These views are confirmed, not merely by direct experiments, but likewise by other changes which fermented liquors undergo on being kept under circumstances favourable for farther chemical action; and the author having proved the existence of a compound of carbon, hydrogen, and nitrogen, and shown that water is decomposed during its formation, he thinks we are enabled to account for many other changes which occur during the decomposition of vegetable matter and the growth of plants: whence he proceeds to show that evidence of the presence of such a compound as the above in fermented liquors is afforded by the changes which take place in consequence of keeping them. In accounting for many of the phenomena of the vinous fermentation, the small excess of oxygen found in all these compounds, which undergo this chemical action, is an essential and indispensable circumstance; a conclusion which is corroborated both by the formation of these new compounds which have been described, and by the generation of the acetic, tartaric, or oxalic acid which is found in fermented liquors at all times, and in quantities varying according to the situations under which these fluids have been kept.—*Athenæum; abridged.*

DR. URE ON THE ELASTIC FORCE AND DENSITY OF STEAM.

THE specific gravity of vapours of different substances is a subject of modern investigation, for which we are primarily indebted to M. Gay-Lussac, whose experimental process consists in measuring the volume of vapour, furnished by a given weight of liquid. Having weighed a globule of glass, like the bulb of a large thermometer, he then filled it, by heating the bulb, and plunging the point of its capillary tube in the liquid to be vapourized, repeating the heating and immersion till it was entirely full. In this state, it was slightly heated, to expel a drop of liquid from the point, which was then hermetically sealed by the flame of a blow-pipe. The bulb was next introduced into a graduated bell-jar, about a foot long, and two inches in diameter, filled with, and inverted in, mercury, contained in a cast-iron trough. This jar was enclosed within a glass cylinder, open at top and bottom, the space between the two being filled with water or oil, whose hydrostatic pressure was supported by the subjacent column of mercury. The trough or pan, with these cylinders, was placed over a little furnace, whereby it was progressively raised to any desired degree of temperature. Eventually, the liquid became sufficiently hot to generate vapour, of such tension as to burst the thin glass bulb. In the case of water, he continued to heat the apparatus till the water contained between the two cylinders entered into steady ebullition; at which instant he measured the space occupied by the steam, noting the height of the mercury in the bell above the level of the bath; by deducting which height from that of the barometer placed in the same apartment, he found the pressure of the internal vapour. He afterwards reduced by calculation the length of the volume of mercury to what it would be at the temperature of melting ice; and made the small correction due to the volume of liquid between the cylinders.

In this research, we must take care that the whole of the liquid of the bulb be reduced into vapour, otherwise we shall fall into great errors. This circumstance would occur, were more liquid introduced into the bulb than would fill the whole space in the bell-jar at the given temperature.

That the densities of vapours vary proportionally with the pressures has also been experimentally determined; or, in other words, vapours when reduced by calculation to the same standard temperature are proportional to the elastic forces. Upon this principle, the column entitled specific gravity of steam, in a certain table, has been constructed; and the column entitled number of cubic feet of steam from one pound of water, has been computed from the preceding column, and the known weight of a cubic foot of water. One volume of water at 32° produces 1694 volumes of steam, at the temperature of 212° Fahr., and a pressure of 29·912 inches of mercury. The number 0·000589, given as the specific gravity of such steam, is obtained by dividing 1000 by 1698, on the supposition that one volume of water produces 1698 volumes of ordinary steam.—*Literary Gazette.*

COMPRESSION OF ATMOSPHERIC AIR.

M. SAVERESSE has patented an application and method of speedily compressing atmospheric air to 150 lb. on the square inch, by means of chemical action, and very little labour. He has successfully applied his invention to the manufacture of ginger-beer and soda-water, which he can produce perfectly in 15 minutes; the material being bottled and corked up with great speed and economy at the same time by means of an apparatus attached to the machine. This consists of a globular generator, from which ascends a cylindrical tube, in which the soda-water or ginger-beer is placed; the air, by ascent, carries the latter from thence into a small vertical cylinder, where the gas is purified; and thence is carried to the large cylinder, which is placed on a swivel, in order to more perfectly mix the liquid with the gas; from whence it is bottled off.—*Railway Magazine.*

RESPIRATION OF DETERIORATED ATMOSPHERES.

MR. C. T. COATHUPE has detailed to the British Association the results of a series of experiments on the Respiration of Deteriorated Atmospheres, which he had instituted to determine whether the injurious effects which have followed the respiration of charcoal vapours had depended on carbonic acid, as was generally thought, or on the specific agency of some other volatile product. The volatile products of the combination of charcoal he stated to be as follows:—

Carbonate of Ammonia,
Hydrochlorate of Ammonia,
Sulphate of Ammonia,
Volatile Empyreumatic Oil,
Carbonic Acid Gas,
Carbonic Oxide,
Oxygen,
Nitrogen,
Aqueous Vapour.

From a number of experiments on the elimination of carbonic acid during respiration, he arrived at the following results:—that 266·666 cubic feet of atmospheric air pass through the lungs of an adult in twenty-four hours, of which 10·666 are converted into carbonic acid, yielding 5·45 ounces of carbon, or 124·628 pounds annually, which will give a total of 147,070 tons of carbon as the annual product of the inhabitants of Great Britain and Ireland! The average amount of carbonic air found in atmospheric air in which animals had expired was found to be, for warm-blooded animals, 12·75 per cent., for the cold-blooded animals, 13·116 per cent. When the animals were removed, on becoming comatose, the average amount of carbonic acid was found to be 10·42 per cent. On confining a taper until its extinction, the quantity of carbonic acid found was 3·046 per cent. From hence it would appear that an atmosphere that has ceased to support combustion can support animal life for some time, which Mr. Coathupe proved by direct experiment.—*Athenæum.*

POISONING BY CHARCOAL.

DR. GOLDING BIRD has read to the British Association a series of

"Observations on Poisoning by the Vapours of burning Charcoal." An opinion has been held, that the vapours of carbonic acid were more injurious when produced by the combustion of coal and charcoal, than from any other source, on account of the admixture of light carburetted hydrogen gas. This opinion he dissented from, as it was well known that, in coal-mines, the fire-damp, as this gas was called, was inhaled with perfect impunity. To ascertain the *modus agendi* of the gas when inhaled, he made numerous experiments, by immersing animals in different mixtures of it and atmospheric air, as well as in the pure gas. In the latter case, the animals died asphyxiated, as when immersed in water or mercury, the spasm of the glottis preventing any portion of it from being inhaled. If not more than 25 per cent. be present, then respiration will go on, and its true poisonous effects take place. As to the amount of this gas necessary to produce fatal effects, Dr. Bird found that, as a general rule, any quantity above $3\frac{1}{2}$ per cent. was capable of producing death. Two opinions prevailed on the nature of these properties: the first was, that the gas acted negatively, as pure nitrogen or hydrogen is known to do, by preventing the due supply of oxygen. To test this opinion, was formed a mixture containing twenty-one parts of oxygen, and seventy-nine of carbonic acid, and death followed instantly from immersion in it; whilst the same result followed when the proportions were reversed, although a taper burned brilliantly in the latter combination; showing, that the burning of a light in any suspected situation is not always a safe test of the absence of danger. The second opinion is, that this gas, when respired, exerts a specific poisonous action on the nervous system. This latter Dr. Bird adopts, from various considerations drawn from his direct experiments, and from the symptoms observed in numerous cases.—*Athenæum.*

GAS FROM GRAPES.

An interesting experiment has been made at Bourdeaux, on the husks of grapes, when pressed, and the lees of the wine, in order to show their use for the purposes of lighting. A pound of the dried husks put into a red-hot retort, gave, in seven minutes, 200 litres of gas, which burnt with an intense light, and free from smoke or smell. A second experiment with the dried lees was equally satisfactory.

NEW MODE OF PREPARING CARBURETTED HYDROGEN GAS.

This invention has gained for M. Selligue the premium of 2000 francs proposed by the Société d'Encouragement of Paris. It consists in obtaining pure hydrogen by decomposing water by means of incandescent charcoal, and then carburizing it by mixture during the simultaneous decomposition of another liquid substance, rich in carbon and hydrogen. Among all known substances, that which appears to answer best is the oil of schist (l'huile de schiste).

The furnace is composed, 1st, of three vertical retorts, communicating, so as to form only one. In a double furnace, there will be six retorts. These are all open at both ends, but closed below by sliding stoppers, *(couvercles rodès)*, so that simple contact and the least pressure are sufficient to shut them firmly. The top of each retort is closed by a head fixed by keyed gudgeons and iron cement. Each head bears itself a stopper, or cover, like those below.

The first retort into which steam is introduced through a tube communicates below, by a tube twice bent, with the second, which connects at top with the third by a similar tube; and this third retort has, below, a vertical tube with branches, by which the gas is conducted to a refrigerator, and thence to the gasometer. This tube dips into a trough of water, to serve as a hydraulic closure. The third retort bears at top a funnel syphon, through which the carburizing substances are introduced. 2nd, Two horizontal tubes, placed in the sides of the vault, serve as boilers to vapourize the water; each communicates at one end with the first retort by an arched tube, and to the other end is attached a funnel syphon, by which the boiler is supplied with water. 3rd, Two furnaces. 4th, A chimney in four parts, uniting at first into two, and then into one, in order to regulate the fire with greater ease.

Operation.—Having filled with charcoal the first two retorts in each of the (double) furnaces, and suspended chains in the two last, in order to increase the surface, the fire is lighted; and when the retorts have attained a cherry red heat, a gentle flow of water and oil is made through the syphons. The water falling into the boilers is instantly evaporated, passes into the first retort, then into the second, where it is deprived of its oxygen; and, reaching the third, the hydrogen alone mingles with and carries along the carbonated hydrogen simultaneously formed from the oil in the last retorts.

The united gases then issue from the lower end of the third retort, and press off through the branches, while the more volatile matters are deposited in the reservoir of water.—*Annales des Mines; Mechanics' Magazine.*

GAS PRODUCED BY A NEW PROCESS.

An experiment in Gas-lighting by the Comte de Val Marino has been made in the presence of several scientific gentlemen. A small gasometer was erected for the purpose, which was connected by tubes with a furnace built of brick, and containing three retorts, one of which was supplied with water from a syphon; another was filled with tar; and both, being decomposed in the third retort, formed the sole materials by which the gas was produced. The principal agent employed to produce the gas was common water combined with tar; but, according to the theory of the inventor of this new species of gas, any sort of bituminous or fatty matter would answer the purpose equally as well as pitch or tar. After the lapse of about half an hour employed in the experiment, the gas was turned into the burners, and a pure and powerful light was produced, perfectly free from smoke or any unpleasant smell. The purity and intenseness of the flame were tested in a very satisfactory manner. The great advantage of this sort of gas over that produced from coal consists, it was said, in the cheapness of the materials employed in its production, the facility with which it is manufactured, and the perfection to which it is at once brought, without the necessity of its undergoing the tedious and expensive process of condensation and purification; for, in this instance, as soon as the preliminaries were completed, the light was produced in a perfect state within a few feet of the gasometer; which, although of inferior size, was said to be capable of affording light for 10 hours to at least 500 lamps or burners. It was also stated, that 1,000 cubic feet of gas manufactured by this process could be supplied to the public for about one-third the price now charged by the coal-gas companies: and it was said to be equally available for domestic use, and more safe than the common gas, inasmuch as small gasometers might at a trifling expense be fixed at the back of grates in private dwellings, from which the gas could be conveyed in India-rubber bags to any part of the house, thereby preventing the many accidents which occur by the use of tubes and pipes. The Comte de Val Marino has taken out a patent for his discovery, and he has improved upon the burners now in use, so as to render the light produced more pure and intense. For this improvement he is also secured by a patent.—*Times; abridged.*

MINERAL SPRING IN NEW SOUTH WALES.

DR. F. LHOTSKY has described in the *Philosophical Magazine* a Spring situated on the Menero Downs, 300 miles from Sydney, and skirting the Australian Alps on their eastern side. The formation about the spring is calcareous, and the immediate neighbourhood of it is formed by an extensive level of travertine, just like that with which the plains of Romagna are covered. The spring formed (after having been enlarged by Dr. Lhotsky), an aperture of 3 feet in diameter, and the water of a nearly constant temperature of 50° Fahr., with the air at a temperature of 98°. The water is perfectly clear, and only disturbed by the continued evolution of carbonic acid gas, with a large quantity of which the water is saturated. This having been subjected to evaporation, yields a salt, which Prof. Daubeny describes as containing muriates, calcareous earth in combination with carbonic acid, a trace of magnesia, and iron. The salt which effloresces upon the travertine level near the spring, contains carbonate of lime, common salt, a little sulphate of soda, and some peroxide of iron. The water has the taste of Seltzer water.

BRINE SPRING IN BAVARIA.

PROF. FORBES, in a paper read to the Royal Society of Edinburgh, on January 7th, gives the following details of the produce of an Intermitting Brine Spring discharging Carbonic Acid Gas, near Kissengen, in Bavaria.

The mineral discharge of the spring consists of from 35 to 40 Bavarian cubic feet per minute. The gas consists, according to Kastner, of almost pure carbonic acid, with a trace of azote. A pound of water is combined with 30½ French cubic inches of gas: but this gives no idea of the quantity which is evolved wholly uncombined. The specific gravity of the water varies from 1·064 to 1·030; and the solid matter in 1000 grains of water compared with that of sea-water by Dr. Murray, (sp. gr. about 1,028, are, Kissengen, 22·37230 gr.; sea-water, 30·309 gr.; carbonic acid in the Kissengen water, 2·06380 grains. The brine is converted into salt by evaporation on artificial hedges of blackthorn, 25 feet high, and 3½ feet thick; of which there are two rows, extending beneath sheds 6262½ feet, or nearly a mile and a quarter in length. After 6 falls in the process, the brine contains 17½ per cent. of salt, instead of 2½. From 2,600 to 2,800 cwt. of salt are thus extracted from the brine, or even 3,000. What a vast charge of water the atmosphere has carried off by this easy process! By the analysis, we see that 14 grains of muriate of soda are combined with 976 or water, or 70 times its weight; hence, besides all loss in manufacturing, 210 millions of pounds of water are annually disposed of; and since the brine is seven times stronger after spontaneous evaporation than before, (rising from 2½ to 17½ per cent.), six-sevenths of the water are driven off during this process. Hence, during that part of the year in which the manufacture continues, (for, during many winter months the frost renders it impossible,) the atmosphere actually absorbs 180 millions of pounds of water in the form of invisible vapour. The volume of this is three millions of cubic feet in round numbers, a quantity which, if uniformly distributed over the area of the thorny stacks through which it percolates, would reach the astonishing depth of 68½ feet, or more than *twice and a half* that of the thorns themselves.—*Abridged from Jameson's Journal.*

ANALYSIS OF HAILSTONES.

M. Girardin, in a letter to M. Arago, read at the French Academy of Sciences, on April 22, gives the result of his analysis of hailstones collected by him in the preceding February. The hailstones were first put into a bottle, washed with distilled water, and weighed 500 grains; the hail readily dissolved, and the liquid resembled water into which a few drops of milk had been let fall; it was turbid and whitish. Gradually there were formed in it a considerable number of white and very light flocks, which soon became one cloudy mass, and deposited at the bottom of the vessel. Next morning the liquor was perfectly limpid.

M. Girardin then details some beautiful experiments with the hail-water, the re-agents employed by him being nitrate of silver, oxalate of ammonia, nitrate of barytes, and nitric acid. It follows, that the hailstones so examined contained a considerable portion of organized and azotized matter, a sensible quantity of lime and sulphuric acid, but no sensible trace of ammonia.

Several chemists have directed their attention to the existence of an organic substance and saline matters in the atmosphere. Their experiments have proved that rain, in falling through the atmosphere, carries with it in solution, into the earth, ammoniacal salts, calcareous salts, and a fleecy matter, which is, without doubt, the origin of the deleterious principles which are designated by the term *miasmata.* Hitherto, however, no one has stated the existence of this organic matter in hailstones.—*Journal de Pharmacie; Philos. Mag.*

METEORIC IRON.

Prof. Shepherd, of South Carolina, has detailed to the British Association an analysis of Meteoric Iron, in which he has detected new elements, viz., chlorine and silicon. There were also exhibited some specimens of native terrestrial iron, the existence of which had been doubted. One had been found in Connecticut, and another in Pennsylvania. The latter was arseniuretted native iron. No traces of nickel or cobalt could be detected in either.

CONSTITUTION OF RESINS.

MR. F. W. JOHNSTON has submitted to the Royal Society three papers on this inquiry, with tabulated results of chemical examinations. The author's general conclusions are the following:

1. Many of the resins may be represented by formulæ, exhibiting their elementary constitution, and the weight of their equivalents, in which 40 C is a constant quantity.
2. There appear to be groups, in which the equivalents, both of carbon and hydrogen, are constant, the oxygen only varying; and others, in which the hydrogen alone varies, the two other elements being constant.

In the third part of the same series of investigations, the author examines the constitution of the resin of Sandarach, of commerce, which he finds to consist of three different kinds of resin, all of which possess acid properties. In like manner, he finds that the resin of the *pinus abies*, or spruce fir, commonly called *Thus*, or ordinary *Frankincense*, consists of two acid resins; the one easily soluble in alcohol, the other sparingly soluble in that menstruum. The gum resin *olibanum*, of commerce, was found to consist of a mixture of at least two gum resins, the resinous ingredient of each of which differs from that of the other in composition and properties.—*Athenæum.*

NEW RESIN.

ON April 20, Mr. Solly read to the Asiatic Society, an account of a New Resin, for specimens of which he was indebted to Dr. Babington. He stated that it was wholly unknown in this country; and the only author he was acquainted with by whom it had been mentioned was Dr. Ainslie, who, in his *Materia Medica* of Hindostan, described it as a substance similar to myrrh, and employed in native medicine, under the name of Cumbi Gum. Mr. Solly described, at length, its properties and nature, and the difference between it and gum myrrh. He stated that it was a resin, and not a gum-resin, and dissimilar to all the resins now known. In the state in which it had been brought over, it was mixed up with a considerable quantity of impurities, amounting to nearly 17 per cent. When purified, it was of a deep amber colour, brittle, easily fusible, and combustible; easily soluble in alcohol, and insoluble in water and caustic potassa. From the virtues attributed to it in India for various purposes, and, amongst others, in several diseases of cattle, he thought it reasonable to expect considerable medicinal powers, and stated that experiments on its properties were in progress.

RESIN OF BENZOIN.

M. BERZELIUS has asserted that the Resin of Benzoin, on distillation, furnishes an oil, which, like that of bitter almonds, is, by long contact with the air, converted into Benzoic acid. Since then, M. Frenéy has shown that this oil is changed into Benzoic acid under the influence of potass. M. Auguste Cahours has been making farther experiments, with the following results: in a pure state this oil is limpid, colourless, a little soluble in water, to which it communicates its odour and its flavour: it is soluble in alcohol and ether in every proportion; its odour is sweet and aromatic, its flavour acrid and burning, its specific gravity greater than that of water, and it boils at about 205°.—*Athenæum.*

BORING ROCKS BY CHEMICAL AGENCY.

M. PRIDEAUX has found that a stream of hydrogen oxygen gas applied to a piece of granite soon produced heat; and that, on the application of cold water, the stone became soft and yielded to the tool. The oxygen might be superseded by common air from a pair of double bellows; and the common coal gas would be found better than hydrogen, because it contained more inflammable matter in a given space; and it might be procured from any neighbouring gas-works, and conveyed down into a mine in a copper vessel. If oxygen gas should be found absolutely necessary, nothing was easier to procure where there was a steam-engine; they had only to get a little iron retort, and, in a county like Cornwall, abounding with manganese, they need never be at a loss for oxygen gas. He did not, however, suppose, in the present state of underground management in our mines, this plan would be adopted; but he was of opinion, should the Mining School be continued for two or three years, there would soon be many young men ready to carry it into effect.—*Proceedings of the Royal Cornwall Polytechnic Society.*

ALKALINE AND EARTHY BODIES IN PLANTS.

ON March 14, was read to the Royal Society a paper by Robert Rigg, Esq., to show that the solid materials which compose the residual matter in the analysis of vegetable substances, and which consist of alkaline and earthy bodies, are actually formed during the process of fermentation; whether that process be artificially excited, by the addition of a small quantity of yeast to fermentable mixtures, or takes place naturally in the course of vegetation, or of spontaneous decomposition. His experiments also show this formation of alkaline and earthy bodies to be always preceded by the absorption of carbonic acid, whether such acid be naturally formed or artificially supplied. This combination takes place to a greater extent during the night than in the day; and in general, the absorption of carbonic acid by the soil is greater in proportion as it is more abundantly produced by the processes of vegetation; and, conversely, it is the least when plants decompose this gas, appropriating its basis to purposes of their own system. Hence Mr. Rigg conceives there to be established in nature, a remarkable compensating provision, which regulates the quantity of carbonic acid in the atmosphere, and renders its proportion constant.—*Proc. Royal Soc.; Philos. Mag.*

GRADUATING GLASS TUBES.

Mr. C. T. Coathupe has detailed to the British Association an improved method of Graduating Glass Tubes for Eudiometrical Purposes. His apparatus for this purpose consists of a truly-bored cylindrical tube of iron, into which an iron piston is accurately fitted. Upon the rod of this piston a screw is cut, with a good pair of dyes, throughout its entire length. The rod is then filed of a triangular form, leaving sufficient of the threads of the screw at the rounded angles for an iron nut to traverse with security and freedom. To the upper extremity of this iron cylinder, a cap of the same metal is screwed, and into this cap is screwed an iron stop-cock. To the stop-cock is attached a glass graduated measure, with a narrow lip, by means of an iron connecting socket. Near the opposite extremity of the cylinder, an iron diaphragm, of about a quarter of an inch in thickness, is inserted, and is fastened in its place by a side screw or pin; and through this diaphragm a triangular-shaped hole is made, through which the piston-rod can slide easily up and down, but without lateral shake. Below the diaphragm, and at the extremity of the cylinder, the nut is inserted, whose action propels or retracts the piston without the possibility of the piston itself deviating from a right line. This nut enters the cylinder to the depth of about half an inch; and around the entering part a deep groove, in the form of the letter V, is turned; into which the pointed ends of three steel screws enter through the exterior of the cylinder at equal distances, in such a manner that the nut can be revolved freely, but cannot be otherwise displaced. From the entering part of the nut a projecting portion forms a shoulder, which is graduated in equal parts: which projecting portion may be of any diameter greater than that of the cylinder. On the exterior of the cylinder an index is fixed, by means of which any number of revolutions of the nut, or any number of equal parts of a revolution, can be ascertained.

To prepare this instrument for use, the piston is to be retracted to its lowest position, and the cylinder is to be filled with mercury, (without air bubbles,) by pouring a sufficient quantity of this metal into the graduating glass that is attached to the stop-cock, and turning the plug for its admission within the cylinder. If, when the cylinder is full, and while some mercury still remains within the graduated glass measure, we turn back the plug of the cock, we get the air-way of the plug filled with mercury: and by pouring off the superfluity, we have the instrument in a proper state to commence graduating any tube for laboratory purposes. Thus, if the tube to be graduated be about one-third of an inch in diameter, and we open the communication between the cylinder and the measure, and propel the piston by one whole turn of the nut, and then close the communication between the cylinder and the measure, by turning the plug of the cock,—we have within the measure a quantity of mercury, which, when poured into the tube to be graduated, will give a tolerably long space for the first division; and such similar spaces may be respectively marked, by repeating the process until the whole tube be divided.

CHEMICAL ABACUS.

Dr. Reid, of Edinburgh, has found this apparatus useful in introducing his pupils to a precise knowledge of the constitution of the more important chemical compounds. It consists of a frame of wood, across which wires are placed, and upon which beads are strung, as in the instrument employed by Chinese clerks, and to be seen in most museums. Each wire corresponds to a chemical element, and the beads to atoms; while the names of the elements are placed on the frame at the extremities of the wires. Dr. Daubeny has suggested an improvement of this instrument, by having the beads of different colours to correspond to the different elements. [The Chemical Abacus may now be seen at most of the philosophical instrument-makers' in the metropolis.]

MODE OF ANALYZING GERMAN SILVER. BY J. C. BOOTH.

The following method of analysis may be successfully practised by any one who possesses a little chemical knowledge. A small piece of about 20 grains is dissolved in nitro-muriatic acid with the assistance of a gentle heat, by which means the metals will be converted into chlorides. If the solution be filtered through a small paper-filter, and a white powder remain after washing with water, it is the chloride of silver, the presence of which metal in the compound is accidental and scarcely appreciable. The acidulated solution is then treated with sulphuretted hydrogen, which separates copper and a little arsenic. The sulphuret of copper is collected on a filter, treated with nitric acid in a gentle heat, till the sulphur appears whitish, then filtered, brought to boiling, precipitated with caustic potass, filtered, and weighed. 100 parts of this precipitate contain 79·83 of metallic copper. To the solution after filtering off the sulphuret of copper, a little nitric acid is added, and the whole heated in order to convert the protoxide into the peroxide of iron. Muriate of ammonia is then added to the same, and a small excess of ammonia, which precipitates only the peroxide of iron. This may be collected on a filter and weighed: 100 parts of it contain 69·34 of metallic iron. The solution is now to be treated with carbonate of soda, and evaporated to dryness; the dry mass is treated with hot water, and the residue washed and dried. This powder, consisting of carbonate of zinc and nickel, is mixed with half its weight of saltpetre, and ignited until the whole is nearly dry. It is transferred to a filter after being powdered in a small mortar, and is then washed two or three times with pure, but dilute, nitric acid, which dissolves the oxide of zinc, and leaves the *peroxide* of nickel. To the zinc solution carbonate of soda is added, the whole evaporated to dryness, treated with hot water, and the remainder after being dried and ignited is weighed: 100 parts contain 80·13 metallic zinc. The peroxide of nickel is dissolved in hydro-chloric acid, precipitated by caustic potassa, filtered off and weighed: 100 parts of it contain 78·71 metallic nickel.

The separation of nickel and zinc is ever attended with difficulty and some uncertainty, but it is rendered much more simple by the method above proposed, and which is not more inaccurate than others in use. Before weighing any of the above oxides, it it decidedly preferable to burn the filter after shaking off as much of the substance as possible into a platinum crucible; to add the ashes, and then subtract their weight from that of the oxide.—*Frank. Jour.; Mech. Mag.*

NEW METHOD OF DETERMINING THE CARBON CONTAINED IN CAST-IRON AND STEEL.

TAKE five grains of cast-iron, reduced to filings when the cast-iron is soft, or pulverized in a mortar when it is brittle; and mix it with sixty to eighty grains of chromate of lead, melted previously. Take away about a third or fourth of this mixture, and put it aside. To the remainder add five grains of chlorate of potass, which contain the quantity of oxygen required to change the iron into peroxide; afterwards introduce the threefold mixture into a tube of glass, similar to those for organic analyses, but which may be much shorter. Add to this the portion of the mixture of cast-iron and chromate of lead, which had been put aside. Lastly, adapt to the tube the common Liebig apparatus, for the analysis of organic substances. The portion of the tube containing the mixture without chlorate is heated; and when it is red hot, begin to heat that part which contains the chlorate, and the fire thus is advanced successively, in proportion as the disengagement of gas diminishes. The cast-iron at first burns completely by the oxygen of the chromate, and only a small quantity of this gas escapes through the tube. Afterwards, the temperature becoming higher, combustion is finished by the chromate of lead, which, in melting, oxidates the last portions of cast-iron. It is convenient to envelop the tube with a sheet of copper, because at the end it is necessary to heat it very strongly in order to obtain a complete fusion of the chromate. The oxidation of the cast-iron is complete, as you may assure yourself by grinding, after the combustion, the matter contained in the tube—not a particle of matter remaining which is attracted by the loadstone. The analysis is so easy that the whole proceeding is finished in less than half an hour. Of the perfect concordance of the results we may judge from the three following analyses, made on the same grey cast-iron obtained by the hot air process:—

1. Five grains have produced 0·582 of carbonic acid.
2. Five ditto ditto 0·585 ditto.
3. Five ditto ditto 0·588 ditto.

Carbon, therefore, 1st, 3·22; 2nd, 3·23; 3rd, 3·25.

When the cast-iron contains sulphur, not a trace of sulphuric acid is disengaged, all the sulphur remaining in the tube in the state of sulphate of lead. With the chromate of lead alone not all the carbon is obtained; the chromate, by losing much oxygen, becomes less fusible, and the oxidation penetrates with difficulty to the centre of the grains of a somewhat thick cast-iron.—*Annales de Chimie.*

TO ASSAY GOLD.

TAKE 6 grains of the gold, and place it in a small crucible, with 15 grains of silver, and from 8 to 12 grains of chloride of silver, according to the supposed impurity of the gold; lastly, add 50 grains of common salt, (chloride of sodium,) reduced to a fine powder, so as to prevent decrepitation; fuse the whole together for five minutes, and allow it to become cold; then take out the metallic button, beat it into a thin plate, and subject it to the action of dilute nitric acid, as in the ordinary mode of parting. By this plan, the tedious process of cupellation is avoided, the lower metals being wholly removed by the chlorine of the chloride of silver, and their place supplied by pure silver.—*Mr. L. Thompson; Philos. Mag.*

SEPARATION OF LIME AND MAGNESIA.

M. DÖBEREINER observes: if anhydrous chloride of magnesium be heated in the air, it absorbs oxygen and gives off chlorine. This decomposition, *i. e.*, the conversion of chloride of magnesium into magnesia, is more quick and complete when chlorate of potash is used instead of air as an oxidizing agent. This property renders the separation of lime and magnesia very easy. A mixture or compound of these two bodies, dolomite, for example, is to be dissolved in hydrochloric acid; evaporate the solution to dryness, and heat the residue in a platina capsule, till it ceases to yield hydrochloric acid, when add gradually to the mass heated to low redness, small portions of chlorate of potash, till the disengagement of chlorine ceases.' The residual mass will then be a mixture of magnesia, chloride of calcium, and chloride of potassium, which may be readily separated by treating the mixture with water, which dissolves the chloride of potassium and of calcium, while the magnesia is left; from the mixture of chloride of potassium and of calcium the lime is precipitated by carbonate of soda.—*Journal de Pharm.; Philos. Mag.*

Or, dissolve the combined earths in dilute nitric or muriatic acid, and precipitate the filtered solution by an excess of carbonate of soda; dry the precipitate, and place it in a coated green glass tube, so disposed that the whole can be heated to a dull red heat; when red, pass a current of well-washed chlorine through the tube for a few minutes: the lime will be converted into chloride of calcium, but the magnesia remain unacted upon. When the whole is cool, remove the mass from the tube, and boil it for a minute or two in water, filter the liquid, and wash the insoluble portion, (which is magnesia,) with water, and precipitate the lime from the mixed liquor by carbonate of soda. The heat should not exceed a dull red; else the mass may become vitrified at the part which touches the tube.—*Mr. L. Lewis; Philos. Mag.*

MANUFACTURE OF SODA.

PROF. GRAHAM has observed that, in the history of the useful application of Chemical Science to the Arts, the past year will be memorable for various improvements connected with the Soda Process. Sulphuric acid, which is the key to so many important chemical products, had been chiefly prepared from the sulphur of Sicily; the supply of which was suddenly much reduced by some fiscal regulations of the Sicilian government. This has led to the invention of several new processes for soda, which possess considerable merit as chemical discoveries. The most interesting is that of M. Gosage, in the neighbourhood of

Birmingham, for the recovery of the sulphur from soda-water; which promise not only a great saving of material, but a benefit of another kind, in abating, or entirely removing the nuisance of the escape of muriatic acid into the atmosphere in the ordinary soda process.—*Proceedings of the British Association.*

NEW SAFETY LAMP.

THE BARON EUGENE DU MESNIL has given to the British Association a description of a safety-lamp, invented by him in 1834. He stated that he had presented it to the French government in 1837, and that it had been now adopted, after a favourable report upon it by M. Ch. Combes.

This lamp consists of a body of flint glass, defended by a dozen of iron bars. The air is admitted by two conical tubes, inserted át the bottom, which are capped with wire gauze, and enter by the side of the flame. The latter rises into a chimney, which has a piece of metal placed in the form of an arch over its top; the chimney, however, being quite open. The consequence of this construction is, that a strong current is constantly passing up the chimney. When carburetted hydrogen passes in, the fact is discovered by numerous small explosions, and the whole glass work is thrown into vibrations, which emit a loud and shrill sound, which may be heard at a very considerable distance.

Prof. Graham stated, that the novelty in Baron du Mesnil's lamp was the circumstance of the chimney being quite open. He considered that the lamp of Davy was left almost perfect by that philosopher, and that all accidents proceeded from carelessness. He alluded to the deleterious effects of the *after-damp*, or carbonic acid left in the atmosphere of a mine after an explosion, which is believed to occasion often greater loss of life among the miners than the original explosion, and often prevented assistance being rendered in case of accidents. In many cases, it was certain that the oxygen of the air was not exhausted by the explosion, although, from the presence of 5 or 10 per cent. of carbonic acid, it was rendered irrespirable. The atmosphere might, therefore, be rendered respirable by withdrawing this carbonic acid, and he suggested a method by which this might be effected. He had found that a mixture of dry slaked lime and pounded Glauber's salts, in equal proportions, has a singular avidity for carbonic acid, and that air might be purified completely from that deleterious gas, by inhaling it through a cushion of not more than an inch in thickness, filled with that mixture, which could be done without difficulty. He suggested the use of an article of this kind by persons who descended into a mine to afford assistance to the sufferers, after an explosion: indeed, wherever the safety-lamp was necessary, and the occurrence of an explosion possible, the possession of this lime-filter would be an additional source of security.—*Proceedings of the British Association.*

IMPROVED MANUFACTURE OF PRUSSIAN BLUE.

THE Gold Isis Medal was last year awarded by the Society of Arts to Mr. Lewis Thompson, of Lambeth, for the following improved mode of manufacturing Prussian blue; extracted from the last publication of the Society's *Transactions:*

"In the usual mode of manufacturing Prussian blue, the requisite carbon and nitrogen are obtained by decomposing animal matter in contact with potash. In this process the potash, being reduced to the metallic state, causes the formation of cyanogen, in consequence of its affinity for that substance. The quantity of nitrogen furnished by a given weight of animal matter is not large, and, in the material employed by manufacturers, seldom perhaps exceeds 8 per cent.; and of this small quantity at least one half appears to be dissipated during the process, thus producing an enormous waste of material, and at the same time increasing the size of the apparatus. Mr. Thompson conceiving that the atmosphere might be made to supply, in a very economical manner, the requisite nitrogen, if allowed to act on a mixture of carbon and potash under favourable circumstances; the experiment proved correct; for the carbonaceous matter employed may be worked over again many times, and is even improved by each operation. To produce Prussian blue:

Take of potash or pearlash, ...	2 parts;
coke, cinders, or coals, . . .	2 parts;
iron turnings, ...	1 part.

Grind the whole together into a coarse powder, and place it in an open crucible or other convenient vessel, and expose the whole for half an hour in an open fire to a full red heat, stirring the mass occasionally. During the process, little jets of purple flame will be observed to arise from the surface of the mixture; when these have almost ceased to appear, which will happen in about the time specified, the whole must be removed from the fire, and allowed to cool; water is now to be added, so as to dissolve the matter soluble in that fluid, and the black matter remaining put aside for another operation. The solution, after being filtered, is to be mixed with one part of copperas, and muriatic acid added in the usual way to brighten the colour of the precipitate. The quantity of Prussian blue produced from a given weight of pearlash or potash is generally about one-fourth of the weight of the pure potash contained in the salt; but the larger the quantity operated upon at one time, the larger is the relative produce. Thus six ounces of pearlash, containing 45 per cent. of alkali, yielded only 295 grains of Prussian blue, whilst one pound of the same pearlash yielded 1355 grains. The Prussian blue here spoken of is the pure perferrocyanate of iron.

"In this process the potash is decomposed by the iron, producing potassium, which, being volatile, rises and combines with the carbon of the coke and with the nitrogen of the atmosphere, the oxygen of which has been removed by passing through the fire, or by the coke or cinders. In the mixture, the cyanide of potassium thus formed is next dissolved in water, and furnishes ferrocyanite of iron on the addition of sulphate of iron, and muriatic acid."

WHITE PRUSSIATE OF IRON.

Mr. R. Phillips, in a series of valuable experiments, has proved the so-called White Prussiate of Iron to be really a combination of ferrocyanide of potassium, with cyanide of iron, or Prussian blue; and that this, by the action of water, becomes resolved into these two salts, of which the ferrocyanide of potassium, being soluble in water, is removed by washing; and the residual cyanide of iron, or Prussian blue, presents the ordinary tint of that beautiful pigment.—*Proc. Brit. Assoc.; Literary Gazette.*

NEW ACID AND COMPOUND.

Mr. L. PLAYFAIR has described to the British Association the true constitution of Chlorochromic Acid, and methods of procuring others of a similar class; together with a new compound called Iodosulphuric Acid. The latter may be obtained by mixing two atomic proportions of iodine with one equivalent of sulphate of lead, and subjecting the mixture to distillation; or, by a more easy method, by sending a stream of pure sulphurous acid through a solution of iodine in pyroxilic spirit to complete saturation. It may be procured from this by distillation, drying over sulphuric acid, &c. It has a powerful acid taste, and acts violently on the cuticle, being analogous in composition with the chlorosulphuric acid of Regnault.—*Literary Gazette.*

VERATRIC ACID.

PROF. SCHRÖTTER has obtained a new and peculiar acid, by treating the seeds of Cevadilla, in the manner directed by Couerbe, with alcohol and sulphuric acid, for preparing veratria: add hydrate of lime to the alcoholic tincture, and distil the alcohol from the filtered liquor. The water liquor remaining with the residue on the separated veratria will then contain the new acid in combination with lime, and it will be requisite only to supersaturate it with sulphuric acid to separate the veratric acid, which, if the liquor be sufficiently concentrated, will crystallize in a few hours. The crystals are completely purified by washing them repeatedly with cold water, dissolving them in boiling alcohol, and treating them with purified animal charcoal. The property of subliming is the only one which this new acid has in common with that which MM. Pelletier and Caventou found in the seeds of Cevadilla, and which ought not, therefore, to be confounded with the acid discovered by Schrötter.— *Journal de Pharm.; Philosophical Magazine.*

CONVERSION OF CHLORATES.

ON Jan. 24, was read to the Royal Society, a paper "On the Conversion of the Chlorates," &c., by Mr. Penny. This communication is insusceptible of analysis: we can only present one or two of the author's results. After mentioning the steps by which Mr. Penny found his equivalents, he gives an account of many experiments with oxygen, nitrate of potassa, nitrogen, chlorate of soda, nitrate of soda, and other substances: he points out the discrepancies which exist between Thomson, Turner, and himself, which appear to be in some instances considerable. The experiment by which he effected the conversion of silver into chlorate, presented, for 100 parts of the former, 157·441 of the latter; and 100 parts of nitrate presented 84·374 of chlorate. Oxygen is the only substance on the equivalents of which chemists are agreed.—*Literary Gazette.*

FLUORIC ACID IN ANIMAL MATTER.

THE existence of fluoric acid as a constituent of human teeth, bones, and urine, has been acknowledged by chemists generally, since the year 1802; when Morichini declared that he had detected it in the teeth, and that he had been led to this examination by the discovery of fluoride in fossil ivory. Dr. G. O. Rees has repeated the experiments upon which the above belief was based by several continental chemists, but with reversed results, save in the case of fossil ivory, wherein he detected the fluoride; but he regards this as an extraneous matter introduced by the partial mineralization of the animal substance: he is convinced that no such constituent exists in recent ivory, the enamel of teeth, human bone, or urine; in fact, that fluoride of calcium should be expunged from the list of the constituents of animal substances. Dr. Rees attributes the fallacy of the continental chemists to their experiments being made in apparatus of bad glass, the peculiar action on which has been erroneously considered to denote the presence of the acid.—*Guy's Hospital Reports*, 1839.

IODINE IN COAL FORMATIONS.

M. BUSSY states, that on examining some mineral specimens obtained from a coal mine in a state of combustion at Commentry, (Allier,) he ascertained the presence of ammonia and iodine: though, in some of these specimens, he was afterwards unable to detect the iodine. The iodine was in the state of hydriodate of ammonia, and the acid had left the alkali. M. Bussy supposes this element to have existed in the bowels of the earth in the state of iodide of potassium, and that it is disengaged in the state of vapour by subterraneous heat. It is well known that ammonia is one of the constant products of the distillation of coal; but the presence of iodine in coal formations is a new fact, worthy of being noticed.—*L'Institut.; Philos. Mag.*

LANTANE.

A NEW metal has been discovered by M. Mosander while submitting the cerite of bastnas to fresh examination. The oxide of cerium, extracted from the cerite by the usual method, contains nearly two-fifths of its weight of the oxide of the new metal, which but little changes the properties of the cerium, and lies, as it were, hidden in it. For this reason, M. Mosander has named it Lantane. It is prepared by calcining the nitrate of cerium mixed with nitrate of lantane. The ceric acid loses its solubility in weak acids, and the oxide of lantane, which has a strong base, may be extracted by nitric acid diluted with 100 parts of water. This oxide is not to be reduced by potassium, but the latter separates, from lantanic chloruret, a grey metallic powder, which oxidates in water, disengaging hydrogen gas, and becoming converted into a white hydrate. Sulphur of lantane may be produced by strongly heating the oxide in the vapour of oxide of carbon. It is of a pale yellow, decomposes water with disengagement of sulphurated hydrogen, and is converted into a hydrate. The oxide of lantane is of a red brick colour, which does not appear to be due to the presence of ceric oxide. In warm water, it is converted into a white hydrate, which turns litmus paper blue, which has previously been made red by an acid. It is rapidly dissolved even by much diluted acids. When employed in excess, it is easily changed into a sub-salt. These salts have an astringent taste, without any sugary flavour. Their crystals are generally rose colour. The sulphate of potass does not precipitate them, unless they are mixed with salts of cerium. Digested in a solution of sal-ammoniac, the oxide is dissolved, sending out the ammonia by degrees. The atomic weight of Lantane is less than that assigned to cerium, or rather to the mixture of the two metals.—*Athenæum.*

SALICINE.

THE memoir of M. Pirria on Salicine, and the products derived from it, is said, by our continental neighbours, to be a remarkable work: the principal points of it are, 1st, the conversion of salicine by contact with acids into sugar and resin; 2nd, the conversion of salicine, by the action of sulphuric acid and bichromate of potass, into a volatile oil, which is a new circumstance in the mode of producing essential oils; 3rd, the exact isomeria of this oil with benzoic acid, under the triple analogy of composition, density of vapour, and atomic weight; 4th, the demonstration that this oil must be looked upon as a hydruret of a ternary base, analogous to benzoile; 5th, the close study of this hydruret, and that of the bodies derived from it, all of which are remarkable for their stability, their easy production, their beautiful crystallization, and the exactness of the phenomena to which they give rise.—*Athenæum.*

PHLORIZINE.

SOME recent experiments on Phlorizine, made by M. Stas, have proved interesting; not only because they furnish a new proof of the means employed by nature in producing certain colouring matters, by causing azote to enter into their constituent principles, but also because they show, that by the reaction of acids, Phlorizine changes into the sugar of grapes, leading to the supposition that the sugary matter of fruits arises from the influence of acids on the gummy or gelatinous parts.—*Athenæum.*

ON THE ALCOHOLIC STRENGTH OF WINES. BY DR. CHRISTISON.

VARIOUS accounts have been given of the alcoholic strength of wines by Mr. Brande, Julia-Fontenelle, and others. The author has been engaged for some time in experiments for determining the proportion of alcohol contained in various wines of commerce, and also the circumstances which occasion a variety in this respect. The present paper is an interim notice of the results.

The method of analysis consisted in the mode by distillation, which was applied with such contrivances for accuracy that nearly the whole spirit and water were distilled over without a trace of empyreuma, and without the loss of more than between 2 and 6 grains in 2000. From the quantity and density of the spirit, the *weight* of absolute alcohol of the density 793·9, as well as the *volume* of proof spirit of the density 920, was calculated from the tables of Richter, founded on those of Gilpin.

The author has been led to the general conclusion that the alcoholic strength of many wines has been overrated by some experimentalists, and gives the following table as the result of the investigations he has hitherto conducted. The first column gives the per-centage of absolute alcohol by weight in the wine, the second the per-centage of proof spirit by volume.

	Alc. p. c. by weight.	P. Sp. p. c. by volume.
Port—Weakest	14·97	30·56
Mean of 7 wines	16·20	33·91
Strongest	17·10	37·27
White Port	14·97	31·31
Sherry—Weakest	13·98	30·84
Mean of 13 wines, excluding those very long kept in cask	15·37	33·59
Sherry—Strongest	16·17	35·12
Mean of 9 wines very long kept in cask in the East Indies	14·72	32·30
Madre da Xeres	16·90	37·06
Madeira, all long in cask in East Indies — Strongest	14·09	30·80
Madeira, all long in cask in East Indies — Weakest	16·90	36·81
Teneriffe, long in cask at Calcutta	13·84	30·21
Cercial	15·45	33·65
Dry Lisbon	16·14	34·71
Shiraz	12·95	28·30
Amontillado	12·63	27·60
Claret, a first growth of 1811	7·72	16·95
Chateau-Latour, first growth 1825	7·78	17·06
Rosan, second growth 1825	7·61	16·74
Ordinary Claret, a superior "vin ordinaire,"	8·99	18·96
Rives Altes	9·31	22·35
Malmsey	12·86	28·37
Rudesheimer, superior quality	8·40	18·44
Rudesheimer, inferior quality	6·90	15·19
Hambacher, superior quality	7·35	16·15
Giles' Edinburgh Ale, before bottling	5·70	12·60
The same Ale, two years in bottle	6·06	13·40
Superior London Porter, four months bottled	5·36	11·91

In addition to certain obvious general conclusions which may be drawn from this table, the author stated, as the result of his experiments, that the alcoholic strength of various samples of the same kind bears no relation whatever to their commercial value, and is often very different from what would be indicated by the taste even of an experienced wine-taster.

Some observations were next made on the effect produced on the alcoholic strength of wines by certain modes of keeping or ripening them, more especially by the method employed in the case of sherry, madeira, and such other wines, which consists of slow evaporation for a series of years through the cask, above all, in hot climates. The researches made by the author on this head are not yet complete; but he is inclined to infer, from the experiments already made, that, for a moderate term of years, the proportion of alcohol increases in the wine, but afterwards, on the contrary, diminishes; and that the period when the wine begins to lose in alcoholic strength is probably that at which it ceases to improve in flavour. The increase which takes place at first in the alcohol of wine undergoing evaporation through the cask, appeared at first view parallel to the fact generally admitted on the authority of Söemering, that spirit becomes stronger when confined in bladder, or in a vessel covered with bladder, in consequence of the water passing out by elective exosmose.

The author, however, on repeating the experiments of Soemering, as related by various writers, (for he could not obtain access to the original account of them,) was unable, by any variation of the process he could devise, to obtain the results indicated by the German anatomist.—*Jameson's Journal.*

SPIRIT FROM THE BILBERRY.

In the province of Luxemburg, in Belgium, wine, brandy, and vinegar have been made exclusively from the fruit of the bilberry, *(vaccinium myrtillus,)* a shrub not hitherto turned to any use. This discovery promises to prove of good account in those northern regions in which the bilberry grows abundantly.—*Polytechnic Journal; abridged.*

MUCILAGE OF THE FUCI ECONOMICALLY APPLIED.

Mr. S. Brown, jun., in a paper on this subject, read before the Scottish Society of Arts, shows that, 1st, the common sea-wares contain great quantities of mucilage; 2nd, this mucilage is easily separable from the other ingredients the wares; 3rd, a solution of mucilage does not become stale or mouldy; 4th, by a very simple process it is converted into gum arabic. For this practical paper entire, *see Jameson's Journal,* No. 52, p. 409.

PROTECTIVE INKS.

The Committee of the Society of Arts for Scotland report that, after many experiments they have found Dr. Traill's Ink to be perfectly indelible and indestructible, if written on proper paper; and that the proper paper is that which is unsized and porous, like common printing-paper. It appears that some banks have adopted this ink for orders, letters of credit, &c. The Committee have also reported that Dr. Veitch's Ink, though indelible, destroys the paper.—*Jameson's Journal.*

EFFECTS OF MUSHROOMS ON THE AIR.

According to Dr. Mariet, Mushrooms produce very different effects upon atmospheric air from those occasioned by green plants under the same circumstances; the air is promptly vitiated, both by absorbing oxygen to form carbonic acid, at the expense of the vegetable carbon, or by the evolution of carbonic acid immediately formed; the effects appear to be the same both day and night.

If fresh Mushrooms be kept in an atmosphere of pure oxygen gas, a large proportion of it disappears in a few hours. One portion combines with the carbon of the vegetable to form carbonic acid, and another is fixed in the plant, and is replaced by azotic gas disengaged from the Mushrooms.

When fresh Mushrooms are placed, for some hours, in an atmosphere of azotic gas, they produce but little effect upon it. A small quantity of carbonic acid is disengaged, and, in some cases, a little azote is absorbed.—*Journal de Pharm.; Philos. Mag.*

PREPARATION OF THE WHITE OXIDE OF ARSENIC, (ARSENIOUS ACID,) IN CORNWALL.

MR. HENWOOD has described to the Cornwall Polytechnic Society, the difficult process of separating tin from copper and iron ores, with which it was often intermixed, and stated that the sulphur and arsenic expelled in the process was for many years rejected as utterly worthless. It was thought, however, that the White Oxide of Arsenic might be extracted from it; and about thirty years since it was successfully effected by the late Dr. Edwards, who established a manufactory for the purpose, near Perranwell, which was long the only one in the United Kingdom. Within four or five years, a second had been erected near Bissoe Bridge, and a third had been more recently set at work near Redruth. The materials were now collected from the burning-houses in all parts of Cornwall, and placed in a reverberatory furnace, having a long flue, and the heat is so slowly increased as to dissipate the sulphur before the arsenic is volatilized. This was assisted by means of small fires, communicating with the flue in different parts, by which means the sulphur was carried far on in it. The temperature was then carefully raised, and arsenic was driven off; but as this required a greater heat than the sulphur, it was the more readily deposited on the sides of the flue. When this had been continued for a sufficient time, perhaps weeks, or even months, the fire was extinguished, the flue was opened, and its contents were removed. The finer and purer parts of the arsenic were found in a crystalline state, near to the fire; and as the distance from it increased, it was more and more mixed with sulphur; this was again placed in the furnace for a repetition of the process. The purer portion now underwent a farther process in cast-iron retorts, in order to render it fit for sale, essentially different from that adopted in Germany.—*Trans. of the Society.*

DETECTION OF ARSENIC.

M. ORFILA has discovered a method of detecting the smallest atoms of Arsenic, even when administered in solution. He employs a lamp, the hydrogen gas of which is produced by a piece of zinc in sulphuric acid. The Arsenic, however small the quantity, when exposed to the flame of this gas, is carried along by it; and if a cold substance be presented to the end of the narrow tube conveying the flames, the Arsenic will be deposited on it like a spot.—*Athenæum.*

M. Orfila has subsequently inquired, whether Arsenic might not exist in all human bodies, independently of poisoning? Whether it might not be introduced into the body by means of chemical tests, and the apparatus used in searching for it in such cases? And, whether it might not also be communicated from the earth in which the corpse might have been interred? He has discovered that these three suppositions are all possible, and not of unfrequent occurrence; but he has also shown that the Arsenic obtained by his boiling process cannot be mistaken for the Arsenic introduced by either of the three above hypothetical causes, and has laid down precise rules for making the distinction. It is considered, that he has thus rendered great service to the study of forensic medicine.—*Literary Gazette.*

NEW METHOD OF DISTINGUISHING ARSENIC FROM ANTIMONY.

In the year 1836, Mr. J. Marsh submitted to the Society of Arts a new process of detecting Arsenic in cases of suspected poisoning, by means of hydrogen gas;* since which has been discovered in the process, a compound, in which antimony combines with hydrogen to form a gas (antimoniuretted hydrogen). This gas gives off metallic crusts, which, to the inexperienced eye, much resemble the metallic substance obtained from arsenical solutions by the same arrangements. Mr. Marsh has discovered a very simple distinguishing test for these bodies. The arrangements being previously made for testing antimony or arsenic, the piece of porcelain or glass whereon the metallic crusts are received, is to have a single drop of distilled water placed on it, and to be then inverted, so that the drop of water is suspended undermost. As the gas issues from the jet, inflame it as usual, but hold the piece of glass, &c., with its drop of water about an inch above the jet, or just above the apex of the cone of flame; the arsenic, by this arrangement, will be oxidized at the same time that hydrogen is undergoing combustion; and coming in contact with the drop of water held above, forms with it a strong or weak solution of strong arsenical acid, according to the quantity of arsenic present, should that substance have been in the mixture submitted to examination. A very minute drop of Hume's test, (the ammoniacal nitrate of silver,) being let fall on the solution so obtained, if arsenic be present, the well-known characteristic lemon yellow-colour produced by this test when used in testing for that substance, is immediately produced, namely, the insoluble arsenite of oxide of silver. Antimony, under these circumstances, from being insoluble, produces no change. Mr. Marsh, when much arsenic has been present in the matter examined, has used a clean glass tube, 6 inches long, and about $\frac{1}{2}$ an inch in diameter: slightly moisten the interior of the tube with distilled water, not allowing the hands or fingers to come in contact with the water; and hold the tube thus prepared vertically over the apex of the jet of burning gas. Thus, a strong solution of the substance may be obtained, and which may be tested by Hume's test, or any other of the usual tests employed for arsenic, &c.—*Communicated to the Philos. Mag.; abridged.*

COLOURS OF STEAM AND OF THE ATMOSPHERE.

Prof. Forbes has read to the Royal Society of Edinburgh two communications, of which the following are abstracts.

The author accidentally remarked, that the colour of the sun seen through vapour issuing from the safety-valve of a locomotive engine is deep red, exactly similar to that which a column of smoke or smoked glass gives to it.

* The details of this important process will be found in the *Arcana of Science*, 1837; pp. 157–163.

He next noticed that this colorific character of steam extended but a short way beyond the orifice, and that it gradually became more opaque, and perfectly white like noon-day clouds, both for transmitted and reflected light. At moderate thicknesses, in this state, its opacity is complete.

These observations have been fully confirmed by direct experiments on high-pressure steam. At the moment of issuing from the steam-cock, it is perfectly transparent and colourless; at some distance from the orifice, it becomes transparent and orange-red; but, still farther off it is white, and merely translucent. These properties were traced in steam from a pressure above that of the atmosphere of 55 lb., down to an excess of only three or four; and, as in all cases the redness of the transmitted light was more or less distinctly seen, (and an excess of 10 or 15 lb. does as well as any higher pressure,) it was concluded that the effect of partial condensation in producing the phenomenon would be rendered visible in great thickness of vapour of the lowest tension.

The great analogy of the colour of steam to that which the clouds assume at sunset, or distant lights in certain conditions of the atmosphere, lead the author to suggest this singular property of condensing vapour as the probable cause of those phenomena, of which no satisfactory explanation could be given whilst this fact remained unknown. The prognostics of weather derived from the colours of the sky also receive elucidation from the fact.

Judging from the similarity of colour of steam and that of nitrous acid gas, and the remarkable power of absorbing certain definite rays of the spectrum discovered in that gas by Sir David Brewster, the author thought it probable that similar lines might be discovered in the spectrum formed by light transmitted through steam; and that these might be found to coincide with the *atmospheric lines* of the spectrum noticed by the same philosopher. The experiment was made with great care, but the expected result has not been hitherto obtained. The general action of steam on the spectrum is to absorb the violet, blue, and yellow rays, finally, leaving only the red and orange, with an imperfect green.

Since a portion of watery vapour in a confined space, originally transparent and colourless, may become, by mere change of temperature, first deep orange-red and transparent; and, finally, white and semiopaque; the author notices another analogy with the singular effect of temperature in deepening the colour of nitrous acid gas, and thinks that these facts may one day throw some farther light on the difficult subject of the mechanical constitution of vapours, and particularly of clouds.

The object of the second communication was to develop an application of the above fact. The discovery that steam in a certain stage of condensation is deeply red coloured for transmitted light, seemed to offer a probable solution of a difficulty which has never yet been fairly met, namely, the red colour of clouds at sunset, and the redness of light transmitted through certain kinds of fogs.

A pretty full history of theories proposed to account for the colours of the atmosphere was first given; it being obtained in almost every case from an examination of the original authorities. These theories were reduced under three general heads, exclusively of that of Göethe, and of most writers before Newton, that the blue colour of the sky results from a mixture of light and shade; and that of Muncke, that that colour is merely *subjective*, or arises from an ocular deception. The remaining theories are:—

1. That the colour of the sky is that transmitted by pure air, and that all the tints it displays are modifications of the reflected and transmitted colours. This is more or less completely the opinion of Mariotte, Bouguer, Euler, Leslie, and Brandes.

2. That the colours of the sky are explicable by floating vapours acting as thin plates do, in reflecting and transmitting complementary colours. This is the theory of Newton and most of his immediate followers, and more lately of Nobili.

3. On the principle of opalescence and of specific absorption, depending on the nature and unknown constitution of floating particles. This head is intended to embrace the various opinions of Melvill, Delaval, Count Maistre, and Sir D. Brewster.

To the last named philosopher, however, the merit is due of having conspicuously turned attention to the important, complex, and hitherto unexplained phenomena of absorption, which he has proved to be totally inconsistent with Newton's theory of the colours of nature, (considered as those of thin plates;) and he has farther demonstrated the inapplicability of it in the case of the colours of the atmosphere, by showing that their constitution is wholly distinct from that which any modification of Newton's theory would assign, by a series of experiments, of which as yet the results only are announced.

Since, then, the constitution of the atmospheric colours analyzed by the prism resembles that produced by absorption, the question is, to what medium are we to refer that absorptive action? Evidently not to pure air, since a distant light is red in a fog, and in clear weather white, or nearly so. The author is disposed to attribute the effect to the presence of vapour in the very act of condensation. This intermediate or colorific stage occurs between the colourless and transparent form of steam wholly uncondensed; and that which may be termed the state of *proximate* condensation in which it is seen to issue from the spout of a tea-kettle, when it is likewise colourless, but semiopaque. During the transition, it was shown in the former paper that steam becomes intensely red, and remains transparent. The absorptive action resembles then, so far, that of the atmosphere observed under certain meteorological conditions; the dark lines and bands noticed by Sir David Brewster in the atmospheric spectra have not been discovered, and so far the analogy is as yet imperfect.*

* Some plausible reasons are assigned why these bands should not have appeared in the experiment as it was made, when steam in every stage of condensation must necessarily have been present; nor does it seem easy to devise a form of experiment free from this objection. A very important observation would be to examine the spectrum produced by a distant artificial light seen through a red fog.

In applying this theory to the colours of sunset in particular, the author quotes many acknowledged facts to prove that the redness of the sky is developed precisely in proportion to the probable existence of vapour in that critical stage of condensation which should render it colorific. And he applies the same reasoning to account for the prognostics of weather, drawn from the redness of the evening and morning sky.—*Jameson's Journal.*

ANALYSES OF THE DIFFERENT ASPHALTS. BY M. P. BERTHIER.

The Seyssel.—There are at Seyssel, (in the department of L'Ain) three kinds of minerals: 1st, the sandy mineral; 2nd, the very fusible calcareous mineral; 3rd, the calcareous mineral of difficult fusion.

The *first* of these melts in boiling water, and becomes detached from the stony matters to which it was adherent. It rises to the surface, or sticks to the sides of the vessel in brown lumps, or forms a transparent coating of a brownish red colour. A rich specimen of it gave

Bituminous oil	·086	} bitumen ·106
Carbon	·020	
Quartzy grains	·690	
Calcareous grains	·204	
	1·000	

In the mass it is much less rich. When purified by hot water, this bitumen is called *la graisse*, grease.

The *second* variety is called at Seyssel *asphaltum*. It may be pulverized and sifted, but the powder spontaneously forms into balls. The specimen analyzed contained ·11 of bitumen, 5·89 of carbonate of lime, without clay, and quite pure.

The *mastic* of Seyssel is prepared by mixing nine parts of *asphaltum* with one of the pure *grease* extracted from the sand.

The *third* variety is a compact limestone, in extremely thin, parallel beds.

It consists of

Bituminous matter	·100
Argil	·020
Sulphate of lime	·012
Carbonate of lime	·868
	1·000

The bituminous mineral of *Belley* is very similar to the preceding. It is found in several communes in very considerable quantities, near the surface of the ground. It is of variable quality. A variable specimen yielded,

Carb. of lime	·824
Carb. of magnesia	·020
Sulphate of lime	·013
Argil	·023
Bitumen	·120
	1·000

Bitumen of Bastennes.—This bitumen flows out from several openings or springs, mixed with water. Analysis of the solid gave

Oily matter	·200	} bitumen.
Carbon	·037	
Fine quartzy sand, mixed with argil	·763	
	1·000	

Bitumen of Cuba.—This is transported to Europe under the name of *Mexican asphalt* or *chapopote.* It is a solid bitumen, which exists in abundance near Havanna. It may be used with great advantage in paving. It consists, like the greater number of natural bitumens, of at least two different substances, the one soluble, and the other insoluble in ether and spirits of turpentine. It is the relative proportion of these substances which imparts to each its peculiar properties.

Bitumen of Monastier (Haute-Loire).—This does not soften in the least in boiling water, and hence cannot be extracted by simple means in the large way. It contains:—

Bituminous oil	·070	} ·105
Carbon	·035	
Water	·045	
Gas and vapours	·040	
Quartz and Mica	·600	
Ferruginous argil	·210	
	1·000	

This bitumen of the Haute-Loire differs essentially from those of Seyssel and Bastennes by its infusibility in boiling water, and its fusibility in alcohol.—*Annales des Mines; Franklin Journal.*

NEW INDESTRUCTIBLE INK.

Dr. Coxe, of Pennsylvania, having gathered a Fungus and placed it on a sheet of white paper, leaving it until the next day, found several drops of an inky fluid, slowly trickling from the inner surface, which had assumed a black appearance; and, by placing the Fungus in a glass, the whole, except the outer skin, liquefied. The colour of the fluid was rather of a deep bistre than black; and, being left in a glass, in a few hours it separated into a solid sediment, with a lighter coloured fluid swimming above. Having afterwards collected a considerable quantity of fluid from the same species, the doctor dried an extract, of a pretty deep black colour, of both parts conjoined, which would otherwise have separated. This, on trial, formed an admirable bistre-like water-colour, well adapted for drawing when mixed with gum.

By diluting this substance with water, an ink was speedily made; and writing with it was exposed to the sun for several months with little change: chlorine and euchlorine gas, muriatic acid, and ammoniacal gases, effected a trifling change; and muriatic acid gas destroyed very considerably the dark tint of the writing.

From several experiments, Dr. Coxe infers that an excellent *India Ink* might be prepared as above for drawing, engravings, and, as an ink, indestructible by any common agents.

Natural History.

ZOOLOGY.

RESEARCHES IN EMBRYOLOGY.

DR. MARTIN BARRY has presented to the Royal Society the *Second Series* of his Researches. The author having in the *First Series** investigated the formation of the Mammiferous Ovum, describes in this second series its incipient development. The knowledge at present supposed to be possessed of the early stages in the development of that ovum, consists chiefly of inferences from observations made on the ovum of the bird. It appearing, therefore, highly desirable to obtain a series of observations in continuous succession on the earliest stages of development of the ovum of the mammal, Dr. Barry has purposely confined his attention to a single species, namely, the rabbit, of which he examined more than 100 individual animals. Besides ova met with in the ovary, apparently impregnated, and destined to be discharged from that organ, he has seen upwards of 300 ova in the fallopian tube and uterus; very few of the latter exceeding half a line in their diameter. The results of these investigations have compelled the author to dissent from some of the leading doctrines of Embryology, which at present prevail, as respects not only the class mammalia, but the animal kingdom at large. — *Philosophical Magazine*, No. 92; which see for the corroborative facts.

ARTIFICIAL INCUBATION.

AN exhibition has been opened in Pall Mall, bearing the classical denomination of "*The Eccaleobion*;" the object of which is the Hatching of Chickens by Heat. By such means, in this exhibition-room alone, it is possible to bring into existence, through winter as well as summer, a hundred birds a-day, or upwards of 40,000 in a year. The exhibition is, however, chiefly to be prized as the means for investigating the process of nature in advancing an organic substance to vitality. Eggs may be broken daily, as they proceed in their development, and examined by the aid of the microscope; thus exposing to view the actual commencement of life, and the gradual formation of those members which life is to animate. In an experiment, about the fourth or fifth day, the first trace of a distinguishable organ appears, where an opaque and cloudy spot had hitherto been witnessed. This is the heart of the bird. By placing the egg conveniently on cotton, in a common wine-glass, with water at 98°, and keeping it to that temperature, it is easy to continue the observation for eight or ten hours. From the heart, fine filaments spread over the surrounding

* A notice of the *First Series* will be found in the Year-Book of Facts, 1839, p. 158.

surface. Anon, circulation begins to appear in them; and soon we are able to distinguish the auricles, veins, and arteries, in full play—in one, yellowish atoms flowing rapidly like sand in an hour-glass; and in the other, assuming a redder colour. Again, a dark speck is observed; and, even before this single broken egg is exhausted, it is ascertained to be the future eye of the chicken. Day after day, similar microscopic inspection will show how the work advances—fibres, brain, intestines, muscles, bones, beak, feathers, are all formed in this wonderful sphere—the yolk, the white, and the shell, contributing their various functions, till about the fourteenth or fifteenth day, when the birds are so far matured in the shells as to be hatched by keeping them moderately warm; the warmth of the human body, or 98° of Fahrenheit, being the standard.

From a description of the machine, which is capable of containing above 2,000 eggs at a time, are detached the following relative facts. Greater heat is required to bring forward, say 1,000 fresh eggs, than to mature 1,000 during the last week of incubation. Thus, a heat is, after a while, generated by the eggs themselves; or a lesser heat is required at the end than at the beginning. Birds in a healthy condition require no assistance to effect their escape from the shell; which operation they perform by making a circular fracture of the shell with their bill, and bursting its integuments by strong muscular exertion. In cases of weakness in the bird, or defective hatching, assistance may be given; but such birds generally die in a few days, or, perhaps, hours. Darkness is also considered favourable to the process. Few eggs, excepting those of rare or foreign birds, are worth the trial of hatching, if more than a month old. Very hot weather destroys vitality in a few days. An egg having been frozen is, of course, also worthless. This machine does not, as is frequently the case with eggs sat upon by the parent bird, ever addle them. This evil is occasioned by the alternation of heat and cold, arising from the hen's unsteady sitting. The warmth imparted by the Eccaleobion is uniform and continued. A flush of fresh cool air passing over them each day, for a short time, is considered beneficial. The chicken, at the time it breaks the shell, is heavier than the whole egg was at first.—*Literary Gazette; abridged.*

DIGESTIVE ORGANS OF INFUSORIA.

M. EHRENBERG considers the separation and isolation of the stomachic vessels of *Infusoria* as surprising only to those who have not observed earth-worms cut to pieces. Mere pieces, he remarks, let them be ever so minute, contract at each extremity in such a manner that but very little of the contained juices escapes, and a similar effect is produced by the contraction of the isolated stomachs of the *Infusoria*. One fact, undoubtedly, is more forcible than all arguments; and M. Dujardin only regrets that that of a venicle containing fragments of *Oscillatoria* has not presented itself several times to the observer; for, with respect to the alleged stomachs without contained aliments, even when they appear slightly coloured, the false comparison with the pieces of earth-worms will not suffice to prove that the globules are not part of the gelatinous substance of the Infusoria; since M. Dujardin has frequently seen these globules coloured, either from their having a tinge of their own, or that this effect was the result of an optical illusion, or of a phenomenon of accidental colours.—*Annales des Sciences, Natur.; An. Nat. Hist.*

MEADOW LEATHER.

An interesting vegetable production, having a deceptive resemblance to white dressed glove-leather, has lately been found on a meadow above the wire-factory at Schwartzenberg, in the Erzgebirge. A green slimy substance grew on the surface of the stagnant waters in the meadow; which, the water being slowly let off, deposited itself on the grass, dried; became quite colourless, and might then be removed in large pieces. The outside of this natural production resembles soft, dressed glove-leather, or fine paper; is shining, smooth to the touch, and of the toughness of common printing (unsized) paper. On the inner side, which was in contact with the water, it has a lively green colour, and we can still distinguish green leaves, which have formed the leather-like pellicle. Dr. Ehrenberg has submitted this meadow-leather to a microscopic examination, and has found it to consist most distinctly of *Conferva capillaris, Conferva punctalis, and Oscillatoria limosa*, forming together a compact felt, bleached by the sun on the upper surface, and including some fallen tree leaves and some blades of grass. Among these *Confervæ* lie scattered a number of siliceous infusoria, chiefly *Fragilariæ* and *Meridion vernale*, including sixteen different sorts, belonging to six genera; besides three sorts of infusoria, with membranous shields, and dried specimens of *Anguillula fluviatilis*.—*Philosophical Magazine.*

POLARIZATION OF LIGHT BY LIVING ANIMALS.

Mr. J. F. Goddard having observed that the scarf-skin of the human subject, sections of human teeth, the finger nails, bones of fishes, &c., possessed the polarizing property, he was led to examine some living objects with his Polariscope, when he discovered that, among many others, the larvæ and pupa of a tipulidan gnat (the *Corethra plumicornis*), possessed the same property, and that in a very eminent degree. Its existence in the different substances above enumerated is exceedingly important; but that it should also exist in living animals is infinitely more so, and opens a new field altogether, disclosing characters that lead to an intimate knowledge of their anatomy, and which cannot possibly be discovered by any other means.

This creature is found in large clear ponds, generally in great abundance when met with; but this is by no means common. Having constructed a water-trough, made with two slips of glass about 1·25 inch wide and two inches long, with very narrow slips of thin glass cemented with Canada balsam between them, at the bottom and sides, thus having it open at one end with about 0·050 of an inch space between in the middle, Mr. Goddard filled it with clear water, in which he placed some of the larvæ; and such was the extraordinary transparency of the creature, as to display, in a most beautiful manner, the whole of its internal structure and organization; and which, when viewed in polarized light, present the most splendid appearances Thus, when they place themselves with their head and tail both in the plane of primitive polarization, or in a plane at right angles to it, they have no action upon the light transmitted through them; but when in a plane inclined 45° to the plane of polarization, the light is depolarized, their whole bodies becoming illuminated in the most brilliant manner, varying in intensity according to their size, and the nature of the different parts and substances; the peculiar interlacing of the muscles marking out regular divisions, which, as the creature changes its position with regard to the plane of polarization, exhibit all the varied hues and brilliant tints that have rendered this important branch of physical optics exceedingly interesting.

And, while thus viewing them, if we place behind a thin plate of sulphate of lime or mica, the change and play of colours, as the creature moves, are greatly increased, and are surpassingly beautiful.

These phenomena in the larvæ of the *Corethra plumicornis* are seen, if possible, in a more splendid manner, in the spawn of many large fishes; but more particularly, in the young fishes themselves, many of which, in their early state, are equally transparent, particularly if of marine production.—*Philosophical Magazine.*

MEXICAN MUMMIES.

A MILLION of mummies, it is stated, have lately been discovered in the environs of Durango, in Mexico. With them were found fragments of finely worked elastic tissues, (probably our modern India-rubber cloth,) and necklaces of a marine shell found at Zacatecas, on the Pacific, where the Columbus of the forefathers of the Indians probably, landed from Hindostan, or from the Malay or Chinese coast, or from their islands in the Indian Ocean.—*Philadelphia Presbyterian; Silliman's Journal.*

WEIGHT OF BLOOD IN THE HUMAN BODY.

THE following account of the quantity of blood in the human frame at the different stages of existence, is given by Dr. Valentin in the *Bulletin Gênêral de Thêrapeutique Mèdicale:*—In the male subject, the blood at birth weighs 0·73 of a kilogramme (the kilogramme is 2lb.); at one year, 2·29; at two years, 2·75; at three, 3·03; at four, 3·46; at five, 3·83; at six, 4·14; at seven, 4·62; at eight, 5·10; at nine, 5·52; at ten, 5·99; at eleven, 6·38; at twelve, 7·11; at thirteen, 8·10; at fourteen, 9·28; at fifteen, 10·64; at sixteen, 12·24; at seventeen, 13·16; at eighteen, 14·04; at nineteen, 14·52; at twenty, 14·90; at twenty-five, 15·66; at thirty, 15·80; at forty, 15·78; at fifty, 15·47; at sixty, 15·02; at seventy, 14·45; at eighty, 14·04.—In females, it is as follows: At birth, kilogramme, 0·59; at one year, 1·88; at two, 2·31; at three, 2·52; at four, 2·87; at five, 3·14; at six, 3·39; at seven, 3·74; at eight, 4·02; at nine, 4·55; at ten, 4·90; at eleven, 5·32; at twelve, 6·19; at thirteen, 7·03; at fourteen, 7·72; at fifteen, 8·37; at sixteen, 9·01; at seventeen, 9·95; at eighteen, 10·77; at twenty, 11·04; at twenty-five, 11·17; at thirty, 11·18; at forty, 11·49; at fifty, 11·85; at sixty, 11·50; at seventy, 10·89; at eighty, 10·45.

NEW MODE OF RESUSCITATION FROM DROWNING.

At the annual meeting of the Bristol Humane Society, the Society's silver medal was presented to Dr. Fairbrother, of Clifton, for his exertions in recovering a boy who had been under the water in the floating harbour about half an hour, another quarter of an hour having elapsed before the doctor could operate on the body. The most remarkable feature in this case is the new mode by which Dr. Fairbrother succeeded in his laudable object: namely, by closing the boy's mouth with his finger, sucking off the foul air from his lungs through the nostrils, and promoting respiration by pressing on the abdominal muscles on the side. The usual method is to inflate the lungs; but it is very seldom that persons are recovered by this method if they have been longer than a few minutes under the water.

LONGEVITY.

(From the First Annual Report of the Registrar-General of England, in 1837-8.*)*

Among the diversities which especially demand attention, and by which there is least danger of being led to false conclusions, are those which relate to Longevity, showing the varying proportions of deaths in old age in different portions of the kingdom. From a few instances of Longevity no inference can be safely drawn; but the fact that, of the deaths in any district, a comparatively large portion is above the age of seventy, is a strong presumption in favour of the healthiness of that district. These proportions are found to vary greatly. In the whole of England and Wales, out of 1000 deaths, 145 have been at the age of seventy and upwards; while in the North Riding and northern part of the West Riding of Yorkshire, and in Durham, excluding the mining districts, the proportion has been as high as 210. In Northumberland, excluding the mining district, Cumberland, Westmoreland, and the north of Lancashire, the proportion has been 198; in Norfolk and Suffolk, 196, in Devonshire, 192, and in Cornwall, 188. In contrast with this evidence of the large proportion of persons who attain to old age in these more thinly-peopled portions of the kingdom, we find results extremely different where the population is densely congregated. In the metropolis and its suburbs the proportion who have died at seventy and upwards has been only 104: and even this proportion is favourable, when compared with that of other large towns; the proportion in Birmingham being 81, in Leeds, 79, and in Liverpool and Manchester only about 63. A comparison of the mining parts of Staffordshire and Shropshire, and of Northumberland and Durham, with the rural districts surrounding each, exhibits great differences in this respect, the former averaging 109, and the latter 76. A very marked diversity also appears in the proportion of deaths of infants in different parts of the country. In the mining parts of Staffordshire and Shropshire, in Leeds and its suburbs, and in the counties of Cambridge and Huntingdon, and the lowest parts of Lincolnshire, the deaths of infants under one year have been more than 270 out of 1000 deaths at all ages; while in the northern counties of England, in Wiltshire, Dorset and Devon, in Herefordshire, and Monmouthshire, and in Wales, the deaths at that age, out of 1000 of all ages, have scarcely exceeded 180.

DENTITION OF MAMMALIA.

Mr. Goodsir has read to the British Association, a paper "On the the Follicular Stage of Dentition in Ruminants, and other Orders of Mammalia." Referring to his previously published paper on the same subject, he has since ascertained that all the permanent teeth, excepting the first molar, which does not succeed a milk tooth, are developed from the internal cavities, and the depending folds of the sacs of composite teeth are formed by the lips of the follicles advancing inwards after the latter closes. Mr. Goodsir farther described the earliest indications of the teeth in the embryo state, and generally the gradual formation of teeth. The author then referred to the distinction which must be drawn between those permanent teeth, which are developed from the primitive, and those which are developed from the secondary groove; and stated that he had been in the habit of dividing the teeth of these animals, the dentition of which he had examined, into three classes, viz.: 1, milk or primitive teeth, developed in a primitive groove, and deciduous; 2, transition teeth, developed in a primitive groove, but permanent; 3, secondary teeth, developed in a secondary groove, and permanent. Mr. Goodsir expressed a hope that other anatomists would verify and extend this line of research, as the results appeared to him not only confirmatory of certain great general laws of organization, but as leading, in his opinion by the only legitimate path, to the determination of the organic system to which the teeth belong, (a subject exciting great interest at present,) and as it might enable us, in investigating the relations of dental tissue to true bone, to avoid the error of confounding what there appears to be a tendency to do, analogy with affinity. The paper concluded with a recapitulation of the principal facts contained in it: 1, in all the mammalia he had examined, the follicular stage of dentition was observed; 2, the pulps and sacs of all the permanent teeth of the cow and sheep, with the exception of the fourth molar, were formed from the minor surfaces of cavities of reserve; 3, the depending folds of the sacs of composite teeth were formed by the folding in of the edges of the follicle towards the base of the contained pulp, the granular body assisting in the formation of these folds; 4, the cow and sheep, (and probably all the other ruminants,) possessed the germs of canines and superior incisives, at an early period of their embryonic existence.—*Literary Gazette and Athenæum.*

STRUCTURE OF THE TEETH.

Mr. Nasmyth has read to the British Association an elaborate paper "On the Structure of the Teeth," embodying the uninterrupted researches of several years, and conclusions on: 1, the covering of the enamel; 2, the structure of the teeth; 3, the structure of the pulp, and its relation to the development of the ivory. In this communication, he endeavours to prove that the pulp, the formative organ of the tooth, is composed of cells; and that the character of the teeth themselves is more or less cellular. From observations which Mr. Nasmyth has made in continuation of this inquiry, "On Epithelium," he has been led to the conclusion that its structure also is cellular. The details of these important papers are reported in the *Literary Gazette*, No. 1183; and *Athenæum*, No. 618.

DEAFNESS.

OBSTRUCTION of the Eustachian tube is a cause of Deafness, though not a very frequent one. "It may arise," says Mr. Curtis, in the Report of the Dispensary for Diseases of the Ear, "from a variety of circumstances—from ulceration, adhesion, stricture, induration, and polypus. In some of such cases, the obstruction, when slight, may be removed by a gargle of cayenne pepper and port wine, and in other instances by injecting air into the tubes; for which purpose, as long ago as 1820, I had an instrument constructed by Mr. Thompson, of Great Windmill-street, Piccadilly. But, in other cases, such as where the passage is not merely obstructed, but *obliterated*, it is manifest no such remedy can be applied, and the disease must be considered incurable. A few years since, when I was in Paris, M. Duleau was kind enough to show me his mode of employing the catheter, and other instruments, for removing obstructions of the Eustachian tube; and I procured a set of instruments for the same purpose from Charriere, and have had recourse to them, occasionally, for the last four years, in slight cases of obstruction. The use of the catheter and air-pump is, however, in my opinion, by no means so simple and harmless in its effects as some of its less experienced advocates would have us believe. Operations of this kind are exceedingly doubtful in their results; and certain recent cases, where death occurred either during or immediately after the employment of this mode of treatment, show, in the most decisive manner, that it may often be productive of disastrous results; in fact, the general sense of the public appears to be growing more and more adverse to operations of any kind, except as aids to, not substitutes for, constitutional treatment. Puncturing the membrana tympani, for example, was formerly an operation in great vogue; but it is now almost universally condemned. Dr. John Tustin Berger, physician to the king of Denmark, also attempted to cure his own deafness by perforating the mastoid cells, but the operation terminated his existence; thus affording a striking warning to all who might afterwards be tempted to follow so rash an example, either upon themselves or upon others."

CILIAGRADA.—LUMINOSITY OF FISH.

AT the late meeting of the British Association, Mr. R. Patterson brought forward a specimen of Ciliagrada, taken on the Irish coast, and which he had named Bolina Hibernica. It led to a question on the Luminosity of this species of fish. Professor Jones stated, that the property was not limited to this species, as several possessed it; nor to life, as the dead also were luminous. Mr. E. Forbes recommended the placing of the animals in a watch-glass, as the best mode for examining them. A subsequent paper, "On the Ciliagrada in our Seas," by Mr. Forbes and Mr. Goodsir (illustrated by drawings of

Alcynoe, Cydippe, and Beroe), led to a conversation respecting the fine tentacula, and other curiously formed organs of these Medusæ, hitherto so unsatisfactorily examined.*—*Literary Gazette.*

SCALES OF FISHES.

M. MANDL supports the opinion of M. Agassiz, that the Scales of Fishes may serve as characters for Classification; and states that these coverings are not to be considered as simply the production of secretion, but consist of a true organized substance. First, he says, that they are composed of an upper and an under layer: then, that the upper layer is composed of longitudinal canals, departing from a centre, which is not always in the middle of the scale; of cellular lines, produced by the union or fusion of cells; of yellow corpuscles, similar to those of bones and cartilages, and, like them, containing salts; of a centre, or focus, which appears to be the rudiment of the scale; of teeth, which, however, only exist on the terminal edge of the *Acanthopterygii:* thirdly, the under layer is formed of fibrous plates, the middle of which are the shortest.—*Athenæum.*

SINGULAR PROPAGATION AMONG THE LOWER ANIMALS.

SIR J. G. DALZELL has communicated to *Jameson's Journal*, No. 51, an interesting investigation of "a Singular Mode of Propagation among the Lower Animals;" whence he deduces: 1st, that two different modes of propagation carry on the race of *Actinia*, one whereby the embryo, a shapeless corpusculum, endowed with locomotion within the parent, is produced symmetrical by the mouth, but then deprived of that faculty, or nearly so; the other whereby a fragment buds externally from the base, thus generating after the fashion of the *hydra tuba.*

* Eschscholtz's work on those of the Pacific Ocean is, perhaps, the best on the subject: the British species of Ciliagrada ascertained, and some of them well known, and not much liked in their jellies by our sea-bathers, are thus enumerated:—

Genus *Cydippe*—Filamentary appendages.
Genus *Alcynoe*—Tentacula round mouth.
Genus *Beroe*—No tentacula or appendages.

1. *Cydippe pileus* (Linnæus).—Rows of cilia, 19 or 20, on the summits of the lobes—filamentary appendages white. St. Andrew's. Mouth of the Thames. (Dr. Grant.)
2. *Cydippe Flemingii* (Forbes.)—(*an Beroe ovatus*, Fleming?)—Rows of cilia 36, on the summits of the lobes—filamentary appendages white. St. Andrew's.
3. *Cydippe lagena* (Forbes).—Rows of cilia about 25, placed in the furrows of the lobes—filaments white, Coast of Ireland.
4. *Cydippe pomiformis* (Patterson).—Rows of cilia about 20—filaments rufous. Coast of Ireland and mouth of Forth.
5. *Alcynoe rotunda* (Forbes and Goodsir).—Ovate rounded, crystalline—tentacula rounded at their extremities—natatory lobes forming half the animal. Kirkwall Bay, Orkney.
6. *Alcynoe Smithii* (Forbes).—Elongato-pyriform, sub-compressed crystalline—natatory lobes not more than a third of the whole length—tentacula acute.
7. *Beroe cucumis* (Otho Fabricius).—No spots on external surface, internal dotted with red points—ciliferous ridges red. Isle of May at the mouth of Forth.
8. *Beroe fulgens* (Macartney).

2nd, That the *Aplidium verrenosa*, a compound ascidia, is originally an inert ovum, next an embryo endowed with an active locomotive faculty, which, in the third stage, is converted to an animal of a form absolutely different, riveted to the same spot.

3rd, That the Zoophytes of certain genera pass through intermediate stages towards perfection, of which that stage exhibiting them endowed with the faculty of locomotion is not the first.

All the preceding names are given provisionally.

BLOOD CORPUSCLES IN THE MAMMALIA.

MR. GEORGE GULLIVER, F.R.S., has communicated to the *Philosophical Magazine*, No. 100, an elaborate series of observations on the Blood Corpuscles, or Red Disks, of the Mamminiferous Animals. We have only space to mention them in fine. Of Australasian animals the corpuscles have the form and size most common in mammals, their diameters varying from 1-4800th to 1-3000th of an inch. These Australasian animals are the *Perameles lagotis, Petaurus Sciurus, Macropus Bennettii, Dasyurus Ursinus*, and *D. Viverrinus*.

In reference to the interesting discovery by M. Mandl of the oval blood corpuscles of the Dromedary, Mr. Gulliver has found the blood disks of the *Auchenia Vicugna, A. Paco*, and *A. Glama*, also very distinctly elliptical. In the *Vicugna* they are rather smaller than in the other species.

In the Musk Deer, (*Tragulus Javanicus,*) Mr. Gulliver observes that the blood disks are smaller than those hitherto described, of any other mammal whatever. In the *Tragulus*, the disks, though very distinct in form, measure, on an average, 1-12,000th of an inch only; but many variations in size are to be seen, from 1-15,000th to 1-9,600th of an inch in diameter.

PULMONIFEROUS MOLLUSCA.

MR. E. FORBES has read to the British Association a Report drawn up at the Society's request, "On the Distribution of the Pulmoniferous Mollusca of Britain, and the Influences which affected that Distribution." In this respect, climate and soil principally operate; the cold northern parts being far less abundantly inhabited than the central and southern regions, where they increase more rapidly, and display superior colours and forms. Rocks have also great influence, and those of a calcareous order are particularly favourable. This able naturalist went at considerable length into the various minutiæ of his subject; and Mr. Lyell enlarged on the expediency of completing observations on the chemical influence of strata, the subaqueous distribution of species, &c. &c.; so that errors in geological deductions, respecting such as still existed, or were extinct, might be avoided.—*Literary Gazette.*

ANIMALCULES.

ON Feb. 1, Dr. Grant read at the Royal Institution, the following paper—"On the recent Discoveries and History of Animalcules;" creatures living, so infinitely small, that the mentioning their size subjects a lecturer, in the minds of many of his hearers, to a charge of inaccuracy, or, perhaps, of worse. Still is it not the less true that animated beings of marvellous minuteness exist; that millions of animalcules live in a single drop of water; and that these creatures are of a complicate structure, and closely allied to animals of a higher class. They are various in species, possessing relatively cerebral ganglia, respiratory, visual, masticating organs, &c., and stomachs, to the number of 150 in one animalcule. They abound in pools, in rivers, extensively in the ocean, and in all waters on the surface of the globe; also in waters in mines, in water-percolating rocks, where no ray of light penetrates. Countless thousands inhabit mud; and dust, clouds of desiccated earth, contain their millions ready to resume their living state. They are capable of a torpid existence in earth dried up by a summer sun; and they hibernate frozen in ice. Poisons, if chemically combined with the water, destroy animalcules, otherwise their particles in any mechanical dissolution are too large to be swallowed; but even in the case of chemical combination, animalcules often revive after imbibing the strongest poisons. These powerful agents, therefore, have little effect upon these creatures, and the ordinary means of destruction none; but a shock of electricity bursts their bodies, and kills them instantaneously. Their increase is astounding. From one individual, in four days, one hundred and forty millions of millions would have existed, sufficient to form two solid cubic feet of siliceous rock. Most of the solid skeletons of *polygastric* animalcules are external siliceous covers which envelop the entire body; they long resist decay, and they exhibit the general form and characters of the species to which they belonged. Both marine and fresh water species abound in the tertiary deposits of all latitudes, forming alone entire strata, or occurring with the remains of other classes; and they are observed, along with the scales of fishes, in the substance of chalk flints, or siliceous poriphera of the newest secondary rocks. Their remains form vast deposits and layers of solid stone. This wonder in some measure ceases when merged in the general view of the tertiary and most of the secondary formations. Testaceous rocks in some instances of more than 130,000,000 of cubic fathoms in bulk are composed entirely of shells, the skeletons of molluscous animals. All the limestone of the globe is formed from animal bodies, and chalk contains myriads upon myriads of cephalopods and others. Animalcules, then, are, as before said, the greater portion of siliceous rock: their nature has been revealed by the microscope, and their internal complicate structure fully developed by that instrument, assisted by the patience and ingenuity of experimenters. Substances considered formerly, indeed almost to our own times, mere gelatinous portions of vegetables or of zoophytes, have been proved, by modern accurate observers, to be distinct animalcules. Colouring matter, carmine put into water, was quickly conveyed into the numerous cavities or stomachs of the animalcules, and thus their whole interior investigated.

The distribution of their species follow, in some respects, the laws of that of the higher classes; but in others present a singular phenomenon. Fresh water species have been identified with those of the sea, and several found in the deserts of Africa have also been discovered in the waters of Prussia. They form the colouring of waters, the luminosity of the sea, &c. Of the polygastric animalcules, the gaillonelle contain a very large portion of iron, and compose the rusty appearance of chalybeate springs.—*Literary Gazette.*

Prof. Ehrenberg has published an important work entitled "The Infusoria, (microscopic animalcules,) as perfect Organisms, a glance into the deeper organic life of Nature:" with an Atlas of 64 coloured plates. In this work, which may be regarded as the summary of Ehrenberg's researches into the *Infusoria,* he arrives at valuable conclusions as to the geographical distribution of the *animalcules,* and establishes two great natural laws: 1, that the animal organization is perfect, in all its principal systems, to the extreme limit of vision assisted by the most powerful microscopes; and, 2, that the microscopic animalcules exercise a very great and direct influence on inorganic nature. One of the inferences drawn from the first law is the great improbability of these *animalcules,* as well as organic bodies in general, being ever produced by spontaneous generation.

In the *Infusoria* themselves, Prof. Ehrenberg has confirmed or first established many very curious qualities and relations, which are highly interesting in a physiological and other points of view, the most important of which we briefly enumerate:

1. Most (probably all) microscopic *animalcules* are highly organized animals. 2. They form, according to their structure, two well defined classes. 3. Their geographical distribution in four parts of the world follows the same laws as that of other animals. 4. They cause extensive volumes of water to be coloured in different ways, and occasion a peculiar phosphorescence of the sea by the light they develop. 5. They form a peculiar sort of living earth; and, as 41,000 millions of them are often within the volume of *one cubic inch,* the absolute number of these *animalcules* is certainly greater than that of all other living creatures taken together; the aggregate volume is even likely to be in favour of the *animalcules.* 6. They possess the greatest power of generation known within the range of organic nature; one individual being able to procreate many millions within a few hours' time. 7. The *animalcules* form indestructible earth, stones, and rocks, by means of their siliceous *testæ;* with an admixture of lime or soda they may serve to prepare glass; they may be used for making floating bricks, which were known to the ancients; they serve as flints, as tripoli, as ochre, for manuring land, and for eating, in the shape of mountain-meal, which fills the stomach with a harmless stay. They are sometimes injurious by killing fish in ponds, in making clear water turbid, and in creating miasma; but that they give rise to the plague, *cholera morbus,* and other pestilential diseases, has not been proved. 8. As far as observation goes, the *animalcules* never sleep. 9. They exist as *entozoa* in men and animals, the *spermatuzoa* not being taken into consideration here. 10. They themselves are infected with lice as well as *entozoa,* and, *on the former, again, other parasites have been observed.* 11. They are, in general, affected by external agents much in the same manner as the larger

organic beings. 12. The microscopic *animalcules* being extremely light, they are elevated by the weakest currents, and often carried into the atmosphere. 13. Those observers, who think they have seen how these minute creatures suddenly spring from inert matter, have altogether overlooked their complicated structure. 14. It has been found possible to refer to certain limits in organic laws, the wonderful and constant changes of form which some of these *animalcules* present. 15. That the organism of these *animalcules* is comparatively powerful, is evinced by the strength of their teeth and apparatus for mastication; they are also possessed of the same mental faculties as other animals. 16. The observation of these microscopic beings has led to a more precise definition of what constitutes an animal, as distinct from plants, in making us better acquainted with the systems of which the latter are destitute.—*W. Weissenborn; Mag. Nat. Hist.*

CORALLINE ANIMALCULES IN CHALK.

PROF. EHRENBERG has ascertained that in the finest powdered Chalk, not merely the inorganic part is separated, but that it remains mixed with a number of well-preserved forms of the minute shells of Coral Animalcules. As powdered chalk is used for paper-hangings, Ehrenberg has also examined these as well as the walls of his chambers, which are simply washed with lime, and even a kind of glazed vellum paper called visiting cards; and obtained the very visible result,—demonstrating the minuteness of division of independent organic life,—that those walls and paper-hangings, and so, doubtless, all similar walls of rooms, houses, and churches, and even glazed visiting cards prepared in the above manner, (of which cards many, however, are made of pure white lead, without any addition of chalk,) present, when magnified 300 diameters, and penetrated with Canada balsam, a delicate mosaic of elegant coralline animalcules, invisible to the naked eye, but, if sufficiently magnified, more beautiful than any painting that covers them.—*Poggendorff's Annalen; An. Nat. Hist.*

ARE THE CLOSTERIÆ ANIMALS OR PLANTS?

EHRENBERG enumerates the following reasons for considering the *Closteriæ* as belonging to the animal kingdom. They enjoy voluntary motion, they have apertures at their extremities, they have projecting permanent organs near the apertures, which are constantly in motion, and they increase by horizontal spontaneous division. Dr. Meyen, who is of the opposite opinion, mentions as the most important observations in favour of their vegetable nature, that their structure is exactly similar to that of the *Conferva*; their formation of seed, and the development of this seed, is like that of the *Confervæ*. The occurrence, moreover, of any turn in the interior of the *Closteriæ* with which they are frequently nearly filled, is a striking proof of their being plants: they have no feet; what Ehrenberg regards as such are molecules having a spontaneous motion, which occur in great number in *Clos. Trabecula*, and quite fill a canal the whole length of the plant. Their function is difficult to determine; but they also occur in very many *Confervæ*, and may perhaps be compared with the *Spermatozoa* of plants.—*An. Nat. Hist.*

GEOMETRY OF SHELLS.

THERE is a mechanical uniformity observable in the description of Shells of the same species, which at once suggests the probability that the generating figure of each increases, and that the spiral chamber of each expands itself, according to some simple geometrical law common to all. To the determination of this law, if any such exist, the operculum lends itself, in certain classes of shells, with remarkable facility. Continually enlarged by the animal, as the construction of its shell advances, so as to fill up its mouth, the operculum measures the progressive widening of the spiral chamber, by the progressive stages of its growth.

The animal, as he advances in the construction of his shell, increases continually his operculum, so as to adjust it to his mouth.

He increases it, however, not by additions made at the same time all around its margin, but by additions made only on one side of it at once. One edge of the operculum thus remains unaltered as it is advanced into each new position, and placed in a newly-formed section of the chamber similar to the last, but greater than it.

That the same edge which fitted a portion of the first less section should be capable of adjustment, so as to fit a portion of the next similar but greater section, supposes a geometrical provision in the curved form of the chamber, of great apparent complication and difficulty. But God hath bestowed upon this humble architect the practical skill of the learned geometrician, and he makes this provision with admirable precision in that curvature of the logarithmic spiral which he gives to the section of the shell. This curvature obtaining, he has only to turn his operculum slightly round in its own place as he advances it into each newly-formed portion of his chamber, to adopt one margin of it to a new and larger surface, and a different curvature, leaving the space to be filled up by increasing the operculum wholly on the outer margin.

Why the Mollusks who inhabit turbinated and discoid shells should, in the progressive increase of their spiral dwellings, affect the peculiar law of the logarithmic spiral, is easily to be understood. Providence has subjected the instinct which shapes out each, to a rigid uniformity of operation.—*Prof. Moseley; Phil. Trans.* Pt. ii., pp. 351, 353, 359.

LAND AND AQUATIC SHELLS.

CERTAIN physiological facts having reference to the growth of the Mollusk, are deducible from the geometrical description of its shell. If it be a *land* shell, its capacity may be supposed, (reasoning from that principle of economy which is an observable law in nature,) to be precisely sufficient for the animal who built it. If it be an *aquatic* shell, it serves the animal at once as a habitation and as a float; enabling it to vary his buoyancy according as it leaves a greater or less portion of the narrow extremity of its chamber unoccupied, and thus to ascend or descend in the water, at will. Now, that its buoyancy, and, therefore, the facility of thus varying its position, may remain the same at every period of its growth, it is necessary that the increment of the capacity of its float should bear a constant ratio to the corresponding increment of its body; a ratio which always assigns a greater amount to the increment of the capacity of the shell than to the corresponding increments of the animal's bulk. Thus, the chamber of the *aquatic* shell is increased, not only, as is the land shell, so that it may contain the greater bulk of Mollusk, but so that more and more of it may be left unoccupied. Now, the capacity of the shell, and the dimensions of the animal began together, and they increase thus in a constant ratio; the whole bulk of the animal bears, therefore, a constant ratio, of greater inequality, to the whole capacity of the shell, in *aquatic* shells: in land shells, it is, probably, equal to it. —*Ibid.*

THE ZOOLOGICAL SOCIETY.

The Annual Report of the Auditors to April, 1839, states the receipts of the Society for the past year at 14,094*l.* 2*s.* 9*d.*, and the expenditure 12,588*l.* 12*s.* 1*d.* The assets consisted of living and preserved collections; uncertain arrears of subscriptions, 900*l.* 15*s.*; invested in Exchequer bills, 209*l.* 6*s.*; in land on the farm at Kingston, 11,000*l.*; capital funded, 11,291*l.* 12*s.* 7*d.*; and cash in the banker's hands, 341*l.* 2*s.* The liabilities were, bills unpaid, 954*l.* 14*s.* 6*d.*; rent unpaid, 820*l.* 1s. 10*d.*; and contracts pending, 304*l.*

The Report of the Council stated that steps had been taken greatly to reduce the permanent expenditure. In 1837, the salaries were 3,548*l.*, but from this year they would be reduced to 2,916*l.*, making a saving in this department of 632*l.*; whilst the expenses of the general establishment, which, on the average of the last few years, were 880*l.*, having been reduced to 541*l.*, would leave a balance of 339*l.* An increase had taken place in the cost of provisions, which was to be ascribed to an increase in the number of animals, and, amongst other circumstances, to the growth of the larger elephant. Every resource of revenue, over which the Council had control, had increased, and an additional income derived from the gardens of 1,720*l.* There were 3,010 members, and thirty-eight candidates for election; the number of Corresponding Members was 126, of whom eight had been elected since the last meeting. The Museum contained 1,228 specimens, of which 760 were characterized species of mammalia; 5,230 birds, to which 113 had been added since last year, and of which 5,000 were named; 1,000 specimens of reptiles; 1,170 fishes; and 83 mounted skeletons. The animals were in good health, and many valuable additions had been made to the library.

METEORIC PAPER.

On Jan. 31, 1687, a great mass of paperlike black substance fell with a violent snowstorm from the atmosphere, near the village of Rauden, in Courland. This meteoric substance, described completely and figured in 1686–1688, was recently again considered by M. V. Grotthus, after a chemical analysis, to be a meteoric mass; but M. V. Berzelius, who also analyzed it, could not discover the nickel said to be contained in it; and Von Grotthus then revoked his opinion. It is mentioned in Chladni's work on Meteors, and appears as an aerophyte in Nees von Esenbeck's valuable appendix to R. Brown's *Botan.*

Schriften. Prof. Ehrenberg has examined this substance, some of which is contained in the Berlin Museum, (also in Chladni's collection,) microscopically. The whole has been found to consist evidently of a compactly matted mass of *Conferva crispata*, traces of a *Nostoc*, and of about twenty-nine well-preserved species of *Infusoria*, of which three only are not mentioned in Ehrenberg's large work on *Infusoria*, although they have since occurred living near Berlin; moreover, of the case of *Daphnia Pulex*? of the twenty-nine species of Infusoria, only eight having siliceous shields, the others having soft or membranous shields. Several of the most beautiful and exceedingly rare *Baccillaria* are frequent in it. These *Infusoria* have now been preserved 152 years. The mass may have been raised by a storm from a Courland marsh and merely carried away, but may also have come from a far distant district, as Carl Ehrenberg has sent from Mexico forms still existing near Berlin. Seeds, leaves of trees, &c., scattered through the mass, were easily visible, on the examination of larger portions. The numerous native *Infusoria*, and the shells of the common *Daphnia Pulex* seem to denote the original locality of the substance to be neither the atmosphere, nor America; but most probably either East Russia or Courland.—*An. Nat. Hist.*

FLANNEL FORMED OF INFUSORIA AND CONFERVÆ.

Prof. Ehrenberg has submitted to the Academy of Sciences of Berlin, a foot and a half square of natural wadding or flannel, consisting of *Infusoria and Confervæ*, which was found to the extent of several hundred square feet, near Sabor, in Siberia, after an inundation.* This substance is analogous to the "Meteoric Paper" and "Meadow Leather," described already, but is far more surprising from its occurrence in such an immense mass. The flannel is chiefly formed of unramified branches of *Conferva rivularis* interwoven with fifteen species of *Infusoria*, and some shells of the water-flea, *Daphnia*. Of the *Infusoria*, eleven belong to the family of *Baccillaria*, and of these six to the siliceous-loricate genera; several *Closterinæ*, &c. Predominating are the *Fragilaria*, *Navicula viridis*, and *Cryptomonas lenticularis*. All the former are known species.—*Berichte der Akademie; An Nat. Hist.*

ZOOLOGICAL COLLECTION.

A valuable collection made in Borneo for the Dutch government, but refused by them, has been bought by the government of Belgium, and the city of Brussels, for 30,000fr. It contains eight skeletons and skins of the oran-outang, skeletons of the rhinoceros, tiger, bear, &c.; a stuffed crocodile, thirty feet long; several fossil remains, and 1,200 birds.—*W. Weissenborn; Mag. Nat. Hist.*

THE GUIANA EXHIBITION.

Mr. R. H. Schomburgk has exhibited in the metropolis a very interesting assemblage of objects in illustration of Ethnography and Natural History, collected by him during three expeditions into the in-

* See pages 179 and 190 of the present volume.

terior of British Guiana. The *Catalogue Raisonnée* includes about 450 items, besides collections of mammalia, birds, reptiles, fishes, mollusca, and insects; the whole illustrating the economy, natural and social, of the Guianese. The animate attractions of the exhibition are three Indians, who were part of Mr. Schomburgk's boat's crew on his last expedition, and who are the first of their tribes ever brought to Europe.

The human varieties in Guiana have already been tolerably well defined. In British Guiana, there are six tribes of natives. The individuals just referred to belong to three different tribes; and, although there exists a great similarity in their manners and customs, they differ in their language. Their respective names are: 1, *Corrienow*; 2, *Saramang*; 3, *Sororeng*.

1. *Corrienow* belongs to the Warrows, who inhabit the coast along the rivers Orinoco, Pomeroon, and Corentyer, and are the Guaranos of the Spaniards. They are excellent boatmen, and famed for the construction of their canoes, which they hollow out of a single trunk of a tree, partly by the axe, partly by fire. Many of the pilots on the river Orinoco, and generally their boats' crews, are Warrows; they are also occasionally met with as sailors in the colonial craft. Corrienow is about five feet in height, and twenty-one years of age: he is very slightly tattooed; he is the least ingenious of the three, having been almost exclusively employed upon his native rivers as a boatman.

2. *Saramang* is a Macusi: his tribe inhabits the vast plains which extend between the river Rupununy, a tributary of the Essequibo and the Rio Branco, which falls into the Rio Negro and Amazons. There were a few settlements of this tribe on the river Essequibo, but they have mostly retreated to the tracts just mentioned. The Macusi are one of the most powerful tribes who inhabit British Guiana, and are more industrious than the generality of Indians. They are noted for making cotton hammocks, which they barter to other tribes, or sell to the colonists. They inhabit the south-western part of the colony which borders on Brazil, and have been, from time immemorial, sufferers, from the atrocious system of carrying them away as slaves by the Brazilians. Saramang is about five feet in height, and twenty-one years of age; his features are *sculptural*, pleasing, and intelligent; with a womanish expression, which is, doubtless, heightened by his feather cap: his features are not tattooed, but occasionally painted in lines; he excels in shooting with the blow-pipe, and is, altogether, the most ingenious of the trio.

3. *Sororeng* is a Paravilhano, or Parawano: his tribe was formerly powerful, and occupied that part of the Rio Branco which lies southward of Fort San Joaquim. They form, at present, only a few settlements on some of the smaller streams which fall into the Rio Branco, and are dispersed among the Rio Negro, and the Amazons. There is much analogy between the language of the Paravilhano and the Macusi. Sororeng is about five feet four inches in height, and is the senior, being thirty years of age; he uses the bow very expertly.

Each Indian is habited in what are technically termed fleshings; that is, a kind of knit shirt, fitting closely to the figure, and of the precise complexion of the individual, who wears the *perizoma*, or waist-cloth, which forms the only garment of the savage Indian. Around the bust of each hangs a necklace, made of peccary teeth, from which, reaching down the back, is a piece of jaguar skin; and from the neck are suspended, upon the chest, two tiger's teeth, which these simple creatures wear as charms, just as persons formerly wore amulets in this country.—*Literary World.*

THE CHIMPANZEE.

LIEUT. SAYERS has communicated to the Zoological Society an interesting account of Bamboo, a Chimpanzee in the Menagerie, which had been purchased about eight months previously from a Mandingo, at Sierra Leone, who related that he had captured him in the Bullom country, having first shot the mother, on which occasions the young ones never fail to remain by their wounded parents. From the habits of this specimen, it is inferred that trees are ascended by the Chimpanzees merely for observation or food, and that they live principally on the ground. Bamboo, at the time of purchase, appeared to be about fourteen months old: the native stated that Chimpanzees do not reach their full growth till between nine and ten years of age; which, if true, brings them extremely near the human species, as the boy or girl of West Africa, at thirteen or fourteen years old, is quite as much a man or woman as those of nineteen or twenty in our more northern clime. Their height, when full grown, is said to be between 4 and 5 feet. The Chimpanzee is, without doubt, to be found in all the countries from the banks of the Gambia in the north, to the kingdom of Congo in the south: and the low shores of the Bullom country, on the northern shores of the river Sierra Leone, are infested by them in numbers quite equal to the commonest monkeys. Lieut. Sayers considers them to be gregarious: for their cries generally indicate the vicinity of a troop. The natives also affirm, that they always travel in strong bodies armed with sticks, which they use with much dexterity. They are very watchful; and the first who discovers the approach of a stranger utters a protracted cry, much resembling that of a human being in the greatest distress. Certain authors affirm that some of the natives on the western coast term these animals, in their language, "Pongos;" but Lieut. Sayers observed that all the natives in the neighbourhood of Sierra Leone, when speaking of this animal, invariably called him "Baboo," a corruption, it is supposed, of our term Baboon.—*Proc. Zoological Soc.; abridged.*

THE SPERMACETI WHALE.

MR. G. T. FOX has communicated to the British Association a paper "On the Bones of a Spermaceti Whale, discovered a few months previously among the rubbish in an old tower at Durham Castle." From an ancient letter of June 20, 1661, in the Surtees' collection, it is shown to have been cast ashore at that time, and *skeletonized* in order to ornament this old tower.

The writer threw an amusing *coup d'œil* over the history of the Whale, as far as known from old writers; the first account of it being that of Clusius, 1605, of one thrown ashore seven years before near Scheveling, where Cuvier supposes its head is still preserved,—for there is an antiquity of the kind still shown there. Since then, only a few instances have occurred of the Spermaceti Whale having been cast ashore on the British coasts.—*Literary Gazette.*

NEW MONKEYS.

Mr. Ogilby has communicated to the Zoological Society a portion of a letter from M. Temminck, relating to two species of Monkeys, *Colobus fuliginosus* and *Papio speciosus;* the former Temminck considers identical with a Bay-Monkey of Pennant, an opinion founded upon its agreement with a coloured drawing now in his possession; this drawing having been taken by Sydenham Edwards from the specimen of the Bay-Monkey formerly in the Leverian Museum, and which is the original of Pennant's description.

The *Macacus speciosus* of M. F. Cuvier is stated by M. Temminck to be founded upon an immature specimen of a species of *Macacus* which inhabits Japan; the habitat of Molucca Islands, given by M. F. Cuvier, being founded upon error. The specimen was originally taken from Japan to Java, where it died; the skin was preserved, and M. Diard having obtained possession of it, sent it to the Paris Museum, and as there was no label attached, M. F. Cuvier imagined it to be a native of the place whence M. Diard had sent it.—*Proc. Zoological Soc.*

Mr. Ogilby has also characterized a new species of Monkey, now living at the Society's Menagerie; thus: Papio Melanotus; *P. cinereo-brunneus; capite, dorso, lumbisque sub-nigris; caudâ brevissimâ, nudâ; facie, auriculisque pallidis.*

This specimen is a young male, said to have been brought from Madras. It has, at first sight, a considerable resemblance to the common Barbary species, *(Papio sylvanus,)* both in general colour and physiognomy; but differs materially in the blackish-brown shade, which covers all the upper part of the head, neck, shoulders, and back. The face and ears are of a pale flesh-colour, not unlike the shade which distinguishes extreme age in the human species; the naked part of the paws is dirty brown, and the temples are slightly tinged with a shade of scarlet, which the keeper states, spread and deepen when the animal is feeding. The tail is about an inch long, very slender, and perfectly naked. The general colour of the sides, under parts of the body, and extremities, is that of a pale olive brown so common among other species of this genus, such as the Bhander, *(P. Rhesus,)* the Maimon, *(P. Nemestrinus,)* &c., and the hairs are equally without annulations. The individual has all the liveliness, good-nature, and grimace of the young Magot, *(P. Inuus and Sylvanus;)* but, like that species, it will, probably, become morose and saturnine as it advances in age and physical development; qualities which, indeed, are common to all the Papios, and pre-eminently distinguish them from the Cercopithecs, Colobs, and Semnopithecs.—*Proc. Zoological Soc.*

REMAINS OF OXEN IN THE BOGS OF IRELAND.

MR. BALL has read to the Royal Irish Academy a paper, in which, having alluded to the occurrence of Fossil Remains of Oxen in Britain, and the existence of the Auroch, or Wild Ox, in some parks in this country, he remarked on the old and generally received opinion, that Ireland could not furnish any evidence of having ever possessed an indigenous Ox; and he stated that a specimen which he had received from the submarine forest, in the Bay of Youghal, seemed to have been the core of a horn of the fossil ox, often found in Britain, and supposed to have been the Urus; but this specimen having been lost, he alluded to it, to direct the attention of the Academy to the subject, in the hope of having his view confirmed. He then showed that the remains of oxen found at considerable depths in bogs in Westmeath, Tyrone, and Longford, belonged to a variety or race, differing very remarkably from any noticed in Cuvier's *Ossemens Fossiles*, or any other work with which he was acquainted. Finally, he expressed a conviction, that Ireland had possessed, at least, one native race of oxen, distinguished by the convexity of the upper part of the forehead, by its great proportionate length, and by the shortness and downward direction of the horns. As this fact seems to have escaped altogether the notice of British and continental naturalists, and as analogy in the case of other Irish mammals justified the view, Mr. Ball urged the great probability of the race in question proving to be one peculiar to Ireland.—*An. Nat. Hist.*

MADNESS IN ELEPHANTS PREVENTED.

THE announcement in the Berlin newspapers of the tragical end of M. Tourniaire's Elephant,* renders it desirable to ascertain some means of preventing similar misfortunes, which have so frequently occurred in Europe. The state of the Elephant which drives it to madness, is termed by the Indians *Mosti*, literally intoxicated by sexual stimulus or by spirituous liquors; and as soon as the keeper of the Elephant observes the symptoms of the Mosti coming on, he places before it a vessel with three pounds of fluid butter, called *Ghie*, which the Elephant swallows, and thus becomes sober. When, on great festivals, Elephants are intoxicated with brandy for the purpose of fighting them, they are rendered sober as soon as desired by the same means. Ghie has moreover the same effect on Dromedaries and Camels, when they are mosti.—*Carl Freihrr von Hugele; Wiegmann's Archiv.; An. Nat. Hist.*

NEW BAT.

THE REV. L. JENYNS has described in the *Annals of Natural History* a White Bat, which has been taken in the Church of Auckland, St. Andrew, and is preserved in the Museum at Durham. Mr. Jenyns infers the white colour to be accidental; but it possesses other characters, those especially in the form of the tragus, which lead

* Poisoned with hydrocyanic acid. The reader will also remember the fate of Mr. Cross' Elephant, which it became necessary to shoot from the same cause.

him to pronounce it distinct from all the bats hitherto met with in this country, or described by continental authors.

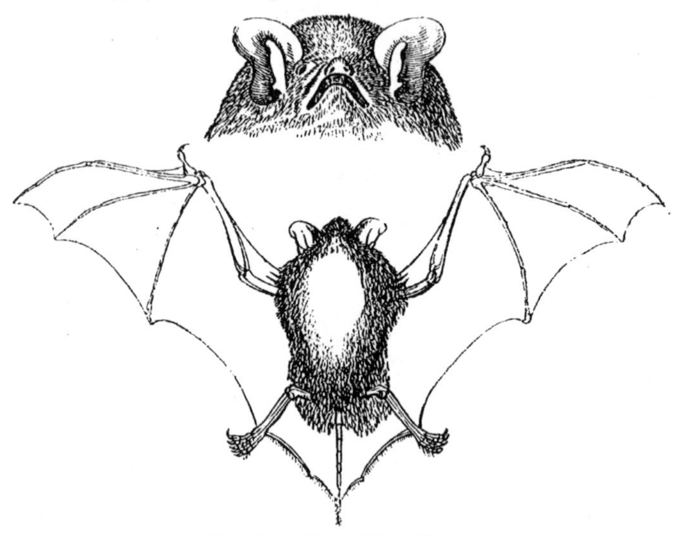

(New Bat; *Vespertilio ædilis.*)

Mr. Jenyns proposes to name this new-bat *Vespertilio ædilis.* In the form of the head, and in its general physiognomy, this bat resembles the *V. mystacinus* more than any other of our British species. The snout is short, but moderately attenuated, and slightly emarginate at the tip between the nostrils. The forehead appears elevated from the erect fur on that part. The face and upper part of the muzzle are hairy. There is some indication of a moustache on the upper lip, with longer hairs interspersed: there are also a few long hairs on the chin. The ears are about the length of the head, widely separate, oval, obtuse at the extremity, bending outwards. On the whole, the auricle very much resembles *V. mystacinus* and *V. emarginatus,* but is not so deeply notched. The *tragus* is of a very peculiar form, and unlike that of any other species. It is not quite half the length of the auricle, if this last be measured in front; but rather more than half if measured behind; its greatest breadth is not quite one-third of its own length: the inner margin is perfectly straight; the outer one arcuate, with a small but rather deep notch a little below the tip which is rounded; there is a somewhat similar notch at bottom, and beneath is a projecting lobe: were it not for the upper notch and rounded apex, the form of the tragus would be nearly that of a small segment of a circle, the broadest part being in the middle. The flying and interfemoral membranes are naked and moderately ample; the latter without any transverse ciliated lines, but dotted irregularly on the under surface with some minute white glands, from which and the margin proceed small bristles. The tail is somewhat shorter than the fore-arm: the feet are larger; the toes, long and bristly, the fur is thick and woolly about the head; elsewhere, of moderate length. The colour, (in this specimen) everywhere of a beautiful silvery white; the membranes are also white, but of a less pure tint. The extent of the flying membrane is ten inches. For the remaining dimensions, comparison with other species, &c., see *An. Nat. Hist.*, No. 15. Altogether, Mr. Jenyns considers this bat to approach nearest to *V. Daubentonii.*

THE VAMPYRE BAT.

A VAMPYRE BAT, stated to be "the first living specimen ever seen in England," has been received from Sumatra at "the Surrey Zoo-

(Vampyre Bat; at "the Surrey Zoological Gardens.")

logical Gardens," and is a very interesting addition to the menagerie. It is of the Indian species, *Vampyrus spectrum;* the *Phyllostoma spectrum* of some authors, *Vampyrus sanguisuga* of others, the *Andira guacu* of Piso, and the *Vespertilio spectrum* of Linnæus. The specimen is a young male: the body is covered with black, and the membrane or wing, in appearance, resembles fine black kid: he is rarely seen at the bottom of his cage, but suspends himself from the roof or bars of the cage, head downwards. The second and third cuts represent the animal in this peculiar position, his wings being also wrapped around his body. The first cut shows him with his wings

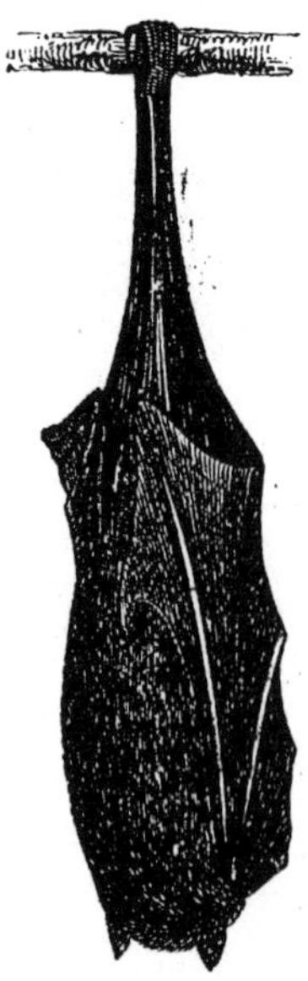

spread, their full exent being nearly two feet. Although this specimen is the Vampyre Bat, to whom so many bloodthirsty traits have been *attributed,* his appearance is by no means, ferocious: he is active yet docile, and the only peculiarity to favour belief in the blood-sucking propensity is his long, pointed tongue. The species has popularly been accused of destroying not only the larger mammifera, but also men when asleep, by sucking their blood: "the truth," says Cuvier, in his *Regne Animal,* "appears to be that the Vampyre inflicts only small wounds, which may, probably, become inflammatory and gangrenous from the influence of climate." In this habit, however, may have originated the celebrated vampyre superstition.—*Literary World.*

MARSUPIAL QUADRUPEDS OF NEW-HOLLAND.

The following table exhibits the relations of these animals in a condensed form, and gives a good idea of what Mr. Ogilby considers to be the most natural and philosophical arrangement of the marsupials. Except the Kangaroos, they are all of nocturnal habits.

	Order	Family	Genera
MARSUPIALIA Marsupiated Mammals	I. SALTIGRADA.. with saltigrade extremities	*Macropidæ* .... and macropoid teeth	*Macropus* *Hypsiprymnus*
		Peramelidæ .... and didelphoid teeth	*Perameles* *Chæropus*
	II. DIGITIGRADA .... with digitigrade extremities and didelphoid teeth		*Myrmecobius* *Phascogale* *Dasyurus* *Thylacinus*
	III. CHEIROGRADA with pedimanous extremities	*Didelphidæ* .... and didelphoid teeth	*Didelphis* *Cheironectes*
		Phalangistidæ...... and macropoid teeth	*Phalangista* *Petaurus* *Phascolarctos*
	IV. PLANTIGRADA .... with plantigrade extremities and rodent teeth.		*Phascolomys*

Magazine of Natural History.

NEW SPECIES OF MERIONES.

DR. HARLAN has read to the Zoological Society the description of a new species of *Meriones (Microcephalus,)* inhabiting the United States of North America. A male and female specimen were taken in 1836, in Philadelphia county: the female at the moment of her capture, carried several young, which adhered to the teats fiercely, notwithstanding the violent efforts and leaps of the parent. The length of the male in the body is three inches; of the tail, four inches; colour above, plumbeous, interspersed with reddish fawn; below, white, similarly but less interspersed, a lateral longitudinal band of reddish fawn separating the sides from the abdomen; tail, sparsely hairy, dark above, white beneath, with a pencil of hairs at the extremity; this member being proportionally longer, and the head much smaller and more elongated, than in *Gerbillus Canadensis.* The female is larger than the male, and of purer white beneath.—*Proc. Zoological Society.*

THE ANT-BEAR.

M. SCHOMBURGK has read to the Zoological Society some very interesting "Remarks on the Greater Ant-bear, *(Myrmecophaga jubata.)*" The chief food of this bear is White Ants, or Termites, which he collects by thrusting his tongue into the tumuli or nests; and the movements of the tongue, alternately protruded and retracted, are so rapid, that it is no longer surprising how so large an animal can satiate his appetite with such minute insects. With the Termites he swallows much of the material of which the Ants' nest is constructed: of this fact M. Schomburgk assured himself by dissection, and he is of opinion that the substance of the nest serves as a corrector. It has been generally thought that the Ant-bear lives exclusively on Ants: this, however, is not the case, for in one which M. Schomburgk dissected, a species of *Iulus* was found; and an adult in M. Schomburgk's possession swallowed with avidity fresh meat hashed up for him. The Ant-bear is very ferocious when attacked; the jaguar, one of the fiercest American animals, being scarcely a match for him. The number of males is supposed to be considerably smaller than that of the females; which circumstance favours the inference that the extinction of the species, like those of the *Edentata* in general, is determined upon.—*Proc. Zoological Soc.; abridged.*

THE MACROSCELIDES.

THIS curious small insect-eater was noticed in the *Year-Book of Facts*, 1839, p. 172. The Macroscelides are exceedingly gentle animals, which never bite, not even when they are tormented. They do not go on the hinder feet like the species of *Dipus*, but always on all fours; and when running, the prolongation of their posterior feet is not at all perceptible. Still, when catching flying or hopping insects, they hide themselves among the dwarf palm, and dart at their prey with a long spring, for which the length of the hinder feet is of great service. Should the pursuer not discover their hiding-place and cut off their retreat under the mass of rock, it will then be necessary to turn over the heavy blocks of stone with iron crow-bars. The very small young of this snouted mouse having been found in the month of February near Tlemsen, Algiers, the time of pairing appears to be during the winter months. When imprisoned, the Macroscelides emit a very peculiar, powerful exhalation. A single Macroscelides which had been confined for some days in a large case, left behind it an odour which the box retained for several weeks. Among themselves these animals appear to be very mild, and not quarrelsome.—*Wiegmann's Archiv.; First part*, 1839; *An. Nat. Hist.*

WOOL OF THE LLAMA.

SOME remarks have been made to the British Association, on the introduction of a species of Auchenia into Britain, for the purpose of obtaining wool, by Mr. W. Danson. Samples and manufactured specimens of Alpaca wool, in imitation of silk (and without dye), as black as jet, were exhibited; and Mr. Danson stated, that the animals producing it ought to be propagated in England, Ireland, Scotland, and Wales; and to the two latter places the Alpaca is well suited, being an inhabitant of the Cordilleras, or mountainous district in Peru. Importations have already taken place to the extent of one million of pounds, and are likely to increase. There are five species of Llamas, of which the Alpaca has fine wool, six to twelve inches long, as shown by the specimens exhibited; the Llamas, the hair of which is very coarse; and the "Vicugna," which has a very short fine wool, more of the beaver cast. The Earl of Derby has propagated the Alpaca in his private menagerie at Knowsley, and Mr. Danson understood that Mr. Stephenson, at Oban, in Scotland, has a few of these animals. The wool of these animals will not enter into competition with the wool of the sheep, but rather with silk. It is capable of the finest manufacture, and is specially suited to the fine shawl trade of Paisley, Glasgow, &c. The yarns spun from it are already sent to France in large quantities, at from 6*s.* to 12*s.* 6*d.* per pound, the price of the raw Alpaca wool being now 2*s.* and 2*s.* 6*d.* per pound.—*Athenæum.*

BIRTH OF A GIRAFFE.

On June 25, Prof. Owen read to the Zoological Society "Some Notes on the Birth of a Male Giraffe at the Gardens," which took place on the 19th. The period of gestation has been, as nearly as possible, ascertained to be fourteen months eighteen days, or fifteen lunar months. The young animal, when born, was perfectly motionless, and apparently dead, or strangulated, its lips and nose being tinged with blood; but after gentle friction had been used for a short time, breathing and motion quickly followed. The mother was in no way depressed or debilitated. It came into the world like other ruminants, with the eyes open, but the hoofs were disproportionately large, and very soft and white at their expanded extremities; the skin was marked as distinctly as in the adult; the horns were represented by stiff and long black hairs, and the mane was well developed. It made many vigorous efforts to stand, raising itself on the fore knees, and was able to support itself on outstretched legs two hours after birth: in ten hours, it had gained sufficient strength to walk. It sucked with avidity warm cow's milk from a bottle, and once or twice uttered low, gentle grunts or bleats, like a fawn or calf. The mother did not show signs of affection or parental care, nor were there any symptoms of her nourishing her offspring; yet once having pushed down the young one when hastily moving away from it, she stood still, and gazed on the prostrate animal with an expression of maternal feeling. The length of the young Cameleopard, from the muzzle to the setting of the tail, was six feet ten inches; and when standing he could reach with his muzzle siz feet. He died on the 28th of June.

HAWKING.

The late Mr. Hoy possessed at his residence, Stoke Nayland, in Suffolk, four Gos-hawks, which he had obtained from the continent for the purpose of hawking. There were three males, and one female; the latter a large and powerful bird, capable of securing, with ease, a full grown hare. Mr. Hoy considered their habits, mode of flight, &c., to be much better suited to an enclosed district than those of the peregrine falcon. When used or taken into the field, the wing of a bird, or the thin end of an ox tail, is generally held in the hand to engage their attention, which they are constantly biting and tearing, without being able to satisfy their appetites, as that would render them unfit for work. They do not require to be hooded, but have bells attached to their legs, (for giving notice of their situation when they alight, which would otherwise be difficult to ascertain,) and a leather strap by which they are held: it is also necessary to have spaniels to hunt up the birds, upon the appearance of which the hawk flies from the hand with incredible swiftness direct at the game, taking it generally at the first attempt; but should he fail, he will perch on some elevated situation, and remain until the game is again started, and is rarely known to miss a second time: when the hawk has captured the game, he is rewarded with a piece of meat, or a pigeon's head, to induce him to give up the prey: if the hawk be allowed to range at pleasure, by whistling he will return with surprising swiftness, when, finding he cannot stop suddenly to settle without striking you with great force, he will glide past, form a circle round you, and alight with ease, and gently, upon the hand.—*Mag. Nat. Hist.*

NEW BRITISH GOOSE.

MR. A. D. BARTLETT has read to the Zoological Society a paper "On a new British species of the genus *Anser*, with remarks on the nearly-allied species." In conclusion, having noticed three nearly-allied species, and described the new one, Mr. Bartlett points out the distinctions between this new species and the Bean Goose, to which it bears the nearest resemblance. First, the average size of the Bean Goose is 33 inches in length, and 64 inches in extent; while the average size of the new species is 28 inches in length, and 60 inches in extent. Secondly, the bill is much smaller, shorter, more contracted towards the tip, and of a different colour. Thirdly, the difference in colour and form of the legs and feet, and in the fleshy character of the foot; and the hind toe being more closely united by its membrane, has, consequently, less freedom of motion. Fourthly, the plumage on the rump and shoulders being more inclined to grey. And lastly, in the form of the sternum, which differs from that of the Bean Goose in shape, and bears a more close resemblance to that of the white fronted goose. Mr. Bartlett has examined, in all, twelve specimens of this new species, four of which were alive: one of them is living in the garden of the Zoological Society, where it has been eight years, without exhibiting any perceptible alteration in its plumage, or in the colour of its legs and feet. The Grey Tail Goose is by far the most rare of the four species here referred to.—*Proc. Zoological Society.*

TALKING CANARY BIRD.

A CANARY BIRD, capable of distinct articulation, has been exhibited in the metropolis. The following are some of its sentences: "Sweet pretty dear." "Sweet pretty dear Dicky." "Mary." "Sweet pretty little Dicky dear;" and often in the course of the day, "Pretty Queen." "Sweet pretty Queen." The bird also imitated the jarring of a wire, and the ringing of a bell: it was three years old, and was bred by a lady who never allowed it to be in the company of other birds. This canary died in the month of October.

NEW FALCON.

M. GENE has read to the Academy of Sciences at Turin, the description of a new species of Falcon, discovered in Sardinia by M. de le Marmora, which has been confounded with the Common Hobby, (*F. Subeteo,*) but from which it differs in its much stronger form, in the colour of the cere, which is blueish; by the form of the cutting edges of the mandible, which are not cut between the base and the tooth; and by the colour of the eggs, which are reddish, spotted, and blotched with brown.—*L'Institut.; An. Nat. Hist.*

RARE FISH.—STERNOPTIX CELEBES.

About the year 1774, Prof. Hermann, of Strasburg, applied the name of *Sternoptix*, (from the apparent folds in the external covering of the breast,) to a very rare osseous fish from around the West India Islands, small in size, truncated in front, narrow and tapering behind, high backed, very compressed, and presenting a triangular pellucid compartment of the region of the tail. From the last-mentioned character, as well as to distinguish it from another genus, from the Azores, (since described by M. Olfers, under the name of *Sternoptix Olfersii*, in which character it is wanting,) the former is known to naturalists as the *Sternoptix diaphana*, or transparent Sternoptix. Owing to the great rarity of this fish—it being hitherto known to no author, excepting through the means of the very incorrect representation of it afforded by Hermann, and the specimen which that naturalist has left in the museum at Strasburg—no opportunity has yet offered for rectifying the mistakes into which Hermann fell, in describing this fish as devoid of a gill-membrane and of a lateral line, and in placing it among the *Apodes* of Linnæus; thus concluding it to be destitute of ventral fins.

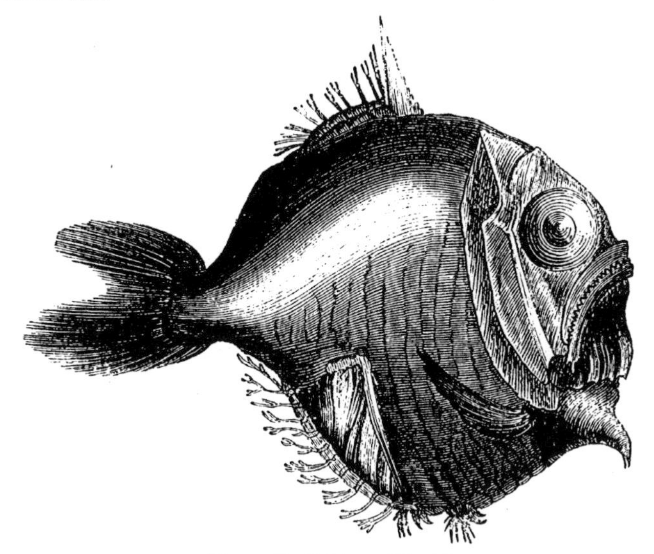

(Sternoptix Celebes: *Fig.* 1.)

The species of *Sternoptix* here figured, however, differs from both the preceding in many even generic characters; such as the situation, character, and number of the teeth, the number of the branchial (or gill) rays, and the components of the different fins: it likewise varies in locality; for, while the two former seem to be confined to the West India Islands, and the warmer parts of the Atlantic, the latter has been hitherto observed only in the Eastern Archipelago. In the present imperfect state of our knowledge respecting this family, the species under consideration is not submitted as a new genus; although (as we have shown) the characters peculiar to it might justify such a course.

The cuts represent the natural size of the *Sternoptix Celebes*, from a specimen caught by Mr. Thomas Kincaid, surgeon, R. N., in the Straits of Macassar, 1° S. lat. 119° E. long., and within thirty miles of the Celebes coast, during calm and clear weather. It is uncertain whether it frequents shoal or deep water; but some fish resembling it were observed swimming about the roots of trees, which had been washed from the coast by the rains, and which trees the fish seemed to have accompanied from the coral reefs near the shore.

(Fig.) 2.

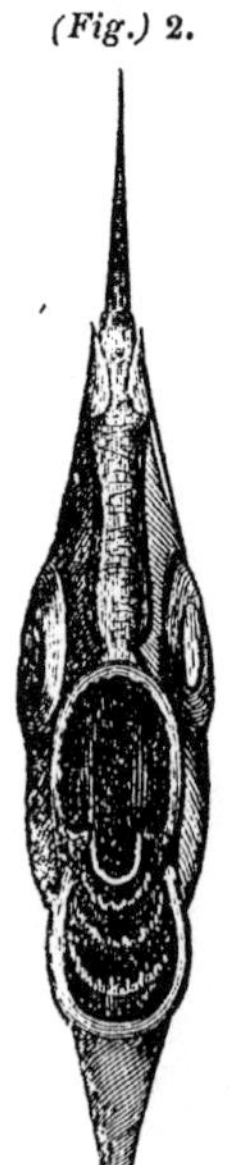

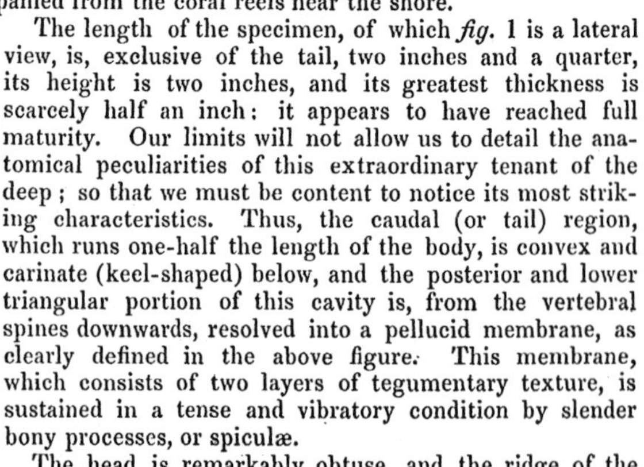

The length of the specimen, of which *fig.* 1 is a lateral view, is, exclusive of the tail, two inches and a quarter, its height is two inches, and its greatest thickness is scarcely half an inch: it appears to have reached full maturity. Our limits will not allow us to detail the anatomical peculiarities of this extraordinary tenant of the deep; so that we must be content to notice its most striking characteristics. Thus, the caudal (or tail) region, which runs one-half the length of the body, is convex and carinate (keel-shaped) below, and the posterior and lower triangular portion of this cavity is, from the vertebral spines downwards, resolved into a pellucid membrane, as clearly defined in the above figure. This membrane, which consists of two layers of tegumentary texture, is sustained in a tense and vibratory condition by slender bony processes, or spiculæ.

The head is remarkably obtuse, and the ridge of the the principal frontal and interparietal bone is distinctly dentated. The eyes are large, salient, and naked; they occupy the middle third of the height of the head, and advance within a line of its anterior boundary. The mouth, which is directed upwards, suddenly descends very obliquely, so as to appear abrupt when viewed in front, (as in *fig.* 2,) and is, therefore, singularly capacious in the vertical direction; while the maxillary (or jaw) bones, (the upper of which slides over the lower) form the superficial boundary of this opening.

The mouth is set with maxillary and palatine teeth; the former being very numerous and minute, and arranged *en crotchets*, three rows in each jaw. (See *fig.* 3.) Each tooth presents the form of two incurvated cones applied base to base, their concavities being directed towards the interior of the mouth. The palatine teeth are much larger, and the existence of them appears to be distinctive of this species. They are five in number, on each side of the mesian plane, and, being arranged *en cardes*, they, on the approximation of the jaws, close after a dove-tailed manner; as shown in *fig.* 2. The branchial arches, (see also *fig.* 2,) are four in number; on the posterior half of the first three of which are placed several slender and curved dental appendages, resembling the teeth of a garden rake; while, on the anterior half of these three arches are placed several tufts of short, straight teeth, arranged *en brosses*. The branchial rays are five in number, naked, attenuated, and curved. (See *fig.* 1.) The dorsal (or back) rays, amount to (one moveable spine and) ten soft rays, each bifurcated at its extremity, the terminated points fimbriated. The anal fin is furnished with thirteen soft rays, connected by a transverse band near the root, and bifid at their extremities. The caudal fin is attached, as in the salmon, to a very fleshy root, being moved by powerful muscles: it is forked, and

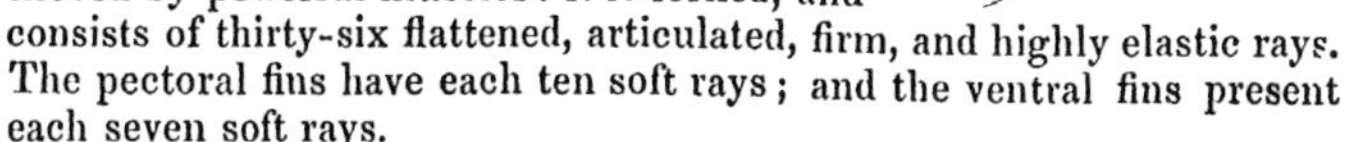

(*Fig.* 3.)

consists of thirty-six flattened, articulated, firm, and highly elastic rays. The pectoral fins have each ten soft rays; and the ventral fins present each seven soft rays.

For the remaining structural peculiarities of this very interesting fish, the reader is referred to the paper by Dr. Handyside, F. R. S. E., communicated to *Prof. Jameson's Edinburgh New Philosophical Journal*, No. 54: the illustrations by M. Willington, of Saltisford, Warwick.

ISINGLASS FROM INDIAN FISH.

Dr. Cantor has examined the Suleah Fish, (the *Polynemus Sele* of Hamilton,) and has found an individual of 2 lb. weight to yield 65 grains of pure isinglass, an article which sells in India at 16 rupees, (1*l.* 12*s.*, per lb.) The air-vessels form the isinglass, requiring only the removal of the vascular membrane that covers them, washing with lime-water, and drying in the sun. As these fishes attain a great size, they promise to furnish an abundant supply of isinglass in India.—*Proc. Zoolog. Soc.*

NEW FISH.

A memoir has been read at the Academy of St. Petersburgh, by M. de Baer, on a remarkable fish found in the White Sea, and called Navaga by the natives. It is a species of Cod, described by Pallas as the *Gadus navaga*, and its length does not exceed ten inches. The transversal apophyses of most of the abdominal vertebræ are of an excessive length, semi-tubular, and are terminated by cavities for air. The five first caudal vertebræ partake of this structure, having on each side of their lower curve, which receives the trunks of the vessels, a hollow prolongation. Although Pallas knew the skeleton, the affinity between these cavities and the swimming bladder seems to have escaped him, as well as Koelreuter. This bladder gives out lateral prolongations, which are hollow, into all these tubular bones; and this structure is peculiar to the species, as those of the genus nearest to it do not present any vestiges of it.—*Athenæum.*

THE WHITE TROUT OF IRELAND,

As a species, is perfectly distinct from the salmon or trout. Mr. Yarrell, in his *History of British Fishes*, vol. ii. p. 37, states that the *Salmo-trutta* of Linnæus, the Sea-trout of Fleming, and *his* Salmon trout, is the White Trout of Ireland; but there is not a close correspondence between the two. In contra-distinction to the common Trout, the flesh of the White Trout is of a richer flavour, and of a deeper orange colour; its skin is much thicker and more oily, its colour bright silvery, with the exception of the back, which is darker: it is destitute of orange spots. It is abundantly taken in salt water, and very seldom in fresh.—*Mag. Nat. Hist.*

CAPTURE OF AN IMMENSE SAW-FISH AT TRINIDAD.

A CORRESPONDENT of the *Magazine of Natural History* describes the taking of an immense Saw-fish on April 15, at Point-à-Pierre. The labour occupied one hundred persons hauling the fish on with ropes before it was exhausted. On endeavouring to raise the fish, it was most desperate, sweeping with its saw from side to side, so that the men were compelled to get strong guy-ropes to save themselves being cut to pieces. At length, a Spaniard got on its back, and, at great risk, cut through the joint of the tail, when animation was completely suspended. It was then found to be 22 feet long, and 8 feet broad, and to weigh nearly five tons. The liver filled a beef-tierce; and in the body were found several eggs, the size of 18 lbs. carronade shot. The head was cut off below the lower jaw, and has been prepared for the Wisbeach Museum: it is believed to be the largest head of a Saw-fish in the world.

NEW OSTRACEA.

A NEW genus of Ostracea has been established by M. Cantraine, of Ghent. It was found in the East by M. Bové, and has been named Carolia, in honour of Prince Charles Bonaparte. It is intermediate between the *Anomiæ* and *Placunæ*. It is rounded, flattened, the valves moderately thick, of a leafy texture, and marked externally with diverging, irregular striæ. A very narrow fissure at the apex divides the dorsal edge. Its greatest diameter is four inches nine lines.—*Athenæum*.

LIVING PROTEI AT PARIS.

PROF. JULES CLOQUET has brought with him from the grottos of Adelsberg, not one *Proteus*, but five, two only of which, however, have reached Paris alive. They have the form of a salamander or lizard, about six or seven inches long; their heads are flattish, and resemble those of an adder. They are either entirely blind, or at least no vestige of eyes can be observed. At the sides of the head are *bronchiæ*, which float about in the water, and are not unlike small coral branches; their feet are very short and flexible; the fore ones having three fingers, the hinder only two. The tail is flattened transversely, is semi-transparent, and endowed with great mobility. Two of these creatures are of a fine white colour tinged with pink; the third is black, and was given to the professor by a chemist of Trieste, who had kept it exposed to a feeble light for several months, thereby causing the change of colour. These reptiles keep themselves, in general, quite tranquil at the bottom of the water, and come from time to time to the surface to breathe, making a slight gurgling noise when they do so. They sink by allowing a portion of air to escape by means of the *bronchiæ*. When the water is disturbed, they quit their apparent state of torpor and move about with the greatest rapidity, sometimes putting their bodies partially out of the water, and then plunging again with the greatest quickness. Some distinguished naturalists intend carefully to examine these reptiles, the habits of which are so little known: they have hitherto been found only in the subterranean waters of Carniola.—*Letter in the Times.*

THE FOUR-HORNED AGAMA.

A Corunha.—Harlan. *Lacerta orbicularis.*—Cuvier.

Of this species of lizard, which has been erroneously described under the names of *spinous toad*, and *horned frog*, specimens have for some time been preserved in spirits in the Zoological Society's Museum, the the British Museum, and the Ashmolean Museum; but we believe that no living specimens had ever been brought to England until last year, when many scientific gentlemen had the pleasure of seeing a very active one in the possession of Mr. Kemp, jun., of Leicester-street, Regent-street; who brought it from Texas, where it was caught among the grass. This specimen was most active during sunshine, when it would run about very rapidly, and catch the flies and other insects within its reach. It was altogether a tame and interesting animal.—*Communicated to the Year-Book of Facts, by James H. Fennell.*

EXISTENCE OF THE TOAD WITHOUT FOOD.

On Sept. 10, 1836, a living Toad was put into the ground, at the depth of eight feet from the surface, in a bed of flinty gravel, with a flower-pot reversed and placed over it. The hole was then filled up and the surface cropped, the spot selected being in a garden. The pit was opened on August 29, 1839, after having been closed three years save ten days; when the Toad was found alive, and used all its exertions to crawl away. It was not a full-grown animal when taken, neither did it appear to have increased in size during its incarceration.

This experiment confirms the statements of Toads existing without food. In the *Mag. Nat. Hist.*, vol. ix., p. 316, is an account of a Toad that was immured, by way of experiment, in a block of stone, for the space of thirty-eight years, and at the end of that period was found alive.—*Mag. Nat. Hist.*

POISON OF TOADS.

In the *Nottingham Herald* it is stated that a dog having worried a Toad, and left it, ere he had gone 300 yards, died from the effects of the poison of the toad. In addition to the above fact, (observes the Editor,) we have had three dogs poisoned by Toads, and we have known instances of pointers and spaniels suffering from their virulent poison.

POISON OF SERPENTS.

DR. CANTOR has read to the Zoological Society a very interesting notice of the *Hamadryas*, a genus of Hooded Serpents, with poisonous fangs and maxillary teeth. The fresh poison of this serpent is a pellucid, tasteless fluid, in consistence like a thin solution of gum-arabic in water; it reddens slightly litmus paper, which is also the case with the fresh poison of the *Cophias viridis*, *Vipera elegans*, *Naja tripudians*, *Bungarus annularis*, and *Bung. cœruleus:* when kept for some time, it acts much stronger upon litmus, but after being kept, it loses considerably, if not entirely, its deleterious effects. From a series of experiments upon living animals, the effects of this poison come nearest to those produced by that of the *Naja tripudians*, although it appears to act less quickly. The shortest period within which this poison proved fatal to a fowl, was fourteen minutes, whilst a dog expired in two hours eighteen minutes after being bitten; the experiments being made at a cold season.—*Proc. Zoological Society.*

SPONGES.

MR. HOGG has perfected his inquiry into the nature of the *Spongilla fluviatilis*, and the *Spongiæ marinæ ;* and endeavours (in addition to what has been already stated,*) to demonstrate the *vegetability* of the River Sponge from the following facts, obtained by many experiments made upon that substance during the two last summers:

1. From the general resemblance of the membrane which invests the soft portion or jelly with the membrane or cuticle of the leaves of many plants.

2. From this gelatinous or soft portion being so similar to the parenchrymatous substance of the more fleshy kinds of leaves; and being, like the latter, chiefly composed of numerous pellucid globules.

3. From the green colouring matter or chromate contained in those globules, on being pressed out, giving a permanent green or yellowish-green colour to white paper, as is the case with the chromate of leaves and plants.

4. From strong acids having the same effect on this Sponge as they are seen to have upon plants when they are macerated in them.

5. From the mode in which numerous bubbles of gas, most probably oxygen, are disengaged from the surface of the living mass of *Spongilla*, when exposed to the brightest solar light, being so extremely analogous to that which is known to occur with the leaves of a plant when immersed in water, and submitted to the direct action of the light of the sun.

Mr. Hogg considers the currents of water in the *Spongilla fluviatilis*, which are the best evidence of their supposed animal nature, to

* See Year-Book of Facts, 1839, p. 193.

be caused by some parasitical insect or other animal, seen generally to inhabit specimens of *Spongilla;* and by means of the animal's respiration, the streams or currents of water are forced to enter into and flow out from the pores or oscules of that structure. But, should these currents occur in specimens free from parasites, Mr. Hogg would consider them to be effected by the same agents as cause the motions or circulations of fluids in vegetables, and most probably by an endosmosis and exosmosis of different fluids, in accordance with the important discoveries of M. Dutrochet.

Mr. Hogg, in conclusion, considers there to be scarcely ever a generic difference between *Spongilla and spongia,* and that both are decidedly vegetable; and he proposes to classify them in a separate order, "*Spongiæ,*" which ought to be placed between the order *Fungi* and that of the *Algæ.—Proc. Linnæan Society; An. Nat. Hist.*

STRUCTURE OF THE VOLVOCINÆ.

PROF. EHRENBERG observes: "all endeavours to acquire some knowledge of the organization of the genus *Volvox* have proved only successful, now that observation has been, at last, directed to the right depth (1833). Formerly, the entire globule was generally regarded as a single verrucose or ciliated animalcule, and its bursting considered as the reproduction of single individuals. But this view leads to wonders and to contradictions: it is evidently erroneous, and the organic relations lie much deeper. Each globule is a hollow *Monadier* (Monadenstock,) of many hundreds, nay, thousands, of minute animalcules; and, within this, several smaller globules are developed; which, however, are not single individuals, but also *Monadiers.* The single animals are those small greenish warts, or points on the surface, and they resemble the Monads. Each animalcule bears precisely the same relation as a single animal of *Gonium pectorale;* it possesses a gelatinous shield, open anteriorly, which when full grown it can leave, and is connected by three to six thread-like tubes with the neighbouring individuals. It is evidently, then, quite erroneous to compare the green bodies of *Gonium* or of *Pandorina* with the larger inner globules of *Volvox;* they are to be compared with the minute outer granules on the surface: and, though *Volvox* is much larger than *Gonium* in its aggregate state; yet the individual animals are much smaller. In these small animalcules, which appear in the form of very minute green warts on the periphery of the *Volvox* globule, and to which little attention has hitherto been paid, Ehrenberg has succeeded in recognising relations of structure which coincide entirely with those peculiar to the family of Monads.

For the structural details of these highly interesting *Infusoria,* the reader is referred to Ehrenberg's work, *Ueber die Infusionsthierchen,* whence this extract is taken; this illustrious naturalist having discovered nutritive organs, mouth, eyes, generative organs, &c.—*An. Nat. Hist.*

NATURE OF POLYPIDOMS.

MR. H. MILNE EDWARDS, in a valuable paper "On the Nature and

Growth of Polypidoms," in the *Annales des Sciences Naturelles*, observes that " the various facts which we have examined seem to prove that the current opinion relative to the nature and to the mode of formation of the Polypidoms is inaccurate; and that these bodies, far from always being external incrustations and without any organic connection with the animals which produce them, are integral parts of these beings, and consist of an organized tissue, the substance of which becomes charged, more or less, with corneous or calcareous matter deposited at its base, and the nutrition of which is effected by intus-susception. In all these animals, there is a tendency in the tegumentary and reproductive portion of the body to harden, but the degree this solidification reaches varies much; and this alone determines the differences which exist between the species distinguished by zoologists under the names of *naked Polypes, Polypes with flexible polypidoms, fleshy Polypes,* and *Polypes with stony polypidoms.* The cartilaginous or stony polypidom of a *Sertularia*, or of a *Zoanthus*, is not, as is usually stated, a habitation which these animals build; it is in some measure their membrane which forms the solid structure of their body; and which, in the same manner as the skeleton of vertebrate animals, assumes, at one time, a membranous form, at another, a cartilaginous texture, and sometimes a condition, in some degree, osseous. —*An. Nat. Hist.*

ON THE NEMATOIDEA.

Dr. Creplin, in many years' personal observations, has found that a *Nematoideun* living singly in a cyst, inclosed on all sides, or enveloped closely in a membrane, *never* possesses sexual organs. Rudolphi has never discovered generative organs in any of the *Nematoidea*. It is true that he mentions in his *Entoz. Hist. Nat.* ii. p. 152, a sexual difference in *Ascaris*, (e mesenterio *Cotti scorpii*,) *angulata*, but he does not prove by his remarks the accuracy of his assertion; and when Zeder, *Naturgeschichte*, 853, 54, talks of an ovarium and probable seminal vessels in his *Capsularia*, he by no means proves that the organs observed possess the functions he ascribes to them.—*Wiegmann's Archiv.* vol. iv. part v.

We may observe that the organization of the incysted microscopic Entozoon, (*Trichina spiralis*, O.) discovered by Mr. Owen in the human muscles, accords with the generalization enumerated by Dr. Creplin.—*An. Nat. Hist.*

CILIOGRADE MEDUSÆ.

Mr. E. Forbes has exhibited to the Royal Physical Society of Edinburgh, drawings and diagrams of the various genera of *Ciliograde Medusæ* inhabiting the seas of Britain. He has also described two new species of *Alcinæ*—a genus observed for the first time last summer, in the northern hemisphere; also, of a new *Beroe*, discovered near the Isle of Man; and he concluded with some observations on the structure and use of Cilia, which naturalists have generally supposed are for motion, but which Mr. Forbes showed could not be so — *An. Nat. Hist.*

THE PORTUGUESE MAN OF WAR.

Mr. Couch, F.L.S., has communicated to the *Magazine of Natural History* an interesting paper "on the Structure and Habits of the *Physalia* (of Cuvier), or Portuguese Man of War; *Holothuria Physalia* (of Linnæus.) In this communication, Mr. Couch shows the errors of the general opinion that the air in the cavity of the body is collected at the will of the animal, and that it can be expelled at pleasure, or through fear of danger: for the accumulation is clearly not received from without, and so far from expelling it at haste, especially in storms, the animal is seen floating on the most turbulent waves, and is frequently thrown ashore in tempests. Its power of floating appears to be independent of air; for Mr. Couch has discharged nine-tenths of the contained air, thereby causing a shrivelling of the external membrane, without bringing it to a state in which it did not swim buoyantly on the water. An examination of the *Physalia*, when in undisturbed liberty, shows the real use of the inflated condition to be not buoyancy alone. The accumulation of air is absolutely necessary as a fulcrum or point of support for the action of the muscular structure: the fulcrum, which in the higher animals is the heaviest portion of their structure, and acts by gravity as well as strength, being in the *Physalia* and other animals of soft texture, no less effective as a moving power, and yet so light as to serve the office of a balloon. Mr. Couch also shows the sticking power of the *Physalia* and several species of *Medusæ* to be voluntary; for while a sailor-boy had his hands peeled by a single specimen, Mr. Couch has handled numerous specimens, swimming at large and out of the water, living and dead, without his being made sensible of unpleasant effect.

THE ARGONAUT SHELL AND ITS INHABITANT.

Prof. Owen has submitted to the Zoological Society a valuable series of specimens of the Paper Nautilus, (*Argonauta Argo,*) consisting of the animals and their shells of various sizes, of *ova* in different stages of development, and of fractured shells in different stages of reparation, all which were commented on by Prof. Owen, to whom they had been transmitted for that purpose by Madame Power; and from which the Professor recapitulated the following evidence, deduced independently of any preconceived theory or statement.

1. The cephalopod of the Argonaut constantly maintains the same relative position in its shell.

2. The young cephalopod manifests the same concordance between the form of its body and that of the shell, and the same perfect adaptation of the one to the other, as do the young of other testaceous mollusks.

3. The young cephalopod entirely fills the cavity of its shell, the fundus of the sac begins to be withdrawn from the apex of the shell only when the *ovarium* begins to enlarge under the sexual stimulus.

4. The shell of the Argonaut corresponds in size with that of its inhabitant, whatever be the difference in the latter in that respect The observations of Poli, of Prevost, and myself, on a series of *Argonauta rufa*, are to the same effect.

5. The shell of the Argonaut possesses all the requisite flexibility and elasticity which the mechanism of respiration and locomotion in the inhabitant requires; it is also permeable to light.

6. The cephalopod inhabiting the Argonaut repairs the fractures of its shell with a material having the same chemical composition as the orificial shell, and differing in mechanical properties only in being a little more opaque.

7. The repairing material is laid on from without the shell, as it should be according to the theory of the function of the membranous arms as calcifying organs.

8. When the embryo of the Argonaut has reached an advanced stage of development *in ovo*, neither the membranous arms nor the shell are developed.

9. The shell of the Argonaut does not present any distinctly defined nucleus.

Finally, Mr. Owen detailed the points which still remain to be elucidated in the natural history of the Argonaut; but in proposing farther experiments, and while admitting that the period of the first formation of the shell still remained to be determined, Mr. Owen stated that he regarded the facts already ascertained to be decisive in proof that the cephalopod of the Argonaut was the true fabricator of its shell.—*Proceedings of the Zoological Society.*

Mr. Rang, in a note submitted to the Academy of Sciences in 1837, details the following newly-observed facts in the economy of the Argonaut:

1. The belief more or less generally entertained since the time of Aristotle, respecting the skilful manœuvres of the poulp of the Argonaut in progressing by the help of sails and oars, on the surface of the water, is false.

2. The arms which are provided with membrane in the poulp have no other function than that of enveloping the shell in which the animal lives, and that for a determinate object.

3. The poulp with its shell progresses in the open sea in the same manner as the other cryptodibranchial cephalopods.

4. When at the bottom of the sea, the poulp creeps upon an infundibuliform disk represented by the junction of the arms at their base, covered with the shell, and the part reputed ventral above; having in this posture the appearance of a gasteropodous mollusk.—*See Mag. Nat. Hist.*, No. xxxv., p. 527.

AGATE SHELL, (ACHATINA ACICULA).

Mr. D. Cooper records an additional instance of this British shell being found in connection with Danish remains; in the *tympanum* of a skull found at Limbury, a hamlet of Luton, Beds., in conjunction with pottery, urns, and a key, supposed of Danish origin. In addition to the perfect chain of *bones*, were found the lower two whorls and a half of a shell, which proved to be the remains of *Achatine Acicula*, (Agate shell,) a species of rare occurrence at the present time in the vicinity of Luton.

COLOUR OF SALT MARSHES.

Prof. Joli, of Montpellier, having carefully investigated the coloration of the salt marshes of the department of the Herault, concludes:—

1. That the *Artemia salina* contributes only in a subordinate manner, and scarcely at all, to this colouring.
2. That it is owing to infusoria animalcules.
3. That the *Hæmatococcus salinus* are only dead infusoria become globular.
4. That the *Protococcus salinus* are the globules which escape from their bodies after death.—*An. Nat. Hist.*

LIMNORIA TEREBRANS.

Dr. Edward Moore, with the view of submitting Kyanized wood to the action of the *Limnoria*, placed the following pieces of wood on the piles of the Pitch-house Jetty, in Plymouth dock-yard, at low water, on Jan. 12, 1838; namely, a piece of American deal 4 in. by 10¾ thick; also a piece of similar dimensions, which had been soaked for two months in a saturated solution of arsenic; and two others, which had been prepared with Kyan's solution. On the 12th of the following August, the pieces having been all under water for seven months, were taken up; when the protected pieces were found to have been acted upon, though not quite to so great extent as the plain piece of deal; but the specimens were dotted with *Balani* and *Flustræ*, and all containing *living Limnoriæ;* and it was evident that, though retarded, the destruction of the wood would, in a few months more, have been equally certain as where none of the above preparations had been employed. Dr. Moore considers it highly improbable that any protection can be afforded in cases of this kind from the employment of soluble substances; for, in the instance of the solution of oxide of arsenic, or of the bi-chloride of mercury, (corrosive sublimate,) which Kyan's solution is known to be, it is evident that any additional quantity of fluid coming in contact with it will dilute it, or re-dissolve any of the salt which might have been deposited in the pores of the wood, by drying; the continual washing of the sea will effectually clear the surface of the wood of any deleterious matter; and although the foremost depredators may perish in making a lodgment in the interior, yet myriads are ready to supply their places, and to maintain the ground already gained, while the continued action of the water will tend to assist them in their efforts: hence he is of opinion that Kyan's solution is not a certain remedy against the destruction of wooden erections in any of the estuaries around our island. Two arches of the wooden bridge at Teignmouth have fallen down, in consequence of the piers having been destroyed by the *Teredo;* so that we have here found another locality for this animal.—*Mag. Nat. Hist.*

NUMBER OF INSECTS ON THE GLOBE.

On the supposition that we are acquainted with one-third of the Coleoptera, one-half of the Lepidoptera, one-fifth of the Hemiptera, one-sixth of Hymenoptera, Neuroptera, and Orthoptera, one-tenth of the Diptera, and one-twentieth of the Parasita, we obtain the following numbers as the absolute amount of the species existing on the globe:

Coleoptera	120,000
Diptera,	100,000
Hymenoptera	72,000
Hemiptera	25,000
Lepidoptera	20,000
Parasita	10,000
Neuroptera	9,000
Orthoptera	6,000
	362,000

Jameson's Journal.

SCALES OF BUTTERFLIES' WINGS SEEN IN THE MICROSCOPE.

THE REV. E. CRAIG has communicated to the *Philosophical Magazine*, an inquiry into the Lines upon the Scales of Butterflies' Wings. Sir David Brewster, Dr. Goring, and Mr. Pritchard, speak of longitudinal lines, of cross striæ, and of two sets of diagonal lines besides, visible with the microscope on each scale of the *P. Brassica.* The above observers admit that the longitudinal and cross lines may be discovered with tolerable facility; but a doubt still exists as to the existence of the diagonal lines, and Sir D. Brewster considers them to be an optical illusion. He regards the whole linear appearances of the scales, not as lines, but a succession of teeth interlocking, arranged in lines: but Mr. Craig having examined these appearances with good instruments, has arrived at very different results from the preceding. He concludes, after a long series of observations, that each scale is a film, regularly ribbed or divided by longitudinal fibres, which are thicker than the rest of the film, like some corded muslins; and that each thinner portion of the film between the ribs is crossed in a slightly curved direction by still finer fibres, which become closer in their arrangements as they reach towards the serrated end of the scale; Mr. Craig too conceives that the dark longitudinal striæ are not formed by a close combination of interlocked teeth, (as Sir D. Brewster supposed,) because there have been seen many lacerated scales, which invariably split and tear up the middle of the dark line; and in every such case, exhibit a smooth edge without any appearance of teeth. The scales tear down these dark ribs in the same way as some leaves easily tear down the nerves. Moreover, it is easy to arrange the lens so as to leave the other parts of the film dark, and the light passing through these longitudinal ribs or striæ in the same way as through the nerves of a leaf. When the instrument has been arranged for the most distinct vision of the longitudinal lines, a delicate adjustment of the object *nearer* to the lens throws the longitudinal lines out of the focus, and brings out the cross striæ distinctly, thus proving them thinner than the longitudinal ones. Now, between these two adjustments exists the range for accurate investigation of the whole phenomena. The cross ribs or striæ may thus be traced running up into the longitudinal, and rising with a curve out of the interstitial, furrow; and in consequence of this cross-ribbing of the scale, each longitudinal line is liable to irregularities of height at different points.

The diagonal lines appear to Mr. Craig, as Sir David Brewster states, to be an optical illusion produced by the arrangement of the cross fibres: a similar effect, on a larger scale, is often observable in printed fabrics for gowns or waistcoats, in which two sets of diagonal lines are seen at a certain distance; but when the pattern is examined closely, they are not there. When the scale is removed to a sufficient distance only for showing the cross fibres, and not the longitudinal fibres, and the light in the position of the scale is a little oblique, then this delusive appearance takes place. The varieties of appearance in a moderate or powerful instrument seem to arise from its commanding at different times the different heights of an uneven surface; and the difference depends upon the distinct vision of the hill, the valley, or the intermediate slope, in a distance, which, between each longitudinal rib, is only the 20,000dth of an inch. The cross striæ are about the 60,000dth of an inch apart, and near the end of the scale still less.

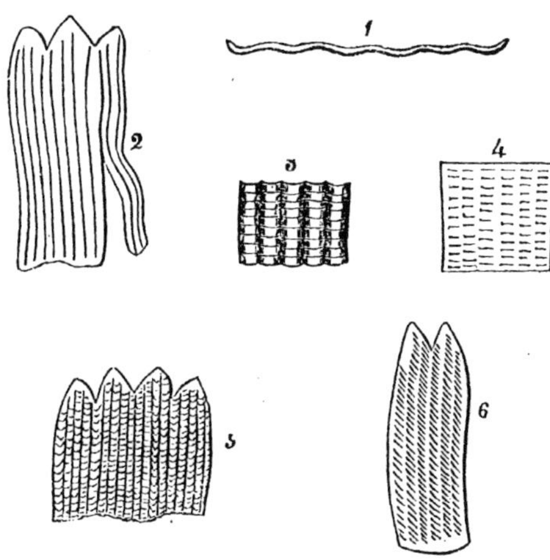

(Scales of Butterflies' Wings; magnified.)

Description of the Engravings.

1. Section of the scales, showing the wavy nature of the ribbed surface.
2. Portion of the scale, showing only the longitudinal lines, and torn along one of the fibres.
3. Portion exhibiting the structure of the film, with the longitudinal and cross striæ.
4. Portion in which the dark longitudinal striæ are thrown out of the focus, and only the cross striæ shown.
5. Portion in which the delusive appearance of the diagonal lines is attempted to be imitated.
6. Portion in which the focus of the lens is so adjusted as to show only a portion of the cross striæ on the rising slope of the furrow.

TISSUE MADE BY INSECTS.

A PIECE of a very fine tissue, a kind of Cloth made by Caterpillars, which was found in Moravia, has been exhibited to the French Academy of Sciences, and proves to be the work of the Caterpillars of the

Yponomeutidæ. A second specimen is attributed to a number of smaller spiders of the genus *Epiere* of Walckenear; for this tissue is somewhat glutinous and very much finer than that of the caterpillars, the thread of the latter not being adhesive like those of the spiders.—*Compte Rendu; An. Nat. Hist.*

MAY-BUGS.

An agriculturist of the department of La Seine-Inférieure has discovered a novel method of destroying May-bugs. He procured a couple of the birds called Kuikimanore from the Sandwich Islands, and taught them to eat these insects. During the season he lets them loose at night, and they return in the morning, after having, according to his calculation, destroyed at least four thousand. These birds have bred under his care, and he has now one hundred couples.—*Athenæum.*

PENSILE NESTS OF BRITISH WASPS.

Mr. Barraud has exhibited to the Entomological Society the small Nest of a Wasp, which had been found near Croydon, built in a sparrow's nest, and attached to the feathers within it. It was considered to be the nest of *Polistes,* a social wasp not hitherto found in this country; or, if not of *Polistes,* certainly of some new species, not yet determined or known.

The nest consists of three cells, with a space about a line wide between each, viz., the rudiments of a basal external one, commenced in a spiral direction, the widest portion of which is about half an inch only. The enveloping one, which gives the form to the nest, and is ovate, about 1½ inch long, and an inch broad at its widest diameter, has a circular aperture at its apex, rather more than 3-8ths of an inch across. Within this case, at the base, there is the commencement of another spiral one, which, at its widest part, laps laterally, scarcely farther than the base of the cells; and within this, in the centre, is placed the tier of cells, originating from a common pedicle; consequently, as usual, the central ones are the most advanced in structure: altogether, there are 15 *perfect hexagons,* the central ones being nearly 4 lines deep, and all somewhat more than 2 lines in diameter.

The nest appears to be constructed of the agglutinated particles of a soft white wood, probably willow, very imperfectly triturated, so as to give it externally a rough granulated appearance. It is sprinkled with black specks, arising, perhaps, from the intermixture of more decayed portions of the wood, and is of very fragile texture.

The nature of the material, its unfinished execution, and the situation in which it was found, constitute its peculiarities; it being considered an accidental deviation from the usual nests of the *Vespa Britannica* of Leach. The situation is very singular: the wasp must have considered the sparrow's nest deserted; or, may it have confided in its means of defence?

There are two British wasps which suspend their nests in exposed situations;—the one above, and the *Vespa Holsatica,* Fab. The nest of the latter is, however, much larger; usually about the size of a man's head, or somewhat smaller. These are of a firm texture, and are attached to shrubs; in the north, they frequently occur in gardens, fixed to goose-berry-bushes.—*W. Shuckard;* in *Mag. Nat. Hist.*

LARGE WASPS' NEST.

A SOCIAL WASPS' NEST, of very large size, has been sent from Ceylon, and deposited in the Museum of the Zoological Society. This nest was found by Lieut. Williams in a talipot tree near Colombo; its apex was secured at the junction of two of the smallest leaves of this magnificent tree, and the bottom of the nest was about 70 feet from the ground, at which elevation the leaves began to shoot. It had been abandoned by the wasps, and its exterior walls were much injured by the monsoon rains and storms, which left the terraces unprotected and unsupported, except by their interior pillars; so that the natives were, in consequence, unable to lower it from such a height without destroying some of the lower terraces. The appearance of the nest, as it hung upwards of 70 feet from the ground, the shaft to it perfectly bare; and the larger leaves, (used by the natives as umbrellas and tents,) waving over it; presented a very singular effect.—*Proc. Zoological Society.*

SCALES OF THE EEL.

DR. BUCKLAND has remarked to the Geological Section of the British Association on the fine provision in the external economy of the Eel, covered with minute scales, and having diffused over them a quantity of slimy mucus, under which, being concealed, they are admirably adapted for that mode of life which consists of imbedding themselves in mud, or penetrating under stones and rock.

SAW-FLIES.

MR. WESTWOOD has lately detailed to the Entomological Society the proceedings of one of the Saw-flies, (*Tenthredinidæ,*) which he had observed in the act of flying over and depositing its eggs in the blossoms of the apple at Hammersmith. In the preceding summer, the entire crop of fruit had failed, in consequence of the interior of the young fruit being infected with the larvæ of a Saw-fly; and he now had no hesitation in considering that the winged insects recently observed, had been produced from some of the previous year's larvæ.—*Literary Gazette.*

PARASITE ON A TORTOISE.

AT a meeting of the Entomological Society, in June, Mr. James H. Fennell exhibited a specimen of a very singular parasitic animal which he had discovered attached to the tail of a land tortoise, and so tenaciously, that it could only be induced to separate from it by actual force. It belonged to the class *Arachnida,* being somewhat like a spider; had eight legs, a flat shelly back of a black colour, bordered with yellow; head and eyes, very small; and a thin short proboscis, through which it, doubtless, sucked the juices of the tortoise.

ACCLIMATION OF SILK IN ENGLAND.

MR. FELKIN has detailed to the British Association an experiment in the Growth of Silk, made in one year at Nottingham: he also exhibited cocoons of the white and gold-coloured products; adding that except in one instance near London, for which the Society of Arts had awarded its premium in the year 1790, no one had succeeded in a similar experiment in England. The white cocoons were the work of worms fed on lettuce-leaves, and then on mulberry leaves; and the gold-coloured, of worms fed entirely on mulberry-leaves, having been hatched a fortnight later in the season, when these could be procured. Of the former, seven out of eight died; of the latter, from 30 to 40 per cent. The eggs were from Italy; and neither the situation in the heart of the town, nor the season, had been propitious. The room was, however, kept at a temperature of between 55° and 70°; and, especially from the worms fed on the mulberry-leaves, the silk was of a good quality. From the above success it was inferred that this method might be introduced with a certainty of prosperity in our colonies, and save us part of an expenditure, increased from three to four millions sterling per annum. General Briggs observed, that in India, the worms were fed on the zizippus jujube. The white mulberry seems to be the best food for supporting the worm, and producing the best material.—*Literary Gazette.*

INSTINCT OF CATERPILLARS.

Mr. A. H. Davis, writing from Adelaide, South Australia, mentions his seeing there an extraordinary procession of Caterpillars. They somewhat resembled in form the caterpillars of the great tiger-moth (*Arctia Caja,*) had a profusion of white hairs, and the body about 2¼ inches in length, of a dark brown colour, with paler lines. These caterpillars were seen crossing a road in single file, each so close to its predecessor as to convey the idea that they were united together, moving like a living cord in a continuous undulating line. At about fifty from the end of the line, Mr. Davis having ejected one from its station, the caterpillar immediately before it suddenly stood still, then the next, and then the next, and so on to the leader; the same result took place to the other extremity. After a pause of a few moments, the first after the break in the line attempted to recover the communication; this was a work of time and difficulty, but the moment it was accomplished by its touching the one before it, this one communicated the fact to the next in advance, and so on till the information reached the leader, when the whole line was again put in motion. On counting the number of caterpillars, he found it to be 154, and the length of the line 27 feet. He next took the one which he had abstracted from the line, and which remained coiled up, across the line; it immediately unrolled itself, and made every attempt to get into the procession; after many endeavours it succeeded, and crawled in, the one below falling into the rear of the interloper. He subsequently took out two caterpillars, about the fiftieth from the head of the procession; by his watch he found the intelligence was conveyed to the leader in thirty seconds, each caterpillar stopping at the signal of the one in its rear; the same effect was observable behind the break, each stopping at a signal from the one in advance;

the leader of the second division then attempted to recover the lost connection. That they are unprovided with the senses of sight and smell appeared evident, since the leader turned right and left, and often in a wrong direction when within half an inch of the one immediately before it: when it at last touched the object of its search, the fact was communicated again by signal, and in thirty seconds the whole line was in quick march, leaving the two unfortunates behind, who remained perfectly quiet, without making any attempt to unroll themselves. Mr. Davis was informed that these caterpillars feed on the *Eucalyptus*, and that when they have completely stripped a tree of its leaves, they congregate on the trunk, and proceed, in procession, to another tree.

NEW SPIDER.

Mr. Blackwall has read to the Linnæan Society a paper describing New Species of Spiders, recently discovered, principally by himself, in the north of England and Wales; no fewer than 53 species having been thus added to the catalogue. The genera to which they chiefly belong are *Drassus Clubiona*, *Lycosa*, *Angelena*, *Theridium*, *Walckenaera*, *Neriene*, and *Linyphia*. Much of Mr. Blackwall's success is to be attributed to the fact of his attention having been chiefly directed to those species which, on account of their diminutive size, require the aid of optical instruments, of a high magnifying power, for their accurate examination.—*An. Nat. Hist.*

TO PREVENT INSECTS CLIMBING UP FRUIT-TREES.

At a late meeting of the Entomological Society, Mr. James H. Fennell communicated the following successful mode of preventing insects ascending the trunks of fruit-trees. Let a piece of Indian rubber be burnt over a gallipot, into which it will gradually drop in the condition of a thick viscid juice, which state it appears it will always retain; for Mr. Fennell has at the present time some which has been melted for upwards of a year, and has been exposed to all weather without undergoing the slightest change. Having melted the Indian rubber, let a piece of cord or worsted be smeared with it, and then tied several times round the trunk. The melted substance is so very sticky, that the insects will be prevented, and generally captured, in their attempts to pass over it. About three pennyworth of Indian rubber is sufficient for the protection of twenty ordinary sized fruit-trees.

BOTANY.

PLANTS GROWN WITHOUT AIR.

Mr. Ward, of Wellclose-square, has succeeded, in what may naturally be called the worst of all London localities, in growing many species of ferns in a superior manner; and amongst them several that had hitherto baffled all the care, convenience, and skill of the gardener. Mr. Ward's success appears to depend on growing Plants in air-tight cases, suffering the moisture which their pores exude to be absorbed again by the roots, while they are preserved from external injuries and sudden changes of temperature. How long plants may be found to submit to this mode of culture is uncertain, but one fact is established—that plants have been imported from New Holland, in such cases, that never before reached Europe alive.

ACTION OF FROST ON PLANTS.

PROF. MORREN has laid before the Academy of Brussels an account of his investigations relative to the Action of Cold on Plants; the results of which are that, however delicate the organization of the plants, not one of their elementary parts is ruptured by the action of the frost, but the functions are entirely deranged: thus, the organs of respiration are filled with water, and those of nutrition with air; so that the natural order is inverted, and death is the consequence.—*An. Nat. Hist.*

FECUNDITY AND DEVELOPMENT OF PLANTS.

PROF. DAUBENY has recapitulated to the Ashmolean Society the new views with respect to the Fecundation and the Development of Plants, which have been brought forward by Brown, Mirbel, Schleiden, and other botanists of the present day. He remarked, that when Linnæus had established, to the satisfaction of naturalists, the doctrine of the sexuality of plants, which formed the basis of his artificial system, he left to his successors two branches of inquiry in a manner untouched: viz., 1st, in what precise method do the stamens operate upon the pistils when they cause fecundation to take place; and, 2ndly, to what extent can we trace an analogy between the mode of fecundation and development, in the case of flowering plants where sexes exist, and in that of cryptogamous ones, where they are not discoverable. The first of these points has been elucidated by the researches of Brown, of A. Brongniart, of Ehrenberg, and others, who have shown that each grain of pollen, when it alights upon the stigma, sends forth a kind of tube, caused by the protrusion of its own internal membrane; and that this pollen tube, as it is called, penetrates the stigma, passes down through the interior of the style, and thus comes into immediate contact with the ovary. The second of these points—as to the analogy subsisting between flowering and cryptogamous plants in these particulars—has been investigated by Mirbel in France, and more lately by Schleiden in Germany. The former observed new cells originating out of those already existing in the case of the *Marchantia;* whilst the latter appears to have shown that a process, the same in kind, takes place within the pollen tubes emitted from flowering plants at the very time they reach the ovary and impregnate it, as well as in the cells of the plant in the subsequent stages of its growth. According to him, each cell, whether it be in the pollen, or in the plant generated from it, contains a portion of starch, which is capable of being converted into nutritious matter at the time it is required. This conversion having taken place, a dark spot is perceived in the coats of the cell, and from this spot a new cell is seen to be protruded.

Accordingly, as this spot is the germ of the cell, it is called by Schleiden the Cytoblast. The new cell, when generated, gives birth in its interior to new cytoblasts, which again generate new cells; and thus a series of cells is produced one within another, until the external one is ruptured, and its contents are enabled to escape, and thus to obtain their natural development. The newly-formed cells are of extreme tenacity; but new matter is afterwards deposited within the interstices of that originally formed, until they gradually acquire firmness and consistency. According to this view of the subject, the reproduction and subsequent development of the more perfect plants, would appear to be conducted fundamentally on the same plan as that of cryptogamous ones: in both instances, the process commences by the generation of a new cell in the interior of one already existing; each cell maintaining, as it were, an individual existence, and differing only in possessing in a higher or lower degree, a principle of independent vitality. From Schleiden's researches, too, another inference would seem to flow, namely, that the embryo exists in the pollen, and not in the ovary; the office of the latter organ being merely that of furnishing to the young individual a receptacle and nourishment. This position, however, is disputed by Mirbel, who adheres to the old doctrine on the subject.—*Athenæum.*

ANOMALOUS FRUCTIFICATION.

Mr. J. Smith has read to the Linnæan Society "A Notice of a Plant which produces perfect Seeds without any apparent action of Pollen on the Stigma." The plant belongs to the natural family of *Euphorbiaceæ*, and has been cultivated for several years in the Royal Botanic Garden at Kew, under the name of *Sapium aquifolium.* It is a native of Moreton Bay, on the east coast of New Holland, where it was discovered in 1829. Shortly after their introduction at Kew, the plants flowered, and they proving to be all females, they were naturally passed over as belonging to a diæcious plant, until Mr. Smith noticed their producing perfect seeds. They have annually flowered and matured their seeds since; and, notwithstanding the most diligent search and constant attention, no male flowers or any pollen-bearing organs have been detected. Young plants have been raised, at different times, from the seeds, and they so clearly resemble their parents that it is scarcely possible even to suspect the access of pollen from any other plants. Mr. Smith considers the plant as the type of a new genus, which he names *Cælebogyne.* It forms an irregularly-branched rigid evergreen shrub, about 3 feet in height, with alternate, petiolate, elliptical, mucronate, coriaceous leaves, having three large spinous teeth on each side, and being furnished with two small subulate persistent stipules.—*An. Nat. Hist.*

FORMATION OF LEAVES.

Dr. H. A. Meeson has read to the Botanical Society a paper "On the Formation of Leaves." He first observed that leaves cannot be expansions of the epidermis, because, if so, they must of necessity be composed entirely of cellular tissue, whereas they are known to abound in vascular tissue. If leaves be expansions of the bark, all modifications of them must be the same; wherefore petals, sepals, stamina, and pistils must be expansions of this substance. But these organs exist in endogens, a class of plants manifestly without bark, and in exogens their texture is different from that of the bark beyond comparison. Dr. M. considers leaves to be the essential part of the plant; they exist in the embryo, and by expanding and unfolding themselves suck up sap through the radicles, and having exposed it to the action of air and light, convert a portion of it into proper juice. A plant is nothing more than a multitude of buds on fixed embryos, which send their roots downwards to form their bark and wood. The leaf should be considered the most essential part of the plant, where all other parts are either directly or indirectly formed; as it is not an expansion of any thing, but a very important organ, having, as it were, a distinct existence of its own.—*An. Nat. Hist.*

VEGETABLE IMPREGNATION.

THE Copley Medal has been awarded by the Royal Society to Mr. R. Brown, for his valuable discoveries of the Organization of the Vegetable Ovule immediately before fecundation; and the direct action of the pollen, manifested by the contact established between it and that point of the ovulum where the embryo first becomes visible. These discoveries are very important, not only in a botanical, but also in a general scientific, point of view, by showing the close analogies of animal and vegetable life.—*Proc. Royal Society.*

STRUCTURE AND FUNCTIONS OF POLLEN.

MR. H. GIRAUD has read to the Botanic Society of Edinburgh, two papers, in which he has pointed out some peculiarities in the external configuration and internal structure of the mature Pollen grain, showing that from the former character no correct indications can be deduced for determining the limits of certain groups of plants; as the figure and surface of the pollen grain often differs widely in individuals of the same genus. The existence of a *third* tunic, or inverting membrane, is shown to occur in the pollen of *Crocus vernus;* and also the presence of minute opaque bodies on the surface of the pollen of *Polemonium cœruleum*, which, when immersed in water, appeared to be possessed of spontaneous motion. The true nature of the furrow which exists in certain spherical and elliptical pollen grains, is pointed out, and shown not to be a slit in the outer membrane, as is the opinion of some vegetable anatomists. The chemical composition of pollen is then described, showing the existence of potassa in the pollen of *Antirrhinum majus;* and of raphides, consisting of phosphate of lime, mixed up with the pollen of *Tradescantia virginica*, and with that of certain species of *Orchidea.* The functions of the pollen are also noticed; and a somewhat elevated temperature is shown to be necessary, in many cases, for the development of the pollen tubes, which condition is supplied by the evolution of caloric in æstivation.

ERGOT IN GRASSES.

MR. QUECKETT has read to the Linnæan Society some "Observations on the Anatomical and Physiological Nature of Ergot in certain Grasses;" from which he infers that the Ergot is a grain diseased by a particular parasitic fungus developing in or about it, whose sporidia find the young state of the grain in a matrix suitable for their growth, and quickly run their race, not entirely depriving it of its vitality, but communicating to it such impressions, which pervert its regular growth, and likewise the healthy formation of its constituents; being at last composed of its diseased materials, which are mixed up with fungic matter, that has developed within it.

The fungus caused to germinate in the way described is quite invisible to the naked eye, seldom measuring beyond the one or two hundredth part of an inch; and from comparisons with British and Foreign genera of Fungaceæ, it has not been found satisfactorily to belong to any as at present constituted; wherefore the author proposes a new genus, with the title *Ergotetæa*, to represent this minute fungus, which will belong to the sub-order *Coniomycetes* of Fries, and to its division Mucedines, very near to the genus *Sepedonium*.

Repeated experiments with the sporidia of the Ergot of Rye, of *Elymus*, and other grasses, induce the author to consider the parasite in each case the same; wherefore he applies the term *abortans*, as the specific name of *Ergotetæa*, to the plant found on the Ergot of Rye; and he believes the parasites found on the other grasses to be the same.

LIGHT FROM PLANTS.

On Dec. 20, Dr. Willshire communicated to the Botanical Society all that was then known with respect to the evolution of Light from Plants; a subject which of late years had attracted some attention on the continent, but which was first observed by the family of Linnæus in *Tropælum majus*. The subject was divided into two portions, viz., light evolved from dead and living vegetable structures; and it appeared that wood rotted in the air never shines, it being requisite to be buried in the earth when the sap is contained within it. It was observed that the effect takes place only in the months of July and August, in warm dry weather, never in damp. Several plants were mentioned as evolving light during the putrefactive stage; among others, mushrooms, potatoes, &c.—*Athenæum.*

COMPARATIVE STRUCTURE OF SUCCULENT PLANTS AND SIGILLARIÆ.

Mr. Link has exhibited to the Berlin Academy some drawings showing the structure of the stem of arborescent Succulent Plants, with reference to the alleged similarity between them and the *Sigillariæ* of a former world. It is certainly remarkable that numerous layers of bark are deposited one on the other, far more so than in all other trees; one consequence of which is their compression of each other into a flattened shape, when the outer bark falls off. The cells, however, of the new layers are flatter than in general. The ligneous bundles pass from the wood to the scars of the leaves; and such a difference in the form of these scars on the outer bark and beneath it, as that observed by Ad. Brongniart in the *Sigillariæ*, was not perceptible. The wood is very thin, even in the thickest stems of succulent plants; the bark and pith very thick; they remain a long while succulent and then rot, so that their preservation among fossil bodies is very improbable. —*From the German; An. Nat. Hist.*

GEOGRAPHICAL DISTRIBUTION OF PALMS.

M. V. MARTINS has published an elaborate treatise on the Geographical Distribution of Palms, which he divides into five groups, viz., *Arecinæ, Lepidocaryinæ, Borassinæ, Coryphinæ, and Coccinæ.* The distribution of palms with which we are at present acquainted is as follows :

	Old World.	New World.	Total.
Arecinæ	53	45	98
Lepidocaryinæ	60	7	67
Borassinæ	11	24	35
Coryphinæ	33	24	57
Coccinæ	2	99	101
	159	199	358

Of these 358 palms, Europe contains 1; New Holland, 6; New Zealand, 1; Oceania, 2; Africa, 13; Asia, 131; and America, 198.

GENTIANEÆ.

DR. GRISEBACH has published an extremely valuable work, entitled *Genera et Species Gentianearum*, &c.; in which the number of species described is 343, or about $\frac{1}{230}$th of the known portion of the vegetable kingdom. These are dispersed over a considerable portion of the world; but no species have been found in several of the isles of the Pacific Ocean, in tropical New Holland; in the islands of Timor, Sumatra, and other of the Polynesian group; nor in the African deserts, the shores of Venezuela, and scarcely in the mountains of the south of Europe. Two hundred and ten species are found in the tropics; 133 are extra-tropical, of which 45 inhabit the southern hemisphere. The New World affords 180 species; the Old 175: but 12 are common to both. The higher mountains of the Andes yield the greater proportion of species (51), then follows tropical Brazil (46), the Himalaya Mountains (41), the United States (33), the Alpine flora of Europe and Siberia (32), Hindoostan (30), the Cape of Good Hope (25). One hundred and thirty-five species flourish at an elevation exceeding 5000 feet above the level of the sea, 230 below that elevation. The maximum of the family may be considered Alpine; nevertheless, the species are rare or altogether wanting in the Alps of Mexico, Java, on the Peak of Teneriffe, and of Sicily.

MORPHOLOGY OF THE ASCIDIÆ.

PROF. MORREN concludes an elaborate inquiry into this subject with the following inferences: 1st, that since all the Ascidimorphous bracts of *Norantea* and of *Marcgravia* are the blades of bracteal leaves joined at their margins so as to form hollow pitchers; 2nd, that since the *Dischidia Rafflesiana* evidently presents leaves with the blade cohering to form an Acidium; 3rd, that since in monstrous states we see blades of leaves become Ascidia, and that petioles are not hollowed to produce this form accidentally, and that when they are winged we do not see their wings cohere at their free margins; 4th, that since the structure of *Sarracenia* proves very decidedly that it is a leaf which forms the Ascidium, retaining the apex of the blade in its coherent state; 5th, that since the Ascidia of *Nepenthes* have already at the lower part a winged petiole, and that the crests of their pitcher are traces of foliaceous blades;—it must be allowed that the Ascidia have, where-ever they have been observed hitherto, a similar organic composition, and that all are metamorphoses of the leaf and particularly of the blade of this organ.—*An. Nat. Hist.*

NEW TREE.

By far the most interesting tree, of which we have an account in the narrative of the recent voyage of Captains King and Fitzroy to the South, is the Alerce, a large *Conifera*, of which the principal forests are in the cordillera opposite to Chiloe. The Spanish settlers conferred this name upon it, no doubt, from some fancied resemblance to the tree of their Arab ancestors, (the *Thuja articulata*;) but, from the description, it would appear rather to be a pine. The earliest description of it is by Captain King; but Captain Fitzroy employed a Mr. Douglas to make an excursion for the purpose of examining the forests, which are now considerably inland, and difficult of access. By his report, there are still trees of large dimensions in the interior; the largest he saw being twenty-two feet and twenty-four feet in girth, at five feet from the ground, though they were unsound. The largest felled, within the last forty years, measured thirty feet, at five feet from the ground, and seventy-six feet to the first branches, furnishing 1,500 planks; the common proportion of the larger trees being from 800 to 900. He gives an account of a landslip which carried down 1,000 trees, a few years since, many of them of large size. Astilloros, or timber-yards, are formed in convenient situations, where the trunks are sawn into lengths of eight or nine feet, and then split, by iron wedges, into planks, which are carried on men's shoulders to the place of embarkation. So straight is the grain, that they split like slates, and are used for roofing, flooring, and many other purposes; they turn blue by exposure to the weather. The wood is brittle, but is not subject to warp or cast. The entire tree makes excellent masts; but the difficulty of transport is such, that, although a very large price was offered, it was impossible to procure one in less than two months. The bark is used for caulking, which purpose it answers while kept under water; but it will not bear the alternation of wet and dry. The timber is not only in general use at Chiloe, but is largely exported to Lima and other places; and, no doubt, a road to the interior forests would repay the projectors, the people being too poor for such undertakings. Far inland, beyond the reach of the Calbruanos, who carry on this laborious business, are said to be trees from thirty feet to forty feet in girth, and eighty feet to ninety feet to the branches, the heads towering forty feet to fifty feet higher. An associated species is called the cypress, which, no doubt, from the description, is different; although Captain King is doubtful on this point. The wood is white, that of the Alerce being red; and it does not split so well as the latter timber.—From an excellent *précis* of the Arboriculture of the Voyage; by Capt. S. E. Cook, R. N.: *Gardener's Magazine.*

SALEP.

PROF. LINDLEY has read to the Linnæan Society "A Note upon the Anatomy of the Roots of Ophrydeæ," in which he shows that Salep, the prepared roots of certain Ophrydeæ, is not a substance consisting principally of starch, as is the common opinion among writers of the present day, but is composed of a balsamic-like matter organized in a peculiar manner. The author briefly explains the error of those who have considered Salep to consist chiefly of starch, by allusion to the mode of its preparation. The tubercles are first parboiled and then dried, the effect of which is to dissolve what starch exists in the cells surrounding the nodules. The dissolved starch flows over the surface of the nodules, from which, when dried, it is undistinguishable; and, consequently, when iodine is applied to Salep, the mass appears to become iodide of starch. If the nodules, however, after this action of iodine, be removed, they are seen to retain their original vitreous lustre.—*An. Nat. Hist.*

EAST INDIAN OPIUM.

Mr. E. Solly has read to the Royal Asiatic Society, a paper on East Indian Opium, which not many years ago was always considered, and described, as being the lowest and most inferior in quality of any in the market. Latterly, however, it has been increasing in reputation, and some sorts of East Indian opium are now classed amongst the best. The quality, however, is very various, and some sorts, from containing a very large proportion of impurities, are of comparatively little value; great influence is also exerted on the quality of the opium by the weather; the presence or absence of clouds, the quantity of dew, the time which the juice is kept before it is dried and formed into cakes, and many other causes. Mr. Solly stated that the Indian opium prepared for the Chinese market contains scarcely any earthy impurities or foreign matters; and he suggested that one of the causes why the Chinese preferred Malwa to many other kinds of opium was, that it contained less caoutchouc and gluten, and therefore yielded a very large per-centage of the watery extract, which is used by the Chinese smokers, and called smokable extract. He detailed a chemical examination he had lately made of some varieties of Indian opium, particularly that from Malwa; from which it appeared that it might enter into competition with some of the best varieties of opium. —*Athenæum; abridged.*

CULTURE OF COTTON.

ON Jan. 19, were read to the Royal Asiatic Society the following communications on the Culture and Economy of Cotton. Dr. Royle read a letter from Mr. Malcolmson, "On the Cotton grown near Pæstum, in the kingdom of Naples;" a small quantity of the seed of which he had forwarded, with a request that it should be sent to the Horticultural Society of Bengal. Mr. Malcolmson entered into details respecting the soil and climate of that part of Italy, and alluded to the remarkable petrifying stream which runs near Pæstum, and which caused a considerable calcareous deposition on the soil which it irrigated, a mixture which had been supposed essential to the cultivation of cotton. There were two kinds of cotton cultivated in the kingdom of Naples, the best of which was grown at Castellamare. Mr. Malcolmson had procured some of the soil from one of the richest cotton-fields of that place, of which he intended to forward an analysis. Dr. Royle stated that he had, at the same time, received a note from the Hon. Fox Strangways, containing an extract from a paper by Professor Tenore, on the very cotton spoken of by Mr. Malcolmson. The extract stated that the usual cotton grown in the kingdom was the *Gossypium herbaceum;* but that the cotton of Castellamare, which had been cultivated from time immemorial in Calabria, was very probably the same as the American cotton described in the "Orto Romano." For the cultivation of this cotton in Castellamare, they were indebted to the French, who had brought it from Calabria. Dr. Royle observed that the cotton was most probably the *Gossypium hirsutum*, or Upland Georgia cotton. As a mail was just about to depart for India, he had been enabled, by the kindness of Mr. W. B. Bayley, to despatch, through the India House, the seed sent by Mr. Malcolmson, and it would reach India time enough for the sowing of the next season.—Two papers were afterwards read, "On the Cultivation of the Bourbon Cotton in the South of India." The first by Mr. Hughes, of Tinnerelly, who had grown that plant largely, and with great success, twenty years ago; and the other by Mr. Heath, who had followed the plan of Mr. Hughes, when commercial resident of Salem and Coimbatore, with equal success. The paper of Mr. Hughes went into minute details on the soil and climate; on the planting, pruning, gathering, and cleansing the cotton; and on its value in the market. At the conclusion of this paper, it was observed by Mr. W. B. Bayley, that so far as he understood, the manufacturers of Glasgow and the north of England considered the defects of Indian cotton generally to arise rather from want of care in gathering and cleansing than from any deficiency of staple; and that, consequently, more attention should be paid to these points than to the introduction of new plants.—*Literary Gazette.*

VEGETABLE TALLOW.

On the 2nd of March, Dr. Royle and Mr. E. Solly read to the Asiatic Society two distinct papers on the Vegetable Tallow Tree (*Valeria Indica*) of the Malabar and Canary coasts. This tree, which has been figured and described by Rheede, is found in the Wynaad and Bednore districts, growing abundantly both in the interior and along the coasts, where it is called the Piney, or Dammar-tree. It grows to a great size, and supplies excellent wood. It also supplies a varnish which is used on the coast in a liquid state: but, when dry, is commercially termed *copal* and *animé*. By boiling the seeds, a fatty matter is obtained, which floats on the surface, becomes solid, and somewhat resembles tallow; being in its most important characters intermediate between wax and tallow, and well adapted in its properties, as a substitute for common tallow, both in the manufacture of candles, and likewise for many other purposes to which the latter substance is now exclusively applied. This vegetable tallow emits no disagreeable smell at any time; therefore, when candles are made of it, they have not that offensive smell which attends common tallow candles. Dr. Babington placed a portion of this vegetable tallow in the hands of a candle-manufacturer, who praised it very highly; he having succeeded in making good candles of it, which came freely from the mould. In 1825, it sold at Mangalore, at twopence-halfpenny per pound. Some brought from India, in January, 1838, sold for £2 4*s.* 6*d.* per hundred weight—nearly the price given for good Russian tallow. Mr. Solly thought that if it could be obtained at such a price as to admit of its being imported as a substitute for common tallow, its valuable and superior properties would soon obtain it a market. Mr. S. Dyer, of the Madras Medical Service, who had long resided at Tellicherry, stated, that the tree will grow readily, even when the branches are put into the ground; and many of the trees were planted on the roadsides in Malabar, about twenty years since, a greater period than is necessary to bring the tree generally to perfection.

MOCHA COFFEE.

The following new and interesting details are condensed from Mr. Cruttenden's Notes on his recent journey from Mokhá (Mocha) to San'á, in Arabia.

The Coffee-plant is usually found growing on the side of a valley or other sheltered situation, the soil which has been gradually washed down from the surrounding heights being that which forms its support. This is afforded by the decomposition of a kind of clay-stone, slightly porphyritic, which occurs irregularly disposed in company with a kind of trap-rock, among which basalt is found to predominate. The clay-stone is only found in the more elevated districts, but the detritus finds a ready way into the lower tract by the numerous and steep gorges that are visible in various directions. As it is thrown up on one side of the valley, it is there carefully protected by stone walls, so as to present the appearance of terraces. The plant requires a moist soil, though much rain is not desirable. It is always found growing in the greatest luxuriance when there is a spring in its vicinity; for, in those plantations where water is scarce, the plant appears dry and withered. The fig, plantain, orange, citron, and indigo, may sometimes be found growing among the coffee. A stream of water from a neighbouring spring is drawn up through the garden, and the roots of each plant are regularly watered every morning and evening. The plant is said to live six years; three of which are requisite for bringing the tree to perfection, for three it bears, and then dies and is rooted up. The bean is gathered twice a year; and one tree, though very small, ought to produce in the two crops, at least ten pounds. The plantation of Dórah, between Mokhá (Mocha) and San'á, yields coffee of very fine quality: it is small, perhaps not covering half an acre, with an embankment of stone round it to prevent the soil from being washed away. Some finer coffee is, however, from 'Uddeïni, the trees of which plantation are very large, or about twelve feet high.

The whole cost of transporting a camel-load of coffee from San'á to Mokhá is forty-four dollars, upon which the merchant clears a profit of three dollars and a half. It is brought into the San'á market in the months of January and December, from the surrounding districts. The nearest place to San'á, where the coffee grows, is Haffrsh, about a short day's journey south-east of San'á. Attempts were made by the last Imán to cultivate the plant in his own garden, but without success, owing, it appears, to the cold. The varieties of coffee are named Sharjí, Uddeïni, Mataríˈ, Harrází, Habbát, Haímí, and Shírází; of these the Sharjí and Habbát are the smallest and best. Keshr (husk), being more in demand at San'á, obtains a higher price. The best is the 'Anezí (Habbát), and is sold at twelve dollars for the hundred pounds; the inferior sorts at four, five, and six dollars for the same quantity. The principal trade of San'á is in coffee; but the merchants are so fearful of trusting their goods to the Turkish government, that they prefer filling their warehouses with it in San'á to sending it to Mokhá.

ASSAM TEA.

The issue of the attempt to cultivate tea in Assam has acquired increased importance since the late suspension of the trade with China, and the difficulties which appear to threaten it. At the same time that an Association was forming in London, one also was established in Calcutta, for carrying out, on a large scale, the cultivation of the tea-plant in Assam, where, from experiments made under the orders of the Indian authorities, its practicability had been ascertained; and, from actual survey, it was proved that the country abounded with the tea-plant, and with every facility for promoting its farther production under approved systems of management, to the greatest extent. A junction of the interests of these two Companies has taken place; the foreign Association being re-constituted under the title of the Bengal Branch Assam Company; the local management of which is to be conducted by a committee of directors, to be elected exclusively in the latter Company. A negotiation has been opened with the Indian authorities for the transfer of their establishment, with the working machinery employed in the experimental cultivation; and letters have been addressed on the part of the Company to the Government superintendents in Assam, requesting all possible information. To one of these it has been replied that there is an "unlimited field for such operations as are contemplated; abundance of tea-plants in a country" said to be "flowing with milk and honey; provisions abundant, and easily procured," and only requiring labour and capital to develop the ample resources possessed.

It was stated, also, in this communication, that, at that time, there were plants in cultivation equal to the production of 100,000 lbs. of tea, if the means of manipulation were provided. Steps had been taken to remedy the deficiency of hands, by procuring families in numbers to proceed to, and settle in, that country; and a correspondence has been opened with Singapore to obtain Chinese artisans conversant with the details of tea preparation. As so large a quantity as 100,000 lbs. had, in so short a space of time, been planted and prepared on the experimental ground, where, on a small scale only, the probationary cultivation had been attempted by the Government, well-grounded expectations may be entertained, that if the cultivation be adequately followed out, a sufficiency of teas will, in the course of a very few years, be produced to render this country entirely independent of the Chinese market.—*Abridged from the Times*, Sept. 28.

It will be remembered that, upon the discovery of the tea-plant being indigenous in Upper Assam, Mr. C. A. Bruce was sent thither to explore the tea country, and was appointed superintendent of its culture. He then proceeded to raise plantations; and, in the year 1838, transmitted to England eight chests of "Assam Tea," each containing 320 lb. It appears, also, that Mr. Bruce has drawn up his second Report, which has been presented by the Tea Committee appointed by the Bengal government.

Notwithstanding the troubles in which the frontier of Assam has been involved, Mr. Bruce has altogether discovered 120 tea-tracts, some of them very extensive, both on the hills and in the plains; whence a sufficient number of seeds and seedlings might be collected, in the course of a few years, to plant off the whole of Assam.

In 1838, on going over one of the hills behind Jaipore, about 300 feet high, Mr. Bruce discovered a tea-tract between two and three miles in length; the trees were mostly as thick as they could grow, and the tea-seeds (smaller than he had seen before,) literally covered the ground: this was in the middle of November, when the trees had abundance of fruit and flower on them. One of the largest trees was two cubits in circumference, and full forty cubits in height. At the foot of the hill was another tea-tract, and doubtless many of the Naga hills are covered with tea. Mr. Bruce crossed the Dacca river at the old fort of Ghergong, and, walking towards the hills, almost immediately came upon tea; and in two days journey he saw thirteen tracts. Farther south-west, the small hills adjoining Gabrew hill were covered with tea-plants. "The flowers of the tea on the hills are of a pleasant, delicate fragrance, unlike the smell of our other tea-plants; but the leaves and fruit appear the same. This would be a delightful place for the manufacture of tea, as the country is well populated, has abundance of grain, and labour is cheap. There is a small stream, called the Jhaugy river, at a distance of two hours walk: it is navigable all the year round for small canoes, which could carry down the tea, and the place is only one and a half days' journey from Jorehaut, the capital of Upper Assam." South-west of Gabrew Purbut, (about two days journey,) is a village inhabited by a race called Norahs, who came from the eastward, where tea abounds. The oldest man in this village told Mr. Bruce, that when his father was a young man, he had emigrated, with many others, and settled at Tipum, opposite Jaipore; that they brought the tea-plant with them, and planted it on the Tipum hill, where it exists to this day; and that when he was about sixteen years of age, he was compelled to leave Tipum, on account of wars and disturbances, and take shelter at the village where he now resides. This man said he was eighty years of age, and his father died a very old man. He was the only man met by Mr. Bruce in his journeys, who could give him any account of the tea-plant; with the exception of an Ahum, who declared that it was Sooka, or the first Kacharry raja of Assam, who brought the tea-plant from Munkum: he said it was written in his Putty, or history. Mr. Bruce found the old Norah man's story true; for the superintendent cleared the tract where it grew thickest, about 300 yards by 300: the old man said his father cut the plant down every third year, that he might get the young leaves.

The Report is accompanied by a map of Muttuck, Singpho, and the country west of the Boree Dibing river, showing all the tea-tracts that have been discovered: they are distributed all over the district. Mr. Bruce does not pretend to say how much tea they would all produce if fully worked. Until lately, he had only two Chinese black tea-makers. These men have twelve native assistants; each Chinaman, with six assistants, can only superintend one locality, and the tea-leaves from the various other tracts, widely separated, must be brought to these two places for manufacture. Hence, additional labourers must be always employed to bring the leaves from so great a distance: the leaves, too, in the journey, soon begin to ferment, and the labour of only preparing them so far in process that they may not spoil by the morning is excessive. The men have often to work very late, and, consequently, the labour is not so well executed; the leaves last gathered are also much larger than they ought to be, for want of being collected and manufactured earlier; consequently the tea is of inferior quality. This is mentioned to show the inconvenience and expense of having so few tea-makers; a disadvantage which may interfere with the success of the experiment. Mr. Bruce considers that it will not become sufficiently forward to be transferred to speculators until a proper number of native tea-manufacturers have been taught to prepare both the black and green sorts; then, under one hundred available tea-manufacturers, it would be worth while to take up the scheme on a large scale. Labourers must be introduced, in the first instance, to give a tone to the Assam opium-eaters; but the great fear is, that these latter would corrupt the new comers. If the cultivation of tea were encouraged, and the poppy put a stop to in Assam, the Assamese would make a splendid set of tea manufacturers and tea-cultivators.

In estimating the extent of the tea-tracts, Mr. Bruce only refers to those patches of plants which grow thickly together; and does not reckon the straggling plants in the forest and jungle. The former are so thick as to impede each other's growth; and, by thinning them, a sufficient number of plants may be found to fill up the patches of jungle between the present tracts. Yet, many tea-tracts have been cut down, in ignorance, by the natives, to make room for the rice field, for fire-wood, and fences. Many of these tracts have sprung up again, more vigorous than before.

Mr. Bruce considers that in Assam, as in China, the hilly tracts produce the *best* Teas. In the lowland, the plants seem to love and court moisture, not from stagnant pools, but running streams. The Kuhung tracts have the water in and around them, and are all in heavy tree-jungles. An extent of 300 yards by 300 will cost from 200 to 300 rupees, clearing; *i. e.*, according to the manner in which the miserable opium-smoking Assamese work. They will not permit their women to come into the Tea-gardens; whereas, females and children might be profitably employed in plucking and sorting leaves. But the gathering is hard work: the standing in one position so many hours occasions swelling in the legs; as the Assamese plants are not like those of China, only three feet high, but double that size, so that one must stand upright to pluck the leaves. The Chinese gather theirs squatting down. The Assamese trees will, probably, become of a smaller and more convenient size after a few years' cultivation; from trimming the plants, taking all their young leaves as soon as they appear, and from the soil being poorer. Transplanting, also, helps to stunt and shorten their growth. The Chinese assured Mr. Bruce, that the China plants, now of Deenjoy, would never have attained half the present perfection under ten years in their own country.

The sun materially affects the leaves; for, as soon as the trees that shade the plants are removed, the leaf loses its fine deep green, and turns yellowish; but it, at length, changes to a healthy green, and becomes thicker than when in the shade. The more the leaves are plucked, the greater number of them are produced. The plants in the sun have flowers and fruit much earlier than those in the shade: flowers and seeds in July, and fruit in November. Some plants, by cold or rain, having lost all their flowers, throw out buds more abundantly than ever. Thus, plants may be seen in flower so late as March, (some of the China plants were in flower in April,) bearing the old and the new seeds, flower-buds, and full-blown flowers all at one and the same time. The rain, also, greatly affects the leaves, for some sorts of Tea cannot be made in a rainy day; for instance, the *Powchong* and *Mingehew*. The leaves for these ought to collected about ten A. M., on a sunny morning, when the dew has evaporated. The *Powchong* can only be manufactured from the leaves of the first crop; but the *Mingehew*, although it requires the same care in making as the other, can yet be made from any crop, provided the morning be sunny. The Chinese dislike gathering leaves on a rainy day for any description of Tea. Some pretend to distinguish the Tea made on a rainy from that made on a sunny day, much in the same manner as they can distinguish the shady from the sunny Teas, by their inferiority. If the large leaves for the black Tea were collected on a rainy day, about seven seers, or fourteen pounds, would be required to make one seer, or two pounds, of Tea; but if collected on a sunny day, about four seers, or eight pounds, of green leaves, would make one seer, or two pounds of tea: so the Chinamen told Mr. Bruce; and from experiment he found their statement correct. The season for tea-making generally commences about the middle of March; the second crop in the middle of May; and the third about the middle of July; but the time varies, according to the rains setting in sooner or later.

We now arrive near the number of tea-plants cultivated in Assam. The China black tea-plants which were brought into Muttack in 1837 amounted to 1609, healthy and sickly; and they mostly flourish as well as if reared in China. Mr. Bruce collected about twenty-four pounds of the China seeds, and sowed some on the little hill of Tipum, in his tea-garden, and others in the nursery-ground at Jaipore; about 3,000 of which have come up, are looking beautiful, and doing very well: but the China seedlings on Tipum hill have been destroyed by some insect.

The Assam and China seedlings are near each other; the latter have a much darker appearance: there may be about 10,000 of them. In June and July, 1837, 17,000 young plants were brought from Muttack, and planted out in thick tree jungles. Six or eight thousand had previously been planted there: many of these died in consequence of the buffaloes constantly breaking in among them; but the rest are doing well, though in jeopardy of the above enemies.

In 1838, 52,000 young tea-plants were brought from about ten miles distance from Jaipore: a great portion of these have been sent to Calcutta, to be forwarded to Madras; should they thrive there, it is Mr. Bruce's opinion that they will never attain the height of the Assamese plants, but be dwarfish, like the Chinese. Transplantation should be done in the rains, when very few, if any, will die; provided, also, that they are removed from one sunny tract to another. Mr. Bruce believes the tea-plant to be so hardy that it would almost live in any soil, if it were only planted in deep shade when taken to it. The roots should be well watered, but not inundated; when they have taken hold, the shade should be removed. From moderately sized plants, removed from the jungle to a garden, a small crop of tea may be gathered next year; from plants raised from seed, a crop may be expected the third year: they reach maturity in six years, and live forty or fifty.

On March 2, Mr. F. C. Brown stated to the Asiatic Society, that he had seen the tea-plant flourishing in the district of Wynaad on the western ghauts of the peninsula of India. In February, 1834, the late Colonel Crewe gave two Chinese tea-plants to Captain F. Minchin, at Manantoddy, the chief place in the Wynaad district. Here the two plants, though small and unhealthy, began in a week or two to improve; and during the rains between June and September, they produced fresh shoots, and produced most healthy plants. On the following year, they were fine and bushy, and were in full bloom when the rains set in, in June. Captain Evans found that a cutting from the tea-plant throve equally well at Manantoddy: he, therefore, infers the soil and climate of the Wynaad district to be adapted to the cultivation of tea; as well as a great portion of the tract of land in south-western India, ceded by Tippoo Sultan; which is as fertile, and of about the same elevation, climate, and temperature.

FORMATION OF INDIGO IN POLYGONUM TINCTORIUM.

PROF. MORREN, in a memoir read before the Academy of Sciences at Brussels, on the culture of, and method of obtaining the Indigo from *Polygonum tinctorium*, observes: "the indigo is contained in the mesophylla of the leaf especially. It is dissolved originally in a liquid which fills the cells, and in which float pure granules of chlorophylle, either inclosing nuclei of cells or bundles of crystals. The formation of the indigo is in connection with the non-development of the fecula, so that the more there is of this substance the less there is of the blue product: whence it follows that the young leaves being less feculiferous than the old ones, are more useful.

The chlorophylle is a formation prior to the fecula, which is developed in separate nuclei in the green granules; but there is nothing to prove that the indigo is influenced by the chlorophylle, or that it is the authocyan, the blue principle of the chlorophylle, which has any connection with the indigo, so that the leaves of a bright and uniform green are those which are the best adapted for the extraction of indigo; for the greener and more healthy a leaf is, the more it contains of the blue principle.—*Bulletin de l'Académie de Bruxelles; An. Nat. Hist.*

THE ASHAR OR ABUK, (ASCLEPIAS PROCERA.)

Dr. Max Koch, a Bavarian traveller, describes this tree as having leaves of a very bright green, and peculiar to the Sennaar. The seeds are enveloped in a fine silk, wherefore it is also called *Asheyr*, (silk-tree). In the plain of Gohr, the natives use that substance for the matches of their guns. The milk-like sap which oozes from the young twigs is collected, and sent to Jerusalem, where the druggists prescribe it for inveterate colds. The flower is poisonous.—*W. Weissenborn; Mag. Nat. Hist.*

NEW BARK.

Dr. Mackay has read to the British Association a communication upon a Bark which he has lately received from South America. It is said to possess febrifuge properties, stated to be equal to those of the best Peruvian bark, for which it has been successfully substituted. Dr. M. submitted specimens of two different oils, obtained from the bark by distillation with water; which, though existing in the same plant, and obtained by the same process, present marked distinctions. The one, being of less specific gravity than the other, floated upon the surface of that which was distilled along with it; while the other, being considerably heavier, sunk to the bottom of the receiver. Both, when fresh, were transparent and colourless; but in a few days they changed to a yellow colour, the heavier oil becoming of a deeper tinge than the other. In smell, the oils differ perceptibly, that of the heavier being fatty and unpleasant, while the odour of the lighter is aromatic and agreeable. In taste, they are equally acrid and disagreeable. The specific gravity of the lighter oil is 0·949, that of the heavier is 1·028, both having been examined several days after their preparation. Upon exposing them to a temperature of 18° F., no effect was produced upon the lighter oil, but in the heavier a great quantity of needle-shape crystals were observed; which, however, speedily dissolved upon the phial being removed out of the febrifuging mixture: the mineral acids decompose them, connecting both oils with fluids of a blood-red colour. In a chemical point of view, the oils referred to are interesting, on account of their presenting such marked distinctions in colour, smell, specific gravity, and the effect of cold upon them; while they are the produce of the same plant, and procured by the same process. —*Birmingham Journal.*

RICE HARVEST IN GERMANY.

A HARVEST of a very uncommon description in northern Europe has been reaped at Blansko, near Brunn in Moravia. After many trials, a Baron Von Reichenbach has, it appears, succeeded in bringing to perfection a field of Rice, which in Germany, even more than with us, is an article of constant and extensive consumption. As the land where the crop has been grown is situated in a cold mountainous region, more than a thousand feet above the sea, and surrounded by forests, where the climate is too severe for the growth of grapes, this success is the more extraordinary. The seed was sown and raised entirely in water; in the first instance in a sort of hot bed, or hot water, for the water was a little warmed whenever, during the spring, the weather was cold enough to render it necessary; and it was then transplanted according to a method practised in Hindostan.—*Athenæum.*

ON MUSHROOMS.

ON June 7, a paper was read to the Botanical Society, by Mr. W. H. White, "On Mushrooms;" the object of which was to stimulate inquiry as to the cause of the deleterious effects some species of Mushrooms have on the human frame in some countries, and not in others. In England, for example, only three species are edible. *Agaricus campestris,* or the common mushroom; *A. pratensia,* or the fairy ring; and *A. Georgii;* whilst in Russia, forty species, almost every variety, are used as articles of food, and many, as delicacies, constantly at the tables of the rich. The mushroom-trade, also, is there considerable; at Moscow alone, to the amount of 200,000 roubles in the year. The peasants, for some months, live almost exclusively on them. The kinds there eaten are considered here, and have been proved to be, by many fatal cases, poisonous. Mr. White believes it not to be dependent on soil or climate, but principally on the culinary process, much salt being used, and also on the care observed not to boil two species together.—*Literary Gazette.*

ON Feb. 15, was read to the Botanical Society a notice of a new species of exotic Polypore, by Professor Kickx, of the University of Gand, translated by Mr. White. The naturalists travelling in America at the expense of the Belgian government, have sent from the island of Cuba various plants, seeds, &c., which have been placed in the Botanic Garden of Gand, and among which is the subject of the present notice. The odour of this exotic polypore is so like that of myrrh, that it has been named the *Polyporus myrrhinus.* The means adopted by Professor Kickx, with invariable success, to determine the diagnostic of vegetable emanation, is, placing the specimen for a few minutes over a vessel containing ammonia; this strengthens without altering the flavour, of which, without this process, the small intensity in many cases would escape detection. The new odorous mushroom was fully and botanically described; also some mosses in the earth, in which the plants before mentioned were sent over.—*Literary Gazette.*

THE ARTICHOKE.

M. VILMORIN, of the Agricultural Society of Lyons, remarks that the Artichoke was known as an esculent plant by the Romans, but neglected in the dark ages, till it again came into notice in the 16th century. Almost all parts of this plant, he says, may be rendered useful. The leaves yield an extract which may be substituted for quinine. The leaves themselves may be cooked and eaten after the fruit is gathered, or used as fodder, mixed with certain grasses; they may be substituted for hops in making beer, and they contain a great proportion of potash.

THE MULBERRY.

M. SERINGE argues, from physiological facts, that pruning the Mulberry at the same time that the leaves are gathered, will produce a handsomer and a longer-lived tree, and a greater abundance of leaves.

THE OAK.

DR. GRENVILLE has presented to the Botanical Society of Edinburgh a series of specimens of *Quercus robur*, exhibiting an extraordinary range of form. From the singular variation exhibited by these specimens in the shape and texture of the leaves, and in the length of the peduncles, Dr. Grenville is of opinion that there is but one species of oak indigenous in Britain.—*An. Nat. Hist.*

VARIETIES OF WHEAT.

IN a conversation which took place at the meeting of the British Association, after the reading of Mr. Webb Hall's paper "On Accelerating the Growth of Wheat," it was stated that a hundred and fifty kinds of wheat were sometimes to be found in one field in England; and that in Spain two hundred kinds were cultivated. If not amenable to the same rule, it should seem that it would be an improvement to keep these varieties as separate and pure as possible, especially in following out the important experiments of Mr. Hall. If we can grow the same corn crop in five months which occupies the ground, at an average, eleven, see what pasturage might be gained in the meantime. —*Literary Gazette.*

NEW MODE OF PROMOTING THE GROWTH OF TREES.

THE Duke of Portland has commenced a new system of strengthening and promoting the Growth of Trees in his Grace's grounds about Welbeck; by putting pigs in the plantations, and confining them within a given space for a certain length of time to root the ground at the foot of the trees, and of course to manure the soil. They are then removed to another part of the plantation, and confined in the same way, and are fed upon potatoes.—*Times.*

Geology.

TO DISTINGUISH TRAP FROM BASALT.

M. H. Braconnot finds that these rocks may be distinguished by submitting them to distillation; when the Trap always yields an empereumatic ammoniacal product, which restores the colour of litmus paper reddened by an acid, whereas Basalt produces no such effect: and he presumes that the organic matter which had existed in the materials of the basalt was destroyed by the volcanic heat by which the rock was produced; whereas he conceives trap to have been formed in water, under the influence of a moderate temperature, insufficient to destroy the organic matter contained in the debris from which it was formed. M. Braconnot has also found that various granites yield ammonia when heated; and the same effect was produced by serpentine and porphyry; but the gneiss of Freiburg, in Saxony, yielded an acid, which appeared to be hydro-fluoric. Many other rocks of various kinds were subsequently found to contain ammonia.—*Annales de Chimie; Philos. Mag.*

USE OF GEOLOGY TO FARMERS.

One of the most obvious sources of advantage to the farmer from an acquaintance with the distribution of mineral masses, is the facility with which, in many instances, the injurious effects of small springs coming to the surface may be obviated. The theory of the earth's internal drainage is so simple, that every man of common sense would be able to drain his lands upon sure principles, or else to know precisely why it cannot be drained, if he were to become so much of a geologist as to learn what rocks existed under his land, at what depth, and in what positions. Springs never issue from stratified masses, except from reservoirs, somehow produced in jointed rocks, and at the level of the overflow of these subterraneous cavities. Faults in the strata very frequently limit these reservoirs, and determine the points of efflux of the water. Let those faults be ascertained, or the edge of the jointed rock be found; the cure of the evil is immediate. But some geological information is needed here; and landed proprietors, who think it less troublesome to employ an agent than to direct such a simple operation, may at least profit by this hint, and choose an agent who knows something of the rocks he is to drain. Another thing, probably of importance to agriculturists, is the discovery of substances at small depths; which, if brought to the surface, would enrich, by a suitable mixture, the soil of their fields. This is very strongly insisted on by Sir H. Davy in his *Essays*, and considering how easy a thing it is for a landowner to ascertain positively the series of strata in his estate, it is somewhat marvellous that so few cases can be quoted, except that of Sir John Johnson, Bart., of Hackness, near Scarborough, in which this easy work has been performed.

Finally, in experiments for the introduction of new systems and modes of management, with respect to cattle and crops, it will be of great consequence to take notice of the qualities of the soil, substrata, and water; for these undoubtedly exercise a real and perhaps decisive influence over the result.—*Mining Review.*

FROZEN WELL.

NEAR Owego, occurs this apparent contradiction of Nature's laws, which is thus described, by a correspondent, in *Silliman's Journal.*

The well is excavated on a table of land, elevated about thirty feet above the bed of the Susquehanna River, and distant from it three-fourths of a mile. The depth of the well, from the surface to the bottom, is said to be seventy-seven feet; but for four or five months in the year, the surface of the water is frozen so solid as to be entirely useless to the inhabitants. On the 23rd of the present month, (Feb.) in company with a friend, I measured the depth, and found it to be sixty-one feet from the surface of the earth to the ice which covers the water in the well, and this ice we found it impossible to break with a heavy iron weight attached to a rope. The sides of the well are nearly covered with masses of ice, which, increasing in the descent, leave but about a foot space (in diameter) at the bottom. A thermometer let down to the bottom, sunk 38° in fifteen minutes, being 68° in the sun, and 30° at the bottom of the well. The well has been dug twenty-one years, and I am informed, by a very credible person who assisted in the excavation, that a man could not endure to work in it more than two hours at a time, even with extra clothing, although in the month of June, and the weather excessively hot. The ice remains until very late in the season, and is often drawn up in the months of June and July. Samuel Mathews drew from the well a large piece of ice on the 25th day of July, 1837, and it is common to find it there on the 4th of July.

The well is situated in the highway, about one mile northwest of the village of Owego, in the town and country of Tioga. There is no other well on that table of land, nor within sixty or eighty rods, and none that presents the same phenomenon. In the excavation, no rock or slate was thrown up; the water is never affected by freshets; and is what is usually denominated "hard," or limestone water. A lighted candle being let down, the flame became agitated and thrown in one direction at the depth of thirty feet, but was quite still, and soon extinguished at the bottom. Feathers, down, or any light substance, when thrown in, sink with a rapid and accelerated motion.

Owego, Feb. 26th, N. Lat. 42° 10′.

Prof. Silliman, in attempting to solve this extraordinary and difficult problem, observes:

At the depth of more than sixty feet, the water ought not to freeze at all, as it should have nearly the same temperature of that film of the earth's crust, which is at this place affected by atmospheric variations, and solar influence, being of course not far from the medium temperature of the climate. Could we suppose that compressed gases, or a greatly compressed atmosphere were escaping from the water, or near it, this would indicate a source of cold; but as there is no such indication in the water, we cannot avail ourselves of this explanation, unless we were to suppose that the escape of compressed gas takes place deep in the earth, in the vicinity of the well, and in proximity to the water that supplies it. Perhaps, this view is countenanced by the blowing of the candle at the depth of thirty feet, blowing it to one side, thus indicating a jet of gas which might rise from the water as low as at its source; and even if it were carbonic acid, it might not extinguish the candle, while descending, as the gas would be much diluted by common air; and still, in the progress of time, an accumulation of carbonic acid gas might take place at the surface of the water, sufficient to extinguish a candle.

GASES IN WELL-DIGGING.

On Nov. 6, was read to the Geological Society, a communication "On the Noxious Gases emitted from the chalk and overlying strata in sinking Wells near London," by Dr. Mitchell. The most abundant deleterious gas in the chalk is the carbonic acid; and it is said to occur in greater quantities in the lower than the upper division of the formation. The distribution of it, however, in that portion of the series is very unequal. Sulphuretted hydrogen and carburetted hydrogen gases sometimes occur where the chalk is covered with sand and London clay, as well as in other situations. In making the Thames Tunnel, they have been both occasionally given out, and some inconvenience has been experienced by the workmen: but in no instance have the effects been fatal. In the districts where sulphuretted hydrogen gas occurs, the discharge increases considerably after long-continued rain, the water forcing it out from the cavities in which it had accumulated. The paper contained several cases of well-diggers having been suffocated from not using proper precautions.—*Athenæum.*

ARTESIAN WELL OF THE NEW RIVER COMPANY.

A paper on this subject has been read before the Institution of Civil Engineers, by Mr. Mylne, engineer of the New River Company. These wells are named Artesian, from their having been first adopted in Artois, called by the Romans, *Artesium.* To test the practicability of this method of procuring water in sufficient quantity for the use of the metropolis, the New River Company caused a well to be sunk at their reservoir in the Hampstead Road; the details of which experiment, with other valuable information, are contained in Mr. Mylne's paper. London is placed in a large hollow or bowl of chalk, generally termed "the London basin;" the superstrata being first sand, and then the deep blue London clay. This geological formation is peculiarly adapted for these wells; and on forcing vertically through the deep stratum of clay to that of sand, water on generally found. This will rise to a height depending, of course, on the elevation of the point at which the sand stratum crops out from under the bed of clay; but the Company's experiments do not promise a sufficient quantity of water for their general purposes. The *modus operandi* pursued at this well is minutely stated by Mr. Mylne; who prefatorily mentions that in various parts of the metropolis, and in other places, wells supplied from sand-springs are so affected by neighbouring wells, or the subsidence of the upper ground from the large quantities of fine sand pumped out, that they have been necessarily abandoned; and many remarkable instances of danger to buildings from this cause, are related in the above paper. The total expense of their experiment was 12,412*l.* 14*s.* 1*d.*—*Railway Magazine.* For an abstract of the above paper, with a sectional engraving, see *Civil Engineer and Architects' Journal*, No. 22.

TEMPERATURE OF ARTESIAN WELLS.

Dr. Paterson has communicated to *Jameson's Journal* several valuable experiments and observations on the temperature of Artesian Wells in Mid-Lothian, Stirlingshire, and Clackmannanshire; a table of which shows how very nearly the results of different localities approximate; and, if we take the average number of these results, 1° for every 48 feet as we descend, we shall find that it comes very near the average fixed upon by the British Association, which is 1° for every 45 feet.

ARTESIAN WELLS AT MORTLAKE AND IN ESSEX.

A successful Artesian operation has lately been performed at Mortlake, within 100 feet of the Thames. First, an auger, 7 inches in diameter, was used in penetrating 20 ft. of superficial detritus, and 200 feet of London clay. An iron tube, 8 inches in diameter, was then driven into the opening, to dam out the land springs, and the percolation of the river. A 4-inch auger was next introduced through the iron tube, and the boring was continued until, the London clay having been perforated to the depth of 240 feet, the sands of the plastic clay were reached, and water of the purest and softest nature was obtained; but the supply was not sufficient, and it did not reach the surface. The work was proceeded with accordingly; and, after 55 feet of alternating beds of sand and clay had been penetrated, the chalk was touched upon. A second tube, 4½ inches in diameter, was then driven into the chalk, to stop out the water of the plastic sands; and through this tube, an auger, 3½ inches in diameter, was introduced, and worked down through 35 feet of hard chalk, abounding with flints. To this succeeded a bed of soft chalk, into which the instrument suddenly penetrated to the depth of 15 feet. On the auger being withdrawn, water gradually rose to the surface and overflowed. The expense of the work did not exceed £300. The general summary of the strata penetrated is as follows: Gravel, 20 feet; London clay, 250; plastic sands and clay, 55; hard chalk with flints, 35; soft chalk, 15; =375 feet.—*Dr. Ure's Dictionary of Arts*, &c.

There is, perhaps, no part of the world, where Artesian wells are more general, or more useful, than in Essex. In the vale of the Lea, they have been bored with the greatest facility, and at a small expense.

In Waltham Abbey, the cost is usually about £16. In the district of Bulpham Fen, 7 miles south from Brentwood, they yield a large supply of water. In the marshes, as well as along the coast, and in the islands of Essex, they have proved of the greatest utility. Formerly, in some seasons, when the ditches became dry, the cattle suffered, the fishes died, and the farmer lost severely on his stock; but, by the aid of Artesian wells, the ditches are now kept full all the year! In Foulness Island, there are no natural springs; and until lately, no water, except atmospheric, collected in the ditches. In hot seasons, this water became putrid, but the inhabitants and the cattle continued to partake of it as long as it lasted; and supplies were then obtained from seven miles distance. Artesian wells now keep the ditches full of fresh and sweet water. Wallisea, Mersea, and other islands, have profited in a similar manner.—*Dr. Mitchell; Proc. Geolog. Soc.*

ARTESIAN WELL AT PARIS.

A Well, upon the Artesian principle, has been, for sometime in progress, at the Abattoir of Grenelle, at Paris.* The sound, or borer, weighs 20,000 lbs.; its height is treble that of the dome of the Hospital of Invalids, and it requires two machines of immense power to put it in motion. In July last, M. Arago stated this well to have reached 483 metres, or 1,584 feet. M. Arago used the thermometer of M. Waferdin; and having provided for the pressure, which at such a depth is equal to 50 atmospheres, six thermometers were successively let down to a depth of 481 metres, care being taken not to lower them until 36 hours after the boring, in order that the heat which this work might have communicated, should have subsided. The thermometers were left in the well 36 hours; and the heat at the above depth was 93¾° Fahrenheit; being about 23 metres for each degree of temperature.

THERMAL SPRINGS.

Prof. Bischof, in his laborious paper on the Natural History of Volcanoes and Earthquakes, observes: "If we take a summary view of all that has been said on the subject of Thermal Springs, we shall find it impossible to avoid recognising a relation between elevations of Plutonic masses, the upraising of Neptunian formations, and thermal springs. Cause and effect have, however, been frequently confounded here. Thermal and mineral springs are seldom, perhaps never, the cause of those effects. Where, however, these effects are observed, where, in consequence, the penetration of meteoric water into the interior of our earth has been rendered possible, and where natural hydraulic tubes have been formed by the upraising of strata, there the phenomena of Thermal and mineral springs were the consequence." The Professor adds that the general aim of his remarks is to show that the degree of heat in Thermal springs depends on the greater or less depth of their origin, consequently, wholly and solely on central heat.

* See Year-book of Facts, 1839, p. 224.

MINERAL SPRING IN NEW SOUTH WALES.

On June 3, was read to the United Service Institution, "An Account of the Mineral Spring on the Rocky Flat, Menero Downs, New South Wales," by Dr. John Lhotsky. This spring is situated about 300 miles from Sydney, at a distance of 12 miles from Kuma, and surrounded by extensive, waste, undulated downs, with long projecting hills. The water of the spring was limpid, and a constant eruption of gas was visible. Its temperature varied from 58° to 60°. "At six P.M., there was lightning to the east, and thunder; wind, S. S. E.; the air became cold to sensation; the spring was still 60°." The water bubbles strongly, with frequent changes in intensity; but as far as Dr. Lhotsky could observe, without any certain succession. Its taste is that of the most valuable mineral waters, Seltzer and Cheltenham. The paper described the geological structure of the neighbourhood of the spring, the appearance of the Travertine rocks, the efflorescence thereon, &c. As New South Wales is becoming more and more the resort of invalids from the adjacent tropical colonies, the existence of such a spring is a valuable discovery, and may lead to a Cheltenham on the Menero Downs.—*Literary Gazette.*

NORTH AMERICAN LAKES.

The following interesting statement respecting the great north-western lakes of North America, appears in the late Report of the "Michigan State Geologist:"—

	Mean length.	Mean breadth.	Area, sq. miles.
Superior	400	80	32,000
Michigan	220	70	22,000
Huron	240	80	20,000
Green Bay	100	20	2,000
Erie	240	40	9,600
Ontario	180	35	6,300
St. Clair	20	14	360
			90,060

The same tabular statement exhibits also the depth and the elevation of each above tide water:

	Mean depth.	Elevation.
Superior	900 ft.	596 ft.
Michigan	1000	578
Huron	1000	578
St. Clair	20	570
Erie	84	565
Ontario	500	232

It is computed that the lakes contain above 14,000 cubic miles of water; a quantity *more than half of all the fresh water on the earth.* The extent of country drained by the lakes, from Niagara to the north-western angle of Superior, including also the area of the lakes themselves, is estimated at 335,515 square miles.—*Athenæum.*

NATURAL EXHALATION OF CARBURETTED HYDROGEN GAS.

In the vale of Cwmdare, near Aberdare, Glamorganshire, is a waterfall, which presents a phenomenon hitherto unnoticed. From the bed of the stream rises an exhalation of gas, which being ignited, continues to burn, with a yellow flame, interspersed with streaks of livid white, orange, purple, and blue. There are more than twelve apertures through which the gas escapes beneath the water, (causing it to rise and bubble); others, on the dry banks, increase daily in size. One of the apertures is considerably larger than the rest, the flame from which burns about 2 feet in height, and $1\frac{1}{2}$ feet in width. The soil consists chiefly of argillaceous schist, or fire-clay, sufficiently hot to burn the hand. Fish caught in the stream, have been boiled upon it. The first impression was that this phenomenon was occasioned by an escape of carburetted hydrogen from a coal level; but it is a considerable distance from any coal mine.—*Athenæum; abd.*

MODIOLÆ ENCLOSED IN LITHODOMI.

MR. CHARLESWORTH has examined a series of specimens of the fossil shells of the genus *Modiolæ*, found in the Bath oolite, enclosed in the shells of the genus *Lithodomus;* submitted by the Rev. H. Jelly. Two of these specimens are engraved in the *Magazine of Natural History*, No. xxxv., p. 552; and Mr. Charlesworth suggests the obvious explanation of supposing that the dead shell of the *Lithodomus* was occupied by a *Modiola*, and the *Modiola* itself subsequently occupied by a smaller individual of its own species; the same thing being repeated in some instances five or six times. The introduction, however, of the *Modiola* in the *adult* state would be opposed by the physical condition in which the *Lithodomus* is placed. Any suggestions bearing upon this curious fact would be acceptable.

THE SILURIAN SYSTEM.

IN the second part of Mr. Murcheson's *Silurian System*, just published, are elaborate engravings of about 350 species of Organic Remains, three-fourths of which are *new* to the scientific world. It is upon this that the chief merit of our author's labours is based; since he demonstrates that, independently of all local or mineral distinctions, these Silurian rocks contain vast quantities of Organic Remains—a *fauna* of their own, totally distinct, except in a very few individual instances, from the fossils of the overlaying systems. It is by the establishment of this fact that he is authorized to claim for his *System* the remarkable individuality and extension of character which justifies its separation from all the earlier deposits, and has enabled other geologists already to identify it in other parts of the earth's surface, of which it constitutes, according to recent information, a not inconsiderable portion.—*Quarterly Review.*

RELATIVE AGES OF THE CRAG OF NORFOLK AND SUFFOLK.

A PAPER from the pen of Mr. Lyell appears in the *Magazine of Natural History*, No. 31; which embodies some results of the highest interest, as bearing upon the Tertiary Geology of Norfolk and the adjoining counties. The district treated of has long been celebrated for the number and beauty of its fossils; but, until a very recent period, no suspicion had been entertained that the fossiliferous beds called "crag" included deposits of distinct geological ages. It is now, however, satisfactorily shown by the application of the per centage test to the very extensive series of crag *Testacea* in the cabinet of Mr. Searle's Wood, that three marine deposits, of different and well marked periods, overlie the chalk and London clay in this part of England. This result confirms the general views upon Tertiary Geology which Mr. Lyell has entertained in opposition to M. Deshayes, who asserts the existence in the tertiary group, of three definite proportions in the per centage of extinct species, and to one of which any member of the series may be referred.*—*Mr. Charlesworth, F. G. S.*

CARBONIFEROUS AND DEVONIAN SYSTEMS OF NORTH-WESTERN GERMANY.

THE following is a correct abstract of a communication from Mr. Murcheson to the British Association; which, in a geological point of view is the most important paper of the meeting, as it unites opinions upon a great principle and much controverted theory. The author states that having, in company with Professor Sedgwick, examined the older rock of north-western Germany and Belgium, it is the intention of his friend and himself to lay before the Geological Society of London a general memoir, (illustrated by numerous fossils,) on the classification of these ancient deposits, showing a succession of the Carboniferous, Devonian, or *Old Red and Silurian Systems.* The present communication bears only upon one point of this analysis, and is offered as a clear and indisputable proof of the geological position of the anthracitic or culm-bearing strata of Devonshire and Cornwall. Transverse sections, in descending order from the productive coal-field of Westphalia on the N.N.E., to the uppermost division of Protogoic rocks on the S.S.W., were explained; and one from Dortmund by Schelke, to the neighbourhood of Limburg and Iserlohn, was specially adduced, in which the various masses of strata are clearly exposed in five natural sections, in the following descending order:—

1. Coal shales, coal, &c. (productive coal-field.)
2. Millstone grit series, with many impressions of large plants, and occasional thin seams of coal.
3. Thinly laminated carbonaceous sandstones and shales, containing many grasses and small plants, together with bands of flat bedded, black, bituminous limestone and shale, charged with Possidoniæ and Gonralites, and attenuating with courses of flinty schist, the Kiesel-schiofer of German geologists.
4. Carboniferous or mountain limestone, of great thickness, and of the usual British mineral characters, loaded with many well-known fossils of the formation.
5. Upper Devonian rocks, consisting of black schists, grey and red sandstones, with occasional calcareous courses, and numerous fossils (the old Grauwacke of German authors.)

The order and sequence of these strata is indicated and maintained along the lower edge of the whole range of the large coal-field of Westphalia, the beds successively rising to the surface at angles

* Constant proportions (3 per cent., 19 per cent., 52 per cent.) in the number of recent species, determine the age of the tertiary strata.—*Deshayes;* translated in *Mag. Nat. Hist.*, vol. i. n. s. p. 12.

varying from 30° to 40°, in perfect conformity, and showing throughout the clearest and most complete transition into each other. It is particularly to the group No. 3, indicated on an exhibited drawing, that the author directs the attention of British geologists; because it is, in all respects, identical with the culm-bearing strata of Devon and Cornwall, first described by Professor Sedgwick and himself, as being a portion of the true Coal-Field, and not belonging to the Grauwacke or older transition rocks, to which they had formerly been referred.

The Westphalian sections establish the geological position of the Bideford culm-strata more clearly than has been done by any stratigraphical evidence in Great Britain, by presenting five masses of unequivocal mountain limestone, rising out from beneath the black limestone and culmiferous schists; and thus the precise age of the latter is demonstrated.

In regard to the rocks of the Devonian system, (old Grauwacke of German authors), which support in mountain masses the carboniferous system above alluded to, the author offered a brief and general sketch, assuring the Section of the Geological Society, that Professor Sedgwick and himself would demonstrate that these rocks fairly represent the British system generally known as the Old Red Sandstone, but to which they had recently applied the term Devonian,—a term which foreign geologists seemed well disposed to adopt, as calculated to prevent that confusion and antiquity which had arisen from the use of the word *Old* Red Sandstone, now that it is ascertained that rocks, for the most part black and slaty, occupy over wide districts the same geological horizon as our *Red* rocks of Herefordshire.

Proofs of the existence of the same order and succession will be hereafter pointed out in the countries of the Hartz and the Fichtelzeberg; as well as upon both banks of the Rhine, in Westphalia, and Nassau, &c.; while a splendid development of the still older Silurian rocks, (both upper and lower,) will be pointed out, chiefly on the left bank of the Rhine; and also in Belgium, and the region around Liege and Namur, already rendered classic ground by the descriptions of D'Omalius, De Halley, and Dumont.—*Literary Gazette.*

SWALLOW-HOLES AND DRAINAGE IN SURREY.

On Feb. 27, was read to the Geological Society, a paper " On the occurrence of Swallow-holes near Farnham, and on the Drainage of the Country at the Western Extremity of the Hog's-back ;" By George Long, Esq.; communicated by Mr. Lyell. Immediately to the north of Farnham, rises a chalk-hill, capped by tertiary strata. No perennial main-springs occur on the face of the hill; but the gullies are, for the greater part of the year, occupied by superficial land-springs, which occasionally become formidable torrents. These rivulets pour down the hill upon the surface of the tertiary clay until they arrive at the chalk, where they are entirely absorbed in Swallow-holes, except during great rains, when a portion of the water flows along channels in the chalk. Seven of these holes, between Clare Park and Farnham Park, were described by Mr. Long. The water absorbed by two of them is supposed to well out in great force at the Bourne Mill stream; and, though soft, where it sinks under ground, it is hard where it re-appears. The Drainage described in the second part of the paper is effected by a stream which passes through a gap at Runfold, the western extremity of the Hog's-back Hill; and flowing northward, traverses the chalk, and carries off the surplus waters of a tract bounded on the north by the Hog's-back, and on the south by a semi-circular range of low hills, extending from Seal on the east by Crooksbury, to Moor Park on the west. This gap in the chalk has hitherto escaped the observation of geologists, but deserves to be recorded among the apertures through the North Downs.—*Literary Gazette.*

ERRATIC BLOCK OF GRANITE.

AN immense erratic block of Granite was floated on the ice during the winter 1837-38, from Finland to the island of Hochland. It weighs about a million pounds, according to the estimate of M. de Baër.—*W. Weissenborn; Mag. Nat. Hist.*

CONVULSION AND LAND-SLIP; NEAR AXMOUTH, DEVON.

THIS convulsion occurred on the south coast of Devonshire, at the distance of two miles east of the mouth of the river Ax, and the town of Seaton, and about one and a half south-east from Axmouth village. It commenced at three o'clock in the morning of Tuesday, the 24th of December, 1839, when a family occupying a farm about half a mile distant was aroused by a crushing and low rumbling sound. Nothing, however, farther occurred until four o'clock on the morning of Christmas-day, when some labourers of the farm, who tenanted two cottages built on the slope of the debris of the undercliff, were awakened by noises similar to that which had been heard the night before. On getting up and endeavouring to open the door, a man who dwelt in one of them, saw that the ground was sinking beneath him, that it was subsiding in terraces towards the sea—that it was gaping with fissures, and that the walls of his dwelling were cracking and tottering as if ready to fall. During the whole of Christmas-day, the disruption continued; making a roaring and grinding noise resembling some kinds of thunder, and causing the earth to tremble at a great distance from the actual disturbance. An immense tract, extending east and west one mile in length, and many hundred feet in width, subsided or sank down so as to form a ravine or chasm more than two hundred feet in depth. Parts of several fields, included in this area, descended with great regularity and precision; so that their surfaces, still bearing their crops, are now at the bottom not much broken up, and only thrown into a slanting position, instead of being level, as they were before. The hedges which divided these fields can be traced on along the fallen portions, as well as across the high country which has remained unmoved. This regularity however, is not universal. Towards the eastern and western extremities of the chasm—particularly toward the former—the devastation has been extraordinary and complete. Columnar masses, resembling vast pinnacles or towers of chalk, are in some places left standing, whilst the more broken and crushed parts have sunk around them: immense banks of flint and broken rock rise in hillocks on every side, whilst the ground is rent and scored in seams many feet wide and deep. An entire orchard is to be seen in one part, which has descended to a level much lower than it before occupied: some of its trees are overthrown and uprooted, whilst many others are still standing, and will bear fruit next season. A wood of forest trees has also been broken up in the same way, the cottages before mentioned are in ruins, and their gardens destroyed: and the devastation around, although dreadful and terrific, is full of beautiful grandeur.

This chasm, however, is but one moiety of the phenomenon. It ranges east and west, and parallel with the sea-shore: and in running through the district, cuts off from the land a portion of the original country measuring one mile in length by half a mile in width. This huge mass, so cut off, has been forced on its foundation many yards in a southerly direction towards the sea, inclined somewhat from its former level, and rent and depressed into terraces. The bed of the sea also, the whole way along in front of it, has been lifted up to the height of forty feet above the surface, to a great distance out from the original line of coast, now forming reefs and islands, inside which are bays and small harbours, into which boats have been, and have found good soundings. These reefs of thrown-up rock are covered with marine productions, such as corallines, sea-weeds, and shells. It is most probable that water, and not fire, has been the cause of this phenomenon. The upper stratum, running through the cliffs, is chalk. This rests on the green-sand formation, much consolidated, and alternated with seams of chert, a species of opaque flint. Beneath this comes a deep bed of loose, sandy marl, or "fox mould," as it is locally termed, and it is this unstable and friable soil that contains the chief causes of the disturbances under consideration; and lastly this stratum is supported by the blue lias, a formation partly composed of beds of tough and impenetrable clay. These being the component strata, let it be borne in mind, in order to the understanding of this explanation, that all the soils above the lias are pervious to water, but the clay in the lowest bed resists it. The rain and other atmospheric moisture which falls on the upper surface, and the springs of water which may tend towards one point, will filter through the chalk and sandstone, and be mainly absorbed in the spongy fox mould. It cannot descend lower, because the clay of the lias resists it. Now, where the edges of these soils are exposed along the cliffs, so as to lay them bare and unsupported, this water will be seen oozing in springs out of the sandy mould immediately above the clay—which it carries away with it—slowly and most imperceptibly perhaps, but surely and inevitably. Such a process, going on through the course of ages must necessarily undermine the superincumbent strata; and when a season occurs more wet than ordinary, and such indeed as England has experienced during the past summer, the catastrophe is hastened on with a sudden crash, even such as we now describe. The precipitate and violent subsidence of such a great mass, had power, by its overwhelming weight, to act laterally, and it was this lateral force which served to thrust upwards the bed of the sea, previously seven fathoms beneath the surface, now into reefs forty feet above.

Several other places on the south coast of England exhibit the remains of similar convulsions, wrought, without doubt in a similar way, but in what age it is impossible to say, as no record of such occurrence exists to inform us. The Undercliff in the Isle of Wight may be instanced as one; the Pinhay Cliffs, only three miles east of Culverhole Point, (the subject of this paper,) as another; and a third on the cliffs between Beer and Branscombe.—*Saturday Magazine; abridged.*

LANDSLIP IN RUSSIA.

On June 18, a remarkable displacement of an entire valley, near the foot of a mountain, took place at the village of Federowk, in Russia; and during seventy-two hours it moved with an undulating motion towards the river Volga. The sinking of the valley is one mile and a half long, and 250 fathoms in breadth. Above 70 houses were damaged or thrown down, but happily no lives were lost.—*Literary Gazette.*

UPRISING OF THE EARTH.

A paper has been read to the Royal Society, "On the Gradual Uprising of the Earth in certain Places," by Mr. Darwin; who follows out the subject taken up some time ago by Mr. Lyell. He refers to observations made in the district of Lochaber, in Scotland; portions of which have been equally raised to an extent of 1,278 feet above the sea. He illustrates and proves his position, by noticing the equable elevation of what had evidently been the margin of ancient waters in the Lochaber district, and also by reference to the deposition of immense blocks of granite on particular parts of many of the hills in that country, as well as in the valleys. These, he considers, could not have been so placed, or rolled into the localities they occupy, by the rushing of the waters, nor by irruptions of the earth. The deposition by the waters is impossible, inasmuch as the rush must have been more impetuous than when the surface of the ground was level. These blocks, in many instances, from the slight elevation of the hills on which they are sometimes found, could not have rolled or fallen from higher ground. M. Darwin describes the probability of their having been transported from situations where their formation had taken place: he describes the positions in which they were found, showing that they must have been removed considerable distances, by an undoubled elevation on the surface of the earth. The subject of the extraordinary parallel roads in Glenroy was discussed at great length in this paper.—*Literary Gazette.*

EARTHQUAKES.

Monmouth.—On Sept. 8, about half-past one o'clock, the shock of an Earthquake was felt generally throughout Monmouthshire and the rest of England. On Sept. 10, one of the finest appearances of aurora borealis ever witnessed in the southern parts of our island, was visible nearly throughout the night, varying its illumination in a thousand brilliant ways.

Scotland.—Several shocks have been felt in Edinburgh, in Perthshire, and in the Highlands.

Scilly Isles.—On Jan. 27, the shock of an Earthquake was felt in St. Mary's, one of the Scilly islands. The tremulous motion of the ground was very slight, and was only felt in the south parts of the island. It was accompanied by a peculiarly harsh, grating sound, which was only of momentary duration, and no particular agitation of the sea was observed.

Messina.—A series of Earthquakes shook the city of Messina during the 27th, 28th, 29th, and 31st of August.

Ava.—A tremendous Earthquake occurred at Amerapoara, between 2 and 3 o'clock, on the morning of the 23rd of March, and extended with equal violence northward as far as Toungnor, and south to Prome. Pagodas, monasteries, brick dwelling-houses, all within the city and on the neighbouring hills, were destroyed, and from 2 to 300 lives lost; but the slightly-built wooden houses escaped. The current of the Irrawaddie was forced upwards for some time; large fissures in the ground, from ten to fifteen feet, formed a deluge of water, and threw up grey earth in great quantities with a sulphurous smell. The towns and villages near the capital are in ruins, and the old city of Ava is stated to be destroyed.—*Foreign Quarterly Review.*

San Salvador.—On Oct. 1, at 2 A.M., a strong shock of an Earthquake was felt at San Salvador, and at 3 A.M., a concussion nearly destroyed the town; and during the day there were altogether 15 smart shocks.

Martinique.—M. Moreau de Jonnès has transmitted to the French Academy of Sciences some particulars of a violent Earthquake at Martinique. On the 11th of January, after a succession of storms, at six in the morning, the wind blew from the north-west, and the whole island was enveloped in clouds and vapours, which entirely hid it from the shipping at sea. Both these circumstances are unusual; for, at this period of the year, the sky is generally very clear, and the north-west wind never blows. The Earthquake consisted of two shocks of unexampled violence, lasting thirty seconds; including a short interval between. They appeared to consist of undulations directed from south to north. The iron grating of the hospital, newly placed, was torn from the stones in which it was inserted, and apparently by electricity. Fort Royal was entirely destroyed, and it was at first believed that the Earthquake proceeded from the long extinct volcanoes of the island. For two hundred years since the French possessed it, these volcanoes have not given any signs of activity, and tradition asserts their tranquillity for several centuries farther back; therefore, M. Moreau de Jonnès believes that the cause is much more extended and general; for, added to these circumstances, it was felt throughout the West India islands, and twenty leagues beyond them at sea. A lieutenant on board the *Recherche* states, that at six in the morning, the ship was shaken in every part by a shock which lasted forty minutes, and the masts bent like bamboos. A few seconds after, a species of vapour rose from the shore, escaping through the crevices of the soil, and then the houses at Fort Royal began to fall. Those on the beach formed clouds of dust, and in the midst of the chaos a frightful cry rose from the lips of thousands of unfortunate sufferers.

All the crews of the vessels, amounting to 500 men, were ashore in ten minutes afterwards; and at the end of some hours, 200 persons, still living, were disengaged from the ruins; and by the evening 400 corpses were found.—*Athenæum.* Martinique was likewise visited by two severe shocks of Earthquake on August 2.

VESUVIUS.

On the morning of Jan. 1., an eruption of Vesuvius took place, which at one time threatened Naples; but the wind shifting, the smoke and ashes were fortunately borne to the shores of Portici. On the 2nd, detonations were heard, and the earth was tremulous under feet. The phenomena continued for another day or two, with considerable violence and dangerous aspects; but, except fears from the fiery glowing of the mountain, the invasion of Naples with cinder-showers, and periods of extreme darkness, no harm was done.—*Literary Gazette.*

SANDWICH ISLAND VOLCANO.

The Volcano of Kirauca, in Hawaii, is described by Count Stzeluki, as superior to all other volcanoes in intensity, grandeur, and extent. The precipitous cliff, forming the N.N.E. wall of the crater he found to be more than 4,000 feet above the sea: it overhangs an area of more than three million yards of half-cooled scoria, and contains more than 300,000 square yards of convulsed torrents of earth in igneous fusion, and gaseous fluids; constantly effervescing, boiling, spouting, rolling in all directions like waves of a disturbed sea, violently beating the edge of the cauldrons, like infuriated surf; and spreading all around its spray, in the form of hair-like glass, which fills the air, and adheres in a flaky and pendulous form, to the distorted and broken masses of the lava. There are five cauldrons, each about 5,700 square yards, almost at the level of the great area; a sixth is encircled by a wall of scoria some fifty yards high, forming the S.S.W. point. Millions of vents, all around the crater, through which the steam escapes, form the security of Hawaii.—*Silliman's Journal.*

SHOWERS OF ASHES.

On Nov. 6, was read to the Geological Society,—"A Notice of Showers of Ashes, which fell on board the *Roxburgh*, off the Cape de Verd Islands, in February last;" by the Rev. W. B. Clarke. On Tuesday, Feb. 4th, the latitude of the ship at noon was 14° 31′ N., longitude, 25° 16′ W. The sky was overcast, and the weather thick, and insufferably oppressive, though the thermometer was only 72°. At three P.M., the wind suddenly lulled into a calm, then rose from the S.W., accompanied by rain, and the air appeared to be filled with dust, which affected the eyes of the passengers and crew. At noon on the 5th of February, the latitude of the *Roxburgh* was 12° 36′ N., longitude 24° 13 W.; the thermometer stood at 72°, and the barometer at 30°, the height which it had maintained during the voyage from England. The volcanic island of Fogo, one of the Cape de Verds, was about 45 miles distant. The weather was clear and fine, but the sails were found to be covered with an impalpable, reddish-brown powder, which, Mr. Clarke states, resembled many of the varieties of ashes ejected from Vesuvius, and evidently was not sand blown from the African deserts.

NEW VOLCANIC ISLANDS.

On Nov. 6, was read to the Geological Society,—A letter from, Mr. Caldecleugh, dated Santiago de Chili, 18th of February, 1839, containing the declaration of the master and part of the crew of the Chilian brig *Thily*, of the discovery, during the evening of the 12th of February, of three Volcanic Islands, about thirty leagues to the east of Juan Fernandez. The island which was first noticed appeared, at the time of its discovery, to be rising out of the sea: it afterwards divided into two pyramids, which crumbled away, but their base remained above the level of a violent surge; and in the course of the same evening, the height of the island was, for a time, again considerably increased. The other two Volcanic Islets bore farther southwards. During the night, the crew of the *Thily* noticed, at intervals, a light in the same direction.—*Athenæum.*

PEAT BOGS.

Dr. Adams has read to the British Association a paper on the great advantage which would result from rendering Peat Bogs serviceable for agricultural purposes. He mentioned that, in some cases, where his plan had been adopted, the bogs were actually more profitable than the neighbouring lands. Another paper was read upon the same subject, in which it was stated that they might be made available as an excellent manure by the use of sulphuric acid, applied to the peat in heaps, mixed with putrifying vegetable matter as a sort of compost.—*Literary Gazette.*

FOSSIL ORGANIC REMAINS.

Mammalia.—M. Lartet announces the discovery of two fossil Carnivora, one of which appears to constitute a sub-genus, intermediate between the badger and the otter, and the second approaching to the dog, differing but little from that gigantic fossil which he has described under the name of Amphicyon. He is of opinion that the latter is the same animal as that of which some remains were found at Epelsheim, and which constitutes the genus Agnethorium of M. Kaup. "There are," says M. Lartet, "a considerable number of fossil mammiferæ found on the borders of the Rhine, which appear to me to be identical with those which are daily brought to light at the foot of the Pyrenees. These affinities are the more interesting, because the intermediate countries, Auvergne, for instance, possessed very different races of animals."—*Athenæum.*

Mammoth.—Dr. Fairbrother has exhibited to the Bristol Philosophical Society the tusk of a Mammoth, which had been found on the line of the Great Western Railway, near the spot where the city boundary crosses the line. It appears that this interesting fossil was discovered in the lower part of a bed of gravel, which reposes upon red sandy and marly beds, forming a part of the New Red Sandstone formation. The gravel is various in size, generally small, much water-worn, and intermixed with very fine gravel and sand; and it possesses so far the appearance of arrangement in layers, as to prove that it was not a sudden and quickly-formed accumulation, but the result of long-continued watery action. It is composed chiefly of the gritty sandstone of the coal measures, called "Pennant," with lias limestone and coarse oolite, all which had evidently passed down the present valley of the Avon. With these are distributed a few flints much worn, which may have found their way from the chalk hills through existing valleys. Pieces of carboniferous limestone, grit-stone, and siliceous iron ore, are also scattered through the bed, particularly in its lower part. As before mentioned, the tusk was found in the lower part of the gravel. Its length is about 5½ feet, its circumference varying from 21 to 10 inches. It is curved in a form nearly circular, and occupies an arc of about 140 degrees. Mr. Stutchbury supposed it to be a part only of the original tusk, which was probably 9 feet in length.

Mastodon.—A Molar, considered to belong to Cuvier's species of the narrow-toothed Mastodon, was dredged up off Eastern Cliff, Suffolk, in June: it is nearly perfect, and weighs 3¾ pounds. Till within the last five years, it was doubted whether the remains of the *Mastodon* had been discovered in England.—*Mag. Nat. Hist.; abridged.*

Glyptodon.—Appended to Sir W. Parish's recently published Account of Buenos Ayres, is an interesting Note drawn up by Mr. Owen, of the College of Surgeons, from a sketch and a tooth of the apparently complete remains of a monstrous fossil animal, entirely new to us. The monster it refers to was found in the bank of a rivulet, near the Rio Matauza, in the Partido of Cannelas, about 20 miles south of the city of Buenos Ayres, in a low marshy place, above five feet below the surface. It appears from the Report sent home with the original drawing, that the entire length of the beast, from the snout to the end of the tail, measures 8½ English feet; the width of the body, 3 feet 4 in.; and its height, 3 feet 6 in. The vertebral column, from the neck to the sacrum, is altogether; the ilia uniting with the vertebral column and sacrum in one single and immoveable piece. Thus, the sketch conveys the idea of a gigantic quadruped of the megatherium or armadillo family, having the internal skeleton and the external dermal bony case in their natural relative positions. The head is covered with a coronal plate, of a form closely corresponding with that which defends the like part in existing armadillos: a long descending process is indicated as being continued from the zygoma, with a slight curve forward: this structure is interesting, as showing that the part in which the megatherium most strikingly resembles the sloth is participated by another species, which indubitably possesses the characteristic armour of the armadillo tribe. The lower jaw, in the peculiarly descending curve of the ramus below and before the angle, also closely resembles that of the megatherium. The armour of the trunk would seem to be more capacious, and to have extended lower down, than in existing armadillos: its structure is described as consisting of polygonal plates, similar to the shelly coverings of all that family; which the animal appears to have resembled in the number of its teeth. Beneath the caudal plate, six hœmapophyses, or chevron bones, are delineated, apparently of disproportionate magnitude; but the indication is interesting, as exhibiting another well-marked feature of the megatherian organization. * * The tooth is slightly curved, with a smooth and polished exterior; its texture resembles that of the tooth of the armadillo, consisting of a central body of ivory, with an external coating of cementum; but the latter is relatively thicker than in the armadillos. The form and structure of the tooth indicates its adaptation to masticate vegetable substances of the softer kind; and the animal must have been provided with claws suitable to the digging up of esculent roots, reeds, &c. The tooth is more complicated in its external form than those of any recent or extinct edentate species hitherto discovered; and seems to indicate a transition from the bruta, or edentata, to the toxodon discovered by Mr. Darwin in the same part of the world. From the regularly fluted or sculptured form of the tooth, it is proposed to name the genus typified by this animal, "Glyptodon," from two Greek words sygnifying *sculpo* and *dens*.

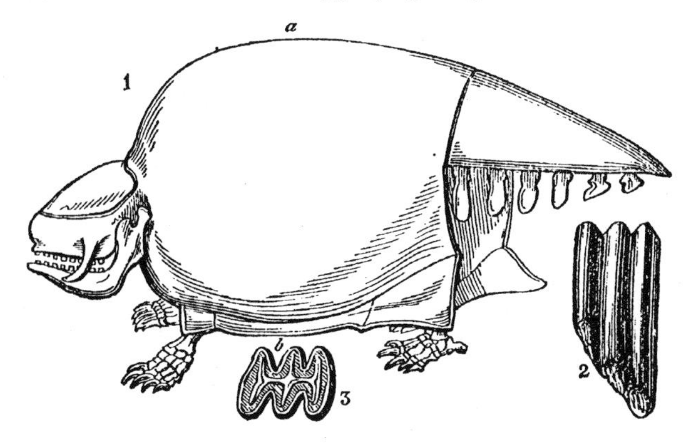

(The Glyptodon.)

Length, 8½ feet; height, from *a* to *b*, 3 feet 6 inches; the feet added conjecturally: Fig. 1. Fig. 2, Side-view of a molar tooth, natural size. Fig. 3, Grinding surface of the same.

Chirotherium Hercules.—On Feb. 27, were read to the Geological Society extracts from two letters; one from Mr. John Taylor, jun., "On the Occurrence of a Slab of Sandstone containing Impressions of Chirotherium Hercules, at the house of Mr. Potts, of Chester;" and the other by Sir Philip Egerton, "On the Peculiarities of the Impressions." When the slab was first laid down, there was no appearance of the remains, which have been gradually developed by the action of the weather. Sir Philip Egerton is of opinion, that the weight of the animal compressed the yielding sand near its foot, and that the print being afterwards filled with the same materials, the stone became nearly homogeneous in composition. The effects of the weather would necessarily remove the softer uncompressed portions, but the denser part, acted upon by the animal's weight, would resist for a time the same operations, and present in relief the outline of the foot. The slab contains the marks of three hind and two fore feet: the latter bearing the same proportion to the former as in the other species.—*Literary Gazette.*

Dinotherium.—Dr. Eichwald has read a memoir before the Imperial Academy of Sciences, at St. Petersburgh, in which he notifies the existence of the Dinotherium in the Crimea, which is very rich in the remains of the ancient world. Most of the country of Kertsch belongs to the recent tertiary formation, characterized by the pisiform iron. In the midst of this iron and heaps of shells, the remains of a large mammalia have been found, which are very heavy, and have passed into a siliceous state; they consist of ribs and vertebræ. A skull has also been met with, which, in some points, approaches that of the Dinotherium.—*Athenæum.*

Footsteps on New Red Sandstone of Shropshire.—Dr. F. O. Ward has exhibited to the British Association specimens, &c., and described the impression of footsteps of animals on the New Red Sandstone of Grinsile, in Shropshire. These phenomena, and the marks of rain, much resemble Dr. Buckland's examples, only that the marks were of three toes, with long nails, pointed forwards; and, except in few instances, where they resembled the dog, there were no footballs visible. Dr. Buckland had some similar sandstone from Dumfrieshire.

Bird on the Continent.—M. Agassiz, mentions a fact of high geological interest. In the slate of Glaris, and in a stratum of the age of the inferior chalk, he has found nearly the entire skeleton of a Bird, of the size of a swallow. This is the first instance of the kind which has occurred to M. Agassiz's knowledge on the continent. Dr. Buckland, however, observes that instances of this kind are not rare in England; and that Dr. Mantell has discovered many animals of the same period in the weald formation of Sussex.—*Proc. Brit. Assoc.; Literary Gazette.*

London Clay.—There has been read to the Geological Society—"A notice of some Organic Bodies procured from the London clay," by Mr. Wetherell; in indurated nodules of clay, obtained from the excavations for the Birmingham Railway, between Euston-square and Kilburn. Some of the specimens consist of flat flabelliform bodies; more or less corrugated on the surface, and covered with minute oviform grains in close contact with each other. Several of the specimens are cylindrical and branched, varying in diameter from half an inch to less than a tenth; and they are likewise wholly or partially covered with the egg-shaped grains. Besides these fossils of a definite form, Mr. Wetherell has procured a multitude of others perfectly amorphous. They are generally of a dark colour, and are for the most part without apparent internal structure, but they occasionally consist of concentric lamellæ. The granules are not unfrequently dispersed through the substance of the specimen, or they are collected in irregular patches on the surface. A series of these bodies, illustrative of the memoir, has been presented to the Society's Museum by Mr. Wetherell.

Mr. Lyell has communicated to the British Association some very interesting details of new fossil remains in the London clay and red crag, in Suffolk. Alluding to the fossil bird, of the size of a swallow, found by M. Agassiz in the Slate of Glaris, he observed that this, like many other discoveries, ought to be a warning to geologists never to erect hypotheses on the evidence of the non-existence of any fossil, simply because they had not happened to meet with them in particular strata or formations. This principle he had endeavoured to enforce years ago, and every year's farther experience had confirmed him in the necessity of attending to it. It now appeared that mammalian remains were found in two strata, among marine deposits, where they were never before thought of;—a caution of a stronger order than they had yet received on the subject of determining the relative age of those various materials of which the crust of the earth was composed. The first specimen was in chalk of the oldest of the tertiary series, full of marine shells (eocene), of which from one to three per cent. only were of known living species. In the red crag of Suffolk, on the contrary, there were thirty per cent. of shells and fishes of the present day (pliocene), so that it might be fairly concluded that a large space of time had intervened between these two formations. He was inclined, however, to change the opinion he had formerly entertained, and to consider this Suffolk crag to belong to the miocene or middle tertiary series. Hitherto, no remains of mammalia had been discovered in this formation, until the first of these specimens was brought under his notice by Mr. Colchester, a well-informed geologist, by whom it was found near Woodbridge. On examining it, together with the many fishes' teeth which accompanied it, he was convinced that it belonged to a mammalian animal; and Professor Owen at once pronounced it to be one of the grinders of a beast of prey allied to the tiger species. On farther examination, it was known to be the left-side jaw-tooth of a leopard. Incited by this discovery, Mr. Wood proceeded to the neighbourhood, where the workmen in the crag-pit had been accustomed for more than twenty years to deliver the teeth they fell in with to an old woman of the village for sale. Among these, were sharks' teeth, and all the rest were of different fishes, but none of mammalia. In another collection, however, of the same kind, made by the Rev. Mr. Moor, of Newbourn, and consisting of the teeth of fish, Mr. Owen detected the molar tooth of a bear; and another, the tooth of some unknown ruminant. These were more or less broken, but still sufficiently perfect to allow of their identification. No doubt they had been found in the large pit at Newbourn, and only one point of doubt rested on the subject. There were many rents or fissures descending from the surface; and it was possible that the teeth might have come out of these rents, which penetrated through the upper red shelly crag into the London clay below. Thus, the remains of the hyena had been found among the marine deposits of the Kentish crag, which had fallen through a similar fissure. They had, therefore, to guard themselves against the possibility that these teeth might have been derived from such an accident, and not belong to the strata with marine deposits below. To guide them in this respect it would be well to see if the colour, condition, and general appearances of the mammalian remains corresponded with those of the fishes among which they were imbedded. In the present instance the resemblance was very great. Mr. L. then referred to the Norwich crag, which held fluvial shells, fresh-water, and marine, intermingled; and supposed that a river might have floated down the carcasses of the quadrupeds, and deposited them with the other remains. In the London clay, where an immense number of fishes' teeth occurred, at Kyson near Woodbridge, one of a mammifer had been found with two fangs —and no fish or reptile ever had two—which on being taken to Mr. Owen he declared it at first sight not to belong to carnivora, for they never had them of this form, and few granivora possessed them. He suspected it to be marsupial, and went to the kangaroos, where he immediately recognised it, and in the North American opossum *(Didelphis Virginiana)* its nearest likeness. The next specimen from the same site, procured by the renewed exertions of Mr. Wood and Mr. Colchester, consisted of the lower jaw of an animal, with four cusps, and containing the last tooth in the jaw and part of that nearest to it. This belonged to the monkey (the *Macacus*), and the macacus radiatus was the living genus the most nearly allied to it. The geological formation where it was found was, first, superincumbent crag; below this, a clay, seventeen feet in thickness; and below the clay, sand ten feet thick, and full of fishes' teeth. At the top of this sand where it was overlaid by clay, these important remains were found. Cuvier had described an opossum in the eocene gypsum near Paris, and he considered them to belong to the same period as the London clay. Mr. Fox had also made similar discoveries in the Isle of Wight; and it was most remarkable to find the quadrumanous organization the nearest approaching to our own informations of so remote an era. He had supposed such possibilities in his published work ; and it was curious that within two years they should have found in France, India, and South America, instances, proofs of their existence. First, a fossil ape, of the miocene period, had been discovered in France; probably a Gibbon, and high in organization so near to human. Next, at the foot of the Himalaya, another had been discovered; but the present was the first instance so far from the equator of quadrumana among eocene remains. In the London clay, crocodiles, creatures and plants of the torrid zone, the bread-fruit, the nautilus, and other natural productions, all showed that the climate had been much hotter than at the subsequent period when the red crag was formed. The whole illustrated the progressive development of animal life; and this single discovery had destroyed the long-maintained theory that those of a high organization had been created just antecedent to the creation of man. M. De la Beche bore testimony to the very great importance of this paper; and observed that the remains of marsupial animals were found in much older rock, of the oolite formation. The discovery of this monkey overset all previous notions connected with the appearance of man upon the earth ; and he was glad to take this opportunity of retracting his previous opinions on the subject.—*Proceedings of the British Association ; Literary Gazette.*

For farther details of these important discoveries, illustrated with engravings, the reader is referred to papers by Messrs. Wood, Owen, and Charlesworth, in the *Magazine of Natural History*, Sept. 1839 ; and to papers by Messrs. Lyell and Owen, in the *Annals of Natural History*, Nov. 1839.

There has been also read to the Society,—A paper, "On the Fossil Remains of a Mammal, a Bird, and a Serpent, from the London Clay." Until a few months since, the highest organized animal remains known to exist in the London clay, were those of reptiles and fishes ; but, during the last summer, there were discovered in the collections of Mr. W. Colchester, of Ipswich, and the Rev. Edward Moore, of Bealings, near Woodbridge, teeth of a quadrumanous animal, of Cheiroptera, plantigrade and digitigrade carnivora, and of a species probably belonging to the marsupial order ; all of which were obtained from the London clay of Suffolk. To this important list, Mr. Owen is now enabled to add the remains of a new and extinct genus of pachydermatous mammals, of a bird, and a serpent. The first of these curious fossil relics was discovered in the cliffs of Studd Hill, near Herne Bay, by Mr. W. Richardson ; and consists of a small mutilated cranium, about the size of that of a hare, containing the molar teeth of the upper jaw nearly perfect, and the sockets of the canines. The molars are seven in number on each side, and resemble more nearly those of the Chæropotamus than of any other known genus of existing and extinct mammalia. The sockets of the canines, or tusks, indicate that these teeth were relatively as large as in the Peccari. The other portions of the head, preserved in the specimen, are then described: the general form of the skull partakes of a character intermediate between that of the hog and the hyrax, but the large size of the eye must have given to the physiognomy of the living animal a resemblance to that of the Rodentia. Mr. Owen has adopted for this new extinct genus the name of Hyotherium, suggested by Mr. Richardson.

The remains of fossil Birds, included in the second part of the paper, consist of two specimens, a sternum with other bones, and a sacrum, both obtained from the London clay at Sheppey. The sternum forms part of the collection of fossils made by the celebrated John Hunter. The sacrum is in Mr. Bowerbank's collection of Sheppey fossils. After minute examination, which we have not space to detail, Mr. Owen finds the Hunterian specimen to bear the greatest number of correspondencies to the skeletons of the Accipitrine species, and closest with the vultures, though of a smaller species than is known to exist at the present day ; and belonging to a group of Accipitrine scavengers, so abundant in the warmer latitudes of the present world. Mr. Bowerbank's specimen consists of ten sacral vertebræ anchylosed together, as is usual in birds with a continuous keel-like spinal ridge ;

and in five of which there is a resemblance to the corresponding part in vultures, in the non-development of the inferior transverse processes. One of the specimens of the extinct species of Serpent, described in this paper, forms likewise part of the collection of fossils left by John Hunter, and consist of about 30 vertebræ; the others, one of which presents a series of 28 vertebræ, are in the cabinet of Mr. Bowerbank. The author considers that all the specimens are referable to the same species; and they were all obtained at Sheppey. The vertebræ are equal in size to those of a *Boa* constrictor 10 feet long. From the agreement, in some points with the Boæ and Pythons, and the absence of all those which might have prevented the living animal from entrapping its prey, and from the length which it may be inferred that the creature attained, Mr. Owen concludes the fossil was not provided with poison-fangs. Serpents of similar dimensions exist in the present day only in tropical regions, and their food consists of cold as well as of warm-blooded animals; he therefore, in conclusion, states, that had there been obtained no evidence of birds and mammals in the London clay, he would have felt persuaded that they must have co-existed with the *Palæophis toliapicus*, which Mr. Owen proposes conditionally to name this interesting fossil.—*Athenæum; abridged.*

Fishes near Manchester.—Mr. Bowman has exhibited to the British Association specimens of fossil Fish found in the vicinity of Manchester; and read a paper from Mr. Binney upon them. There were scales and teeth of Megalichthus from the coal shales, in the Manchester coal-field, lying above the millstone grit; and remains of *Diplodus, Ctenoptychus, Holoptychus,* and *Palæoniscus,* up to the fresh-water limestone formation of Ardwick. The shells of *Cypris,* &c., among which they were found, seem to prove a calm and gradual deposit; and the condition of the fish indicated their sudden destruction, which Mr. Binney attributed to water charged with the gas from decayed vegetation. None of the fossils had been detected in the coal, though they approached to the very edge of it.—*Literary Gazette.*

Fishes in Yorkshire and Lancashire.—There has been read to the Geological Society,—a paper "On the Fossil Fishes of the Yorkshire and Lancashire Coal-Fields," by Mr. W. C. Williamson. Within the last four years, the coal measures of these counties have assumed a zoological importance which previously they were not supposed to possess. In Lancashire, ichthyolites have been lately found to pervade the whole of the series from the Ardwick limestone to the millstone grit, and in Yorkshire they have also been obtained in great abundance. On comparing the specimens procured at Middleton colliery, near Leeds, with the fossil fishes of Lancashire, the author detected the following as common to both coal-fields, viz.—*Diplodus gibbosus, Ctenoptychus pectinatus, Megalicthus Hibbertii, Gyracanthus formosus*; also remains of apparently species of *Holoptychus* and *Platysomus:* but he has obtained some ichthyolites in the Yorkshire field which he has not seen in the Lancashire; and he is of opinion that the latter deposits are characterized by the greater prevalence of Lepidoid fishes, and the former by Sauroid. These remains, except in the case of the Ardwick limestone, always occur in highly bituminous shale; and they are most abundant where it is finely grained, and in general where plants are least numerous. This distinction in the relative abundance of ichthyolites and vegetables, Mr. Wilkinson conceives may throw some additional light upon the circumstances under which the coal formations were accumulated. The fishes are found chiefly in the roof of the coal, rarely in the seam itself, and not often in its floor.—*Athenæum.*

Infusoria in Ireland.—Dr. Drummond has communicated to the *Magazine of Natural History* a notice of fossil Infusoria received from Newcastle, at the base of the Mourne Mountains, County Down; and consisting of a very light, white, earthy substance, which had been found in considerable quantity in the above neighbourhood. This Dr. Drummond found to be the same kind of substance as Prof. Bailey had found in a bog at West Point, in America, and which consisted of the siliceous remains of organized microscopic beings, either animal or vegetable.* Dr. Drummond is not aware that fossil Infusoria have hitherto been found in the British Islands.

The substance referred to is, when dry, of the whiteness of chalk, but becomes brownish when wet; it is as light as carbonate of magnesia, which it much resembles, but it is not acted on by nitric, muriatic, or sulphuric acids, and is indestructible by fire. The specimen was a compact mass, of the shape, and nearly the size, of an ordinary building brick; it could easily be rubbed down into powder, and had a coarse and somewhat fibrous fracture; when a portion was rubbed between the finger and thumb, it had no grittiness, but felt like an impalpable powder, and when it was then blown into the air, it flew about almost like wood ashes.

On examining many times small portions of the fossil mixture with a little water, on a slip of glass, the whole was found to be composed of the bodies, of which long linear *spicula* formed, at least, four-fifths. Occasionally confervoid fragments were seen, and frequently minute annular portions. There was no admixture whatever of unorganized matter, and no medium of cement whatever.

The spicular bodies are joints of the *Diatoma elongata*, (*Engl. Flora*, vol. v., pt. i., p. 406.) This species grows abundantly in a small drain of clear water in the grounds of the Royal Belfast Institution, and its joints in the microscope are seen to be precisely similar to the spicular bodies. When the loricated *Infusoria* are burned to ashes, the latter are found to be their siliceous coverings unchanged; and the same thing occurs in the *Diatoma*, as was discovered by De Brebisson and Prof. Bailey. On burning the *Diatoma* to a red heat, when cold, it was found to be unchanged in form, appearance, and sharpness of outline. The *Navicula tripunctata* was found to be equally unaffected by heat, as also some other Infusoria.

The deposit here described is evidently of the same description as that found in the New World, by Prof. Bailey, and analogous to that

* See the abstract of a paper on Fossil Infusoria discovered in peat-earth at West Point, in *Year-Book of Facts*, 1839, p. 232.

found in several places of the Old World; viz., the *Kieselguhr* of Frauzenbad, and the deposit in peat-bog near the same place, the *Bergmehl* of Santa Fiora, &c., which are formed of fossil Infusoria remains.

Berghmehl.—Dr. Trail has analyzed a Berghmehl from the north of Sweden, and found it to be composed of the minute shields of *Infusoria*, about one-thousandth of an inch in size, consisting chiefly of siliceous earth and alumina.—*Proc. Wernerian Society.*

Brazil.—M. Lund, of Lagoat-Santa, has made a voluminous report to the French Academy of Sciences concerning the fossil mammiferæ of that country, from which the following is an extract: Of this class of animals he has found 75 distinct species, belonging to 43 genera, most of which abound in caverns. The part of the country studied by him lies between the Rio das Velhas, one of the tributaries of the Rio San Francisco, and the Rio Paraopeba. This district forms a plain 2,000 feet above the level of the sea, and is traversed by a chain of mountains only from 300 to 700 feet high. This chain is composed of secondary limestone, stratified in horizontal layers, having all the characters of the zechstein and hochleu kalkstein of the Germans. It is entirely hollowed out into caverns, and crossed by cracks in every direction, and its interior more or less filled with a red earth, identical with that which forms the superficial stratum of the country, and which is from 10 to 15 feet thick. This is often so ferruginous, that its particles of iron are transformed into a pisolithic mineral, like that of the Jura. This earth has undergone some modifications in the caverns, for it contains angular or rolled fragments of the calcareous rock, particles of lime deposited by the water which filters through the cracks, and it is impregnated with saltpetre. The fossils lie in this earth, and are disposed pell-mell in the middle of it. They are all fragile, of white fracture, often petrified, adhere closely to the sockets in which they lie, frequently present calcareous spath, are broken, crushed, or otherwise mutilated, and bear marks of teeth, showing that they have been carried there by the ferocious animals which inhabited those caverns, and also by a species of diurnal bird. *Athenæum.*

Vegetable Skeletons.—Mr. Bowman has read to the British Association a paper on some skeletons of fossil Vegetables, found by Mr. Binney, in white impalpable powder, under a peat-bog near Gainsborough, in a stratum four to six inches in thickness, and covering several acres. It remained unchanged by the sulphuric, hydrochloric, and nitric acids, and by heat, and was concluded to be pure silica, in a state of extremely minute subdivision. On submitting it to the highest power of the compound microscope, it was found to consist of a mass of transparent squares and parallelograms of different relative proportions, whose edges were perfectly sharp and smooth, and the areas often traced with very delicate parallel lines. On comparing these with the forms of some existing *Confervæ*, Mr. Bowman found the resemblance so strong, that he entertained no doubt they were the fragments of parasitical plants of that order, either identical with, or nearly allied to the tribe *Diatomaceæ*, which grow abundantly on other *Algæ*, both marine and fresh water, but are so minute, that individually they are invisible to the naked eye. They are the counterparts of the fossil *Infusoria* of Ehrenberg, and occupy the same place in the vegetable kingdom as those do in the animal.—*Athenæum.*

SAND-PIPES IN THE CHALK NEAR NORWICH.

MR. LYELL, V. P. G. S., has communicated to the *Philosophical Magazine*, No. 96, an important paper "On the tubular Cavities filled with Gravel and Sand called 'Sand-pipes,' in the chalk near Norwich:" which communication was of such interest as to afford the principal topic of discussion and observation during nearly two days at the late meeting of the British Association.

The white chalk with flints in the neighbourhood of Norwich, is covered with a mass of variable thickness of irony sand and gravel, with some intermixture of red clay, the sand passing occasionally into a ferruginous sandstone. The surface of the chalk, when the gravel is removed, presents sharp ridges, deep furrows, and pits, and protuberances larger at the summit than the base. In a word, it is impossible to conceive that so soft a rock as chalk could have acquired such an outline simply by ordinary denudation, or could have retained it, if once acquired, during the accumulation of the mechanical deposit now superimposed. It is equally difficult to refer to any known mode of denudation those deep and narrow hollows, filled with sand and gravel, which are the same as those called in France "*puits naturels,*" and which form the subject of this paper.

The localities of these Sand-pipes are three; 1. at Eaton, about two miles west of Norwich, where they are very symmetrical, having the form of inverted cones, which, at their upper extremity, vary in width from a few inches to more than four yards, while at their lower, they taper down to a fine point: *see fig.* 1. The smaller ones, usually about 1 foot in diameter, seldom penetrate to the depth of more than 12 feet, while the larger are sometimes more than 60 feet deep.

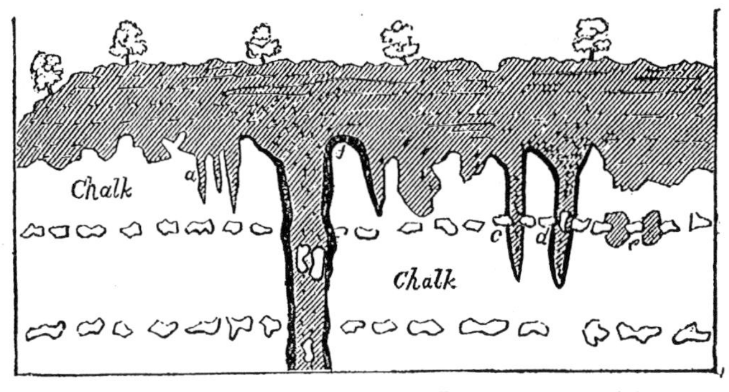

(FIG. 1.—Sand-pipes in the Chalk at Eaton, near Norwich.)

They are mostly perpendicular in direction, and nearly circular in shape; although they often appear of an oval form when cut through in the precipices surrounding the Eaton quarries, because the plane of intersection is there, in reality, oblique, and inclined at an angle of about 80 degrees with the horizon. Several sand-pipes often approach very near to each other without any tendency to unite; see *a*.

These pipes are filled with three kinds of materials: 1st, sand and pebbles; 2ndly, loose unrounded chalk flints; 3rdly, fine ochreous sandy clay, not impervious to water. The pebbles are chiefly of black flint, but a few are of white quartz. With these are sometimes seen unrounded fragments of sandstone, with a cement of oxide of iron; the whole agreeing with the contents of the deposit incumbent on the chalk, which at Eaton is about 20 feet thick. The clay is also similar to the finer portion of that found in the gravel above. As a general rule, the sand and pebbles occupy the central parts of the pipe, while the sides and bottom are lined with clay. Neither organic nor inorganic calcareous matter occurs in any part of the pipes where the clay is in contact with the chalk. Large unrounded nodules of flint, with their original white coating *(b b)*, are dispersed singly and at various depths in the larger pipes. The smaller pipes are frequently crossed by horizontal layers of siliceous nodules, as at *c*, *d*, *e*, *fig.* 1., which still remain in *situ*, not having been removed with the chalk in which they must have been originally imbedded. Single flints, forming part of these layers, sometimes appear in the middle of a small pipe, as at *d*, *fig.* 1, surrounded and supported by sand, so that, at first sight, it is not easy to imagine how it can have retained its position during the substitution of the sand and gravel for the original chalk. But, these flints are usually of large size and irregular shape, and may be still supported at one extremity by the chalky matrix. Neither a loose nodule of flint, nor a heap of nodules, has ever been observed at the bottom of a Sand-pipe at Eaton. The middle of each pipe is generally filled with sand and gravel, and the outside and bottom with clay, which is sometimes horizontal over the chalk, as at *f*, *fig.* 1.

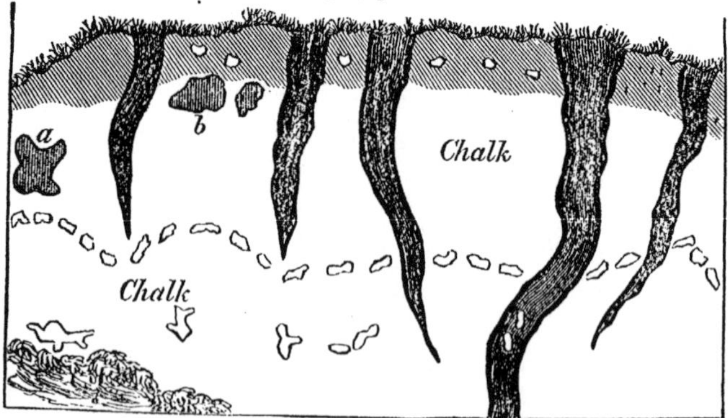

(FIG. 2.—Tortuous Sand-pipes in the Chalk at Heigham, near Norwich.)

The age of the gravel and sand undoubtedly belongs to the Norwich crag; as there are not only casts of marine testacea, a characteristic of that formation, in the ferruginous sandstone at Eaton, but also some shells of the genera *Mya, Mactrum, Cardium*, and *Mytilus*, in which the calcareous matter is still preserved. One remarkable fact is, that the sides of these pipes are lined almost continuously with a fine clay, nothing similar to the materials, the gravels and pebbles, which fill up their centres, and appear evidently to have been precipitated from above, slowly and gradually after (probably at a much later period) the openings in the chalk have been made to receive them. The chalk on the edges of the pipes is in a most discomposed state and slightly discoloured, of a yellowish hue, as if from clay and sand, and extending to the distance of from one to five feet.

The details of the second locality have been communicated to Mr. Lyell by Mr. Wigham. These pipes, (*fig.* 2,) occur at Heigham, in the suburbs of Norwich, and are tortuous in their descent. The pit represented is 30 feet deep, and the chalk barely covered with vegetable soil. Its upper portion, 3 or 4 feet deep, is mixed with sand and gravel; *a b* are cavities in the chalk, from 10 inches to 2 feet in diameter, and are elbows of tortuous sand-pipes, the other parts of which have been removed in excavating the pit.

The layer of chalk flints, *c*, not being horizontal, implies the chalk to have been disturbed.

At Thorpe is a Sand-pipe, (*see fig.* 3,) 20 feet in diameter, which on entering the chalk, is filled with gravel, sand, clay, stones, and

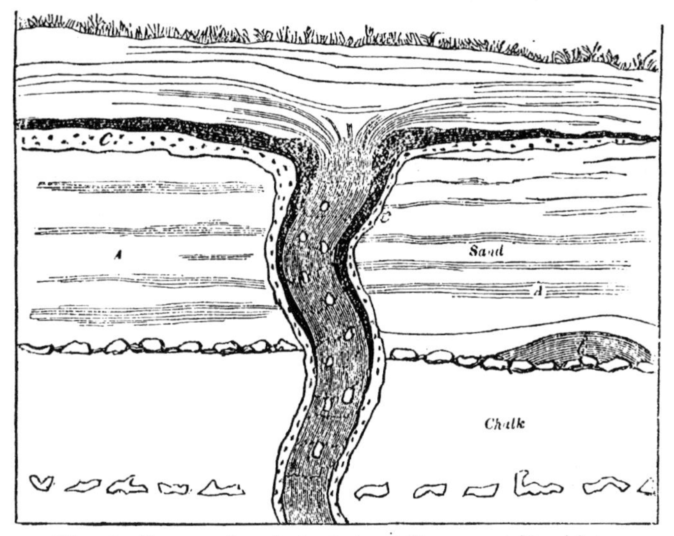

(FIG. 3.—Upper portion of a Sand-pipe at Thorpe, near Norwich.)

chalk-flints. It penetrates the rough 35 feet of chalk, tapering gradually. Its course is regular through 10 feet of sandy strata, A A, overlaying the chalk, some beds of which, as at *b*, are rich in the shells of the Norwich crag. A layer of light coloured sandy clay, *c*, *fig.* 3, (indicated by dots) lines the sides of the pipe vertically, and horizontally beyond the opening. The dark bed, *d*, *fig.* 3, is an indurated layer of sand coloured by oxide of iron, which contains casts of marine shells, not only where horizontal, but where vertical, as far as the entry into the chalk.

At the junction of the chalk and overlaying sand occurs a layer of large flints, which have suffered slightly from attrition.

Mr. Lyell then proceeds to the origin of these Sand-pipes, and concludes: 1. That the chalk has been removed by the corroding action of water charged with acid, in which the siliceous nodules, being insoluble, were left in *situ*, in the smaller pipes after the calcareous matrix had been dissolved. 2. That, from the dispersion of the flints through the widest pipes, the excavation and filling of the pipes were gradual and contemporaneous processes; for, had the tubes been first hollowed out, flints must have fallen to the bottom from intersected layers in the chalk above. 3. As a corollary of the above propositions, we must hold that the strata of the Norwich crag had been already deposited upon the chalk before the excavation of the Sand-pipes; and this is farther confirmed by the manner in which the layers in loose gravel of the pipe *d*, *fig.* 1, and the dark sand with casts of shells, *e*, *fig.* 3, have sunk into the pipe.

Mr. Lyell thus rejects all sudden and violent agency, whether for the erosion or filling of the cavities; and concludes that land in this country must have emerged from the sea after the deposition of the Norwich crag, and yet at a period anterior to that of an aqueous denudation which has removed large portions of a deposit overlaying the chalk, and which supplied the contents of the sand-pipes. But, as we know of no denuding agency capable of excavating great valleys in a flat country like Norfolk, except the power of the ocean, operating either at the time of the submergence of land or that of its emergence from the waters, we must infer from all the facts and reasonings set forth, that land, consisting of chalk covered by crag, was first laid dry before the origin of the sand-pipes, and then submerged again before it was finally raised and brought into its present situation.

Astronomical and Meteorological Phenomena.

NEW COMET.

At 28 minutes after 3 in the morning of December 9, (civil reckoning at Berlin,) Encke observed this Comet at the Royal Observatory at Berlin, where he found its right ascension to be 13 hours, 42 minutes, and 44 seconds; its southern declination being 11 minutes and 30 seconds.

At 31 minutes, 13 seconds, after 6 in the morning of December 10, (civil reckoning at Altona,) Professor Schumacher, at the Observatory of Altona, determined its right ascension to be 13 hours, 43 minutes, 45 seconds, and its north declination 8 minutes and 18 seconds; whilst at 2 minutes 42 seconds after 6 in the morning of the 11th, its right ascension, as observed by the same astronomer, was 13 hours, 53 minutes, 19 seconds, and 27 hundredths; and its north declination was 27 minutes 57 seconds, and 7 tenths.

On the 15th, it was observed at Hamburgh Observatory by Mr. Rumker, when at 24 minutes, 55 seconds, and 38 hundredths after 4 in the morning, (civil reckoning at Hamburgh,) its right ascension was 14 hours, 31 minutes, 59 seconds, 49 hundredths; and its north declination, 1 degree, 39 minutes, 33 seconds, and 49 hundredths.—*Times, Dec.* 24, 1839.

This Comet was also seen at the Cambridge Observatory on the mornings of December 28, 29, 30, 1839; and January 2 and 3, 1840. The apparent right ascensions and declinations, (exclusive of corrections for parallax,) were found on those days to be nearly as follows, at the subjoined times from Greenwich, mean midnight:—

Time from Midnight.		Right ascension.			Declination.		
h.	m.	h.	m.	s.	deg.	m.	s.
5	59,83	16	29	50	3	21	23N.
5	52,74	16	37	37	3	18	31
6	9,54	16	45	15	3	14	20
7	14,45	17	7	3	2	53	40
5	54,94	17	13	29	2	44	20

This comet is a brighter and more considerable object than that of Encke, which appeared towards the close of 1838. It has a well-defined nucleus, which is either solid, or consists of very condensed matter. The tail is well seen in a telescope of very low power, extending in the direction from the sun, making an angle of about 54 degrees, with a circle through the pole.—*Observatory, Cambridge, Friday Morning, Jan.* 10, 1840.

SPOTS ON THE SUN'S DISC.

On October 2, Count Decuppis observed an unusual number of Spots on the Sun's disc, and on a quarter before 9 on that day, perceived a small black spot entirely free from penumbra, and of entirely spherical form, which had advanced upon the disc, describing an arc of about seven minutes. Reiterated observations convinced him that it had, in the mean time, advanced towards the sun's limb to the extent of two minutes and 30 seconds. At three minutes after 9, when M. Decuppis attempted to make a new observation, the spot had disappeared.

A Correspondent of the *Dundee Advertiser*, in September, notes:—

"The surface of the sun has of late presented a very striking and diversified aspect, when contemplated through a powerful telescope. The spots of all descriptions by which its disc is divaricated have been more numerous, and some of them much larger than have been observed for several years past. On Monday, the 2d inst., almost the whole surface of this luminary seemed to be diversified with large and small spots of every description peculiar to the sun. A cluster containing four or five very large spots, and about thirty or forty smaller ones, disappeared from the western part of the disc on Tuesday, but a very great number of both large and small spots still remain. At present (Sept. 4) there is a very large cluster approaching the centre of the disc, which consists of about eighteen large spots, the smallest of them not much less than the size of the earth, and some of them much larger. Besides these, there are within the compass of the same cluster above eighty smaller spots, which can be distinctly counted by means of an achromatic telescope magnifying 120 times, making about 100 spots in all within the limits of one cluster. The smallest of these spots cannot be less than from 500 to 900 miles in diameter. One of the spots which lately passed off from the western margin of the disc, measured about the 1-30th part of the sun's diameter, and consequently, must have been about 30,000 miles in diameter, or nearly four times the diameter of the earth; and, if it is to be considered as a solid body, it must be above sixty times larger than the earth. It contained an area of more than 700,000,000 miles. Besides the cluster noticed above, there are five other clusters nearer the western edge of the disc, containing several large, and a number of smaller spots, amounting in all to about seventy or eighty; so there are at present nearly 200 spots, great and small, diversifying the surface of this luminary. The largest cluster will likely remain for about eight days longer before it disappears from the western limb. Some of the other spots will disappear in the course of three or four days. There are indications of other clusters about to appear on the eastern limb. Each of the larger spots has a dark nucleus, surrounded with a penumbra, or fainter shade, nearly of the same shape as the nucleus. Some of the nuclei appear nearly round, others elliptical, others conical, and some of them are divided in the middle by a bright streak. When these spots are near the margin of the sun, they appear surrounded with a mottled appearance, such as is seen on some parts of the lunar disc, evidently indicating elevations and depressions, or, in other words, mountains and vales of very great magnitude. These mottled appearances generally precede the appearance of spots on the eastern limb, and plainly show that there is a very great diversity of surface and scenery on this magnificent orb; and that changes and operations of inconceivable magnitude are continually going forward—probably for the purpose of preserving this central body in a proper state for diffusing light and heat, and other influence, to surrounding worlds. Four or five of the larger spots may be distinctly seen by means of an opera glass which magnifies about two or three times, and even by the naked eye, provided a coloured glass is interposed between the eye and the sun, or a common plain glass smoked with the flame of a candle."

THE BLUE SUN.

M. BABINET observes: "In studying the phenomena of meteorological optics, I have not neglected those colours which the sun and moon occasionally assume, of an exceedingly dull tint, and without any surrounding rings. The phenomenon of a red sun may be attributed to a defect in the transparency of the atmosphere, arising from vapours, or to any other cause; for the fundamental interval of interferences being much greater for the red than for the blue and violet, these latter are first extinguished, and the obstacles to their transmission are comparatively much greater than to their transmission of the red, as is the case with the reflection from a glass that has been merely smoothed, which always begins with the red. Upon this, I may remark, that it is very doubtful if the reddish brown tinge of smoked rock-crystal is owing to a true colour, and not to the exclusion of the lower colours of the spectrum induced by the imperfect transparency of the foreign matter. Another phenomenon, which is much more rare and curious than the red sun, is the *blue sun;* when this luminary is seen of a fine blue tint, though somewhat mixed with white. Our scientific repositories contain some instances of this, and I have myself seen two. It is evident, that the yellow hue, which is much less remarkable on account of its analogy with the white, must also frequently occur, whilst the violet, owing to the difficulty with which it traverses imperfectly diaphanous media, will but seldom appear. I attribute these colours to the interference of the rays which have traversed the vesicles of water or vapour, with those which have traversed air only. The phenomenon simply implies, that the part of each traversed vesicle is not too thick, a supposition which is easily admitted *a priori.* It is completely of the same nature with that which you have yourself observed in mica or gypsum plates of different thicknesses, and in which the two neighbouring rays which traverse the different thicknesses of the mica or gypsum interfere, and so produce the coloured rays (an experiment which has twice been reimported during the last year from England); they are likewise the phenomena known under the term of the *mixed plates* of Dr. Young.

To produce, then, a blue sun, the red, yellow, and even the violet sun, I have taken (see *Société Philomatique,* 1827,) two plane circular glasses, separated by a layer of mixed water and air, of oil and air, and finally of oil and water; and by suitably approximating the glasses, I have made the flame of a lamp to be seen through them of an uniform red tint, and of a blue and violet tint, at pleasure. The enfeebled image of the sun reflected by water assumes the same colour ; and the moon still more strikingly, and by direct vision. Hence, then, I imagine that nothing requires to be added to the explanation, and the reproduction of the meteorological phenomenon."—*Communicated to Jameson's Journal.*

JUPITER'S SATELLITES.

AN irregularity in the first satellite of Jupiter has for some time attracted the attention of M. Boguslawski, the director of the Observatory at Breslau. He observed it on the 14th of April, 1838, and on the 1st of May he made farther observations, when he found that its lustre, which, generally speaking, is greater than that of the second, appears to be much weaker when it quits the disc of the planet after its passage across, especially when its shadow is seen upon Jupiter. By means of the heliometer, he kept the second satellite always by the side of the first, and at the same distance from the disc, in order to ascertain if the light of the planet were the cause of this diminution. For several hours after leaving the disc at 12h. 54′ 26″. 2, sidereal time, the first satellite was evidently paler than the second: it then slowly began to resume its light, but which at 15′ 18″ had not yet attained that of the second. The next day it had quite resumed its lustre, and again surpassed that of the second.—*Athenæum.*

AUGUST AND SEPTEMBER ASTEROIDS.

THE periodical return of these meteors in August, during the past year, has been noted by several observers.*

Sir James South has addressed to *The Times*, from the Observatory, Kensington, a letter detailing his observations; the substance of which is as follows:—"The evening of the 10th of August was fine, and for a considerable time, the sky was cloudless. A celestial globe being brought out on the lawn for the purpose of tracing the tracks (of the looked-for stars,) and a loud beating clock being set with that at the transit instrument, by which the instant of disappearance of any of these bodies might be noted, between twenty-two minutes after nine, and two minutes after midnight, 165 shooting stars were not only seen, but their flights amongst the fixed stars, and their disappearances, to the nearest tenth of a second, registered. Between five minutes after midnight, and twenty-nine minutes after one in the morning, 150 were seen. Clouds, which continued till daylight, prevented farther observations.

Of these, the principal part resembled stars of the sixth magnitude stealing from one part of the heavens to another. Many were as bright as stars of the first magnitude. Several had a brilliancy many times surpassing that of the planet Venus; whilst some few, apparently of a discal form, were not unlike the planet Jupiter, as seen with a magnifying power of fifty or sixty. These, as well as those of the two preceding classes, prior to their disappearance, frequently burst into thousands of intensely luminous points. Not the

* For notices of the Asteroids of 1838, see Year-book of Facts, 1839, p. 249.

slightest report could be heard. The directions which these fugitives took were very various, as was the extent of arcs they traversed; generally, they took their course from the zenith towards the horizon, but, in several instances, they passed from horizon to zenith: some appeared when within ten or fifteen degrees of the horizon, and disappeared in it. Every part of the visible heavens teemed with them; the constellations, however, of Cassiopeia and Perseus were most prolific. Instantly on the explosion, and at the spot where it occurred, was formed a circular luminous cloud of one to four or even five degrees diameter, and which remained visible for a second and a half before it gradually disappeared."

A Correspondent of the *Philosophical Magazine* observes, that "the average number per hour in the half hemisphere to which the writer attended, was 44, exceeding considerably the average of last year at Geneva." He adds, that a circumstance particularly worthy of notice last year is, that "several of the shooting stars appeared to move upwards, whereas no instance of this was remarked last year at Geneva."

The following statement of the great fall of stars on the 10th of August, is from the *Prussian State Gazette*:—

"The sky has been again particularly propitious for observing another fall of stars. On many days and nights preceding the 10th, the heavens had been so covered that we could not observe when the uncommonly frequent fall of stars commenced. On August 10, however, as early as dusk, an extraordinary fall of stars began. It was not, however, sufficient to count the numbers that fell; it was desirable also to measure the time of their appearance, and of the continuance of their fall, according to Franzmann's instrument, which beats thirds of seconds, and moreover to ascertain their relative light and apparent course in the heavens; and all these observations could commence only at 26 minutes past 9, when the observers, fifteen in number, were assembled. Four observers registered the time of each appearance, according to two clocks, till 14 minutes past 3, when dawn put a stop to the observations: they noticed 1,008 falling stars, not including numbers which must have been overlooked. The courses of only 977 have, therefore, been marked upon the star-maps, with all the circumstances relative to them. The following result is as near the exact truth as possible:—Five stars appeared as bright as Venus, 14 as Jupiter, 238 as stars of the first magnitude; 354 were noted of the second, and 257 of the third, magnitude; 101 were reckoned smaller still, and the size of eight was omitted in the hurry; 273 exhibited themselves with tails. It is useless to mention the apparent paths of the stars, inasmuch as they varied according to the places of observation. Three observers saw on the following night 323 falling stars, whilst the sky was partly covered. In the night of August 12, an observer counted 103 more, from 10 o'clock to 45 minutes past 1. Therefore, the annual periodical return of an uncommon fall of stars towards the 10th of August is once more confirmed, as well as that the passage of this host of meteors near the earth lasts several days.

"VON BOGUSLAWSKI."

"Breslau, August 14, 1839."

Between the hours of ten, September 3, and three the next morning, was observed one of the most magnificent displays of falling stars and northern lights, witnessed of late years. The primary indication of the phenomenon was at about ten minutes before ten, when, apparently, a light crimson vapour rose from the north, and gradually extended to the centre of the heavens; till, by ten o'clock or a quarter past, the whole, from east to west, was one vast sheet of light, resembling that occasioned by a terrific fire. At one time, the light seemed to fall, and then rose with intense brightness mingled with volumes of smoke. These appearances lasted upwards of two hours; and towards morning the spectacle assumed more grandeur. At two o'clock the whole of London was illuminated with noonday brightness, and the atmosphere was remarkably clear. The south, at this time, although unclouded, was very dark; but the innumerable stars shone brilliantly. The opposite quarter of the heavens was extremely clear; the light was very vivid, with a continued succession of meteors, varying in splendour. They apparently formed in the centre of the heavens, and spread till they seemed to burst, when the effect was electrical; myriads of small stars shot swiftly over the horizon towards the earth; they seemed to burst also, and throw a dark crimson vapour over the entire hemisphere. At half past two o'clock, the spectacle changed to darkness, which, on dispersing, displayed a luminous rainbow in the zenith, and round the ridge of darkness that overhung the south. Then from it radiated columns of silvery light, which increased, and intermingled with crimson vapour, stars darted in all directions, and so continued until four o'clock when all died away.—*Abridged from the London Newspapers.*

WHEWELL'S ANEMOMETER.

On May 12, Mr. Whewell read to the Cambridge Philosophical Society, a note respecting the working of his Anemometer, since his memoir on that subject, read May 1, 1837. The Anemometer had since that time been in action at the Society's house, and at the Cambridge observatory; but, in consequence of the instrument being several times repaired and improved, the observations were frequently interrupted. The observations for July and August, 1838, were, however, represented to the Society by comparative diagrams; from which, it appeared, that the form of the line representing the course of the wind, as registered by the instrument at the two places, is nearly identical, thus proving the consistency of different instruments of this construction with one another. The scale of the two instruments appeared to be different, nearly in the ratio of 2 to 1; but no direct comparison of scales had been attempted. It was stated also, that during 1837 and 1838, observations had been made with Mr. Whewell's Anemometer at Edinburgh, by Mr. Rankine, and expressed in a diagram, according to the method recommended by Mr. Whewell in the 14th volume of the *Edinburgh Transactions.* Observations with this instrument have also been made at Plymouth, and reduced by Mr. Southwood, (of St. Peter's College,) by whom also the diagrams for Cambridge were constructed. Mr. Whewell stated, in conclusion, that there is every reason to believe, from the results hitherto obtained, that if any person with sufficient leisure were to take up this subject, it would reward him by leading to the knowledge of important meteorological facts and laws. —*Cambridge Chronicle.*

WIND TELEGRAPH.

LIEUTENANT WATSON, of Liverpool, has contrived a telegraph, or means of indicating the state of the wind and barometer at Holyhead or Bidston. The instrument consists of a large circle, 12 feet diameter, with the points of a compass marked thereon. "It has two hands, like a clock, the longer one showing the point of the wind at Holyhead, the shorter one at the north-west light-ship, or Bidston. At the top of the mast is an iron rod, on which a ball works. When the ball is at the top of the rod, it indicates a light breeze: when in the middle, moderate; and when seen at the bottom, blowing very fresh. The space between this rod and the circle is for the barometer; the pointer on the *right*-hand side, or *north*, shows the position of the barometer at Holyhead, *on that day*; and the one on the *left*, or *south*, the position the preceding day. The whole is very simple, and easily understood."—*Railway Magazine.*

ATMOSPHERIC ELECTRICITY.

IN March, Mr. Andrew Crosse delivered, at Taunton, a lecture on Atmospheric Electricity, illustrated by a number of beautiful experiments. He illuminated 400 feet of iron chain, hung in festoons about the room, the whole extent being brilliantly lighted at the same instant by the passage through it of the spark from the battery; and he melted several feet of wire. Mr. Cross next detailed the results of many experiments on thunder-clouds and mists. By means of a wire apparatus suspended in his park at Bromfield, he has discovered that a driving fog sweeps in masses, alternately negatively and positively electrified; and once the accumulation of electricity in a fog was so great that there was an incessant stream of sparks from his conductor, each of which would have struck an elephant dead in an instant.—*Times.*

METEORIC STONE AT THE CAPE OF GOOD HOPE.

ON March 21, was read to the Royal Society the following account of the fall of a Meteoric Stone in the Cold Bokkeveld, Cape of Good Hope, by T. Maclear, Esq.; in a letter to Sir J. F. W. Herschel. The appearance attending the fall of this aerolite, which happened at half-past nine o'clock in the morning of the 13th of October, 1838, was that of a meteor of a silvery hue, traversing the atmosphere for a distance of about sixty miles, and then exploding with a loud noise, like that from artillery, which was heard over an area of more than seventy miles in diameter—the air at the time being calm and sultry. The fragments were widely dispersed, and were at first so soft as to admit of being cut with a knife; but they afterwards spontaneously hardened. The entire mass of the aerolite is estimated at about five cubic feet.

Next was read Dr. Faraday's "Chemical Account" of the above

Meteor. The stone is stated as being soft, porous, and hygrometric; having, when dry, the specific gravity of 2·94, and possessing a very small degree of magnetic power, irregularly dispersed through it. One hundred parts of the stone in its natural state, were found to consist of the following constituents: namely,—

Water	6·5	Alumina	5·22
Sulphur......	4·24	Lime	1·64
Silica	28·9	Oxide of Nickel....	·82
Protoxide of Iron	33·22	Oxide of Cromium..	·7
Magnesia	19·2	Cobalt and Soda, a trace.	

Athenæum.

FALL OF A METEORITE IN MISSOURI.

On the afternoon of Feb. 13, 1839, a Meteor exploded near the settlement of Little Piney, Missouri, (lat. 37° 55′ N.; lon. 92° 5′ W.) and cast down to the earth one stony mass or more in that vicinity. Although the sky was clear, and the sun, of course, shining at the time, the Meteor was plainly seen by persons in Potosi Caledonia, and other towns near which it passed. At Caledonia, which is about nine miles south-westerly from Potosi, the Meteor passed a little north, and at the latter place, a little to the south of the zenith. Its course was almost precisely to the west. The most eastern spot at which it was seen is about fifteen miles west of St. Genieve, (or about lat. 37$\frac{5}{6}$° N.; long. 90° W.); the most western is Little Piney, near which it exploded, with three reports in quick succession. Although the ground was covered with 3 or 4 inches of snow, there was found a meteoric stone, about as large as a man's head, partly embedded in the earth.

The total weight of all the fragments collected is 973 grains. The specific gravity of one of the small fragments is 3·5; but different portions of the stone may vary slightly in this respect, as they contain more or less of the metallic matter. The stone crumbles under a moderate blow: two of the fragments retain portions of the crust or exterior coating, which is the fifteenth of an inch thick, and bears evidence of intense ignition and partial fusion. It is black, with a wrinkled or cellular surface, and is traversed with seams. The whole mass is studded with metallic particles, and rusty spots, with occasional small spheroidal concretions. The little metallic masses, (doubtless, of nickeliferous iron, are attracted by the magnet, and are generally permeated by the earthy matter. They are mostly of an iron-white colour, but several are yellow and slightly irridescent. One of these minute masses being removed from the stone, it was by the hammer at once extended into a thin lamina, and was evidently malleable.—*Abridged from Silliman's Journal.*

GREAT STORM OF JANUARY.

Mr. F. Osler has read to the British Association a few remarks on the Great Storm of the 6th and 7th of January, 1839. In addition to the records obtained by the anemometer at Birmingham and at Plymouth, Mr. Osler has collected information concerning this storm from many parts of the British isles. A careful analysis of these strongly leads him to the opinion that this was a small but violent rotatory storm, moving forward at the rate of about thirty to thirty-five miles per hour. The tendency of this eddy, or violent whirling of the air, would, of course, be to produce a vacuum in the centres. The air that formed the eddy being constantly thrown off in a slight degree spirally upwards, and dispersed on the upper portion of the atmosphere, the effect of this would, therefore, be to produce a strong current upwards. Now, supposing this large eddy to be perfectly stationary, there would be a rapid rush of air towards it from all sides, which would be drawn up and thrown off through this rotating circle, and dispersed with amazing rapidity above: but, as it is moving on with great velocity, the air that is in the advance of the storm is not sensibly affected until the whirl is close to it; while in the rear, the motion of the air is greatly increased: first, by the tendency of the air to rush into the great vortex of the storm; and, secondly, by the motion onward of the vortex itself. This vortex, or revolving column, would increase in size upwards, so as somewhat to resemble a funnel; it would, in fact, be similar in its shape and action to an immense water-spout; whether it was vertical or not is entirely a matter of conjecture, but Mr. Osler considers it probable that it would incline in the direction that the storm was moving, namely, to the N.E.; and that it was an upper current that carried it in that direction. The greatest intensity of the storm in England was evidently across Lancashire and Yorkshire: wherefore, Mr. Osler conceives, that the nucleus of the hurricane passed in a N.E. direction over these two counties. Towards the sides, however, a little current set in a S. and even slightly in a S.E. direction, on the S. side of the vortex, and in a N.W. and westerly direction on the N. side; but the main rush was behind. Our anemometer shows that we first felt a fresh S. wind with a slight bearing of E. in it, which very shortly became more westerly, increasing considerably in violence; and it then moved round to the S.W. and became quite a hurricane, and continued so, very violent at first, but decreasing in strength during the remainder of the day: at Plymouth it commenced as a S.W. and then very gradually moved round a little more westward.—*Literary Gazette; abridged.*

NEW OBSERVATORIES.

Captain James Ross is commissioned to plant, in his Antarctic Expedition, three Magnetical and Meteorological Observatories, at St. Helena, the Cape, and Van Diemen's land. The commander himself especially wishes to observe at Kerguelen's Land, New Zealand, and other stations on the land and ice; and he regards these as only part of a system of observations, simultaneous or combined, stretching from one side of the earth to another, undertaken or promised, through the whole extent of the British empire, from Montreal to Madras; and blending in co-operation with chains of observatories established, or on the point of being established, by other nations in the four quarters of the world.—*President's Address; Proc. British Association.*

INFLUENCE OF THE MOON ON THE BAROMETER.

Mr. Snow Harris has reported to the British Association his investigations into the supposed Influence of the Moon on the Barometer; and with this view has reduced about 4000 of the observations, so as to show the pressure at the time of the moon's southing, and for each hour before and after; but he cannot discover any differences which can be supposed to arise from the moon's influence. He is, therefore, disposed to agree with the conclusion lately arrived at by Mr. Lubbock, from a discussion of the Barometric Observations at the Royal Society—*viz.*, that no lunar irregularity is observable from this method of discussing the observations—that, if at any time established, must prove extremely small.—*Literary Gazette.*

IMPORTANT ATMOSPHERICAL LAWS.

Prof. Kaemtz considers the following two laws amongst the most important in the whole range of Meteorology:—

1. If two neighbouring parts of the earth have unequal temperatures, we find that, in the upper regions of the atmosphere, there are winds which blow from the warmer to the colder part; while, near the surface of the earth, there are winds from the colder to the warmer portion.

2. When a tract of the earth is unusually heated, or if it is distinguished from the neighbouring regions by a high temperature, the barometer sinks; but if, on the other hand, its temperature is unusually low, the pressure of the atmosphere increases.—*Schumacher's Jahrbruchfür*, 1838; *Jameson's Journal.*

CAUSES OF WATER-SPOUTS.

Prof. Orsted, in considering the Water-Spout, has endeavoured to arrive at its proximate causes from observed and recorded effects; and he has ascertained that a whirlwind which begins in the higher regions of the air, and becomes expanded as it descends, constitutes the essential element of the phenomenon. He reasonably assumes the occurrence in the higher regions of the atmosphere of a whirlwind produced by two currents of air following parallel courses, but flowing in opposite directions. They must often occur while the air beneath is perfectly tranquil, as we know from experience: and we also know that the opposite currents produced by the inequality of the temperature over the land and sea, often extend upwards to a great height, and are there in great commotion, while all beneath is tranquil.—*From Schumacher's Jahrbruchfür*, 1838; *Jameson's Journal.*

LUNAR RAINBOW.

On October 20, between 7 and 8 o'clock, p.m., there was a splendid appearance of this rather unfrequent atmospheric phenomenon witnessed at Edinburgh. There was a brilliant arch of white, outside one of darker shade, forming about three quarters of a large circle.—*Caledonian Mercury.*

METEOROLOGICAL SUMMARY OF 1839.

(Communicated by Dr. Armstrong, *the Retreat, South Lambeth.)*

MONTHS.	TEMPERATURE.					ATMOSPHERIC VARIATIONS.			HYGROMETER.		MODIFICATIONS OF CLOUD.						
	Fahrenheit.		MEAN.														
	Max.	Min.	Reaumur.	Centigrade.	De Lisle.	Mean pressure in Inches.	Prevailing Currents.	Solar Variation.	Mean.	Rain in Inches.	Cirrus.	Cirro-stratus.	Cumulus.	Cirro-cumulus.	Cumulo-stratus.	Nimbus.	Stratus.
January ..	50	30	3·75	4·5	143	29·90	NW. SW.	9·76	34·2	1·428	:	*	*	:	:	:	
February	55	27	9·9	5	142·5	30·40	SW. NE. NW.	10·48	34·10	1·615	:	*	:	*	*	:	
March....	59	21	3·75	4·5	143	29·90	NE. SW. .W.	10·92	38·12	1·84		*	:	:	:	:	
April . ..	75	28	8·1	10	135	29 99	NE. SW.	13·83	44	1·45	:	*	:	:	:	:	
May	74	29	8·4	10·3	134	30·15	NE. SW.	18·38	45	1·725	*	*	*	:	*	*	
June	79	40	12	14·5	127	29·97	NE. SW.	17·28	49·5	1·93	*	*	:	*	*	:	:
July	78	45	12·9	15·	125·5	30·00	SW.	18·32	54	2·577	:	*	:	:	:	*	
August ..	84	41	12·3	16·9	124·5	29·80	SW.	16·47	56	2·425	:	*	*	*	*	*	
September	74	46	11·2	14·2	128·5	30·12	SW.	14·37	52	4·795	:	*	*	:	:	*	
October ..	76	38	11·5	14·75	128	30·10	SW. NE.	11·61	44	1·695	:	*	:	:	:	:	:
November	61	30	1·2	2·5	147	29·80	SW. E.	7·81	39	4·000	:	*	:	:	:	*	
December	58	25	4	5·2	142	29·82	SE. SW.	7·47	37	2·580	:	*	:	:	*	*	

Number of Days for the greater part rainy 43.
—————————— fair 69.

Number of Days fair throughout, but cloudy 227
—————————— cloudless 4

Aurora Borealis slightly exhibited, on February 8, 11; May 6; remarkable on March 14, 19. High wind on January 1, 3, (disastrous hurricane 7) 13; February 16, 23; April 2, 8, 18; May 9, 10, 11; June 20, 22, 23, 24, 29; July 18, 19; September 1, 2, 7, 8; November 1; December 23, 24, 27. Highest tides on January 1; March 16, 17, 18, 19, 21; May 10; July 29, 31; August 27, 29; September 22 to 28; October 23 to 27; November 20, 21, 22, 27; December 21 to 26. The November and December tides were the highest in the Thames, owing chiefly to the land flood meeting the tidal wave. Thunder and Lightning on May 7; June 13, 17, 18, 19, 26; July 7, 17, 27; August 6, with hail. Snow of 1838 on the ground till January 4. Snow on 6, 30, 31; March 6, 7; April 5, 7; May 14; Dense Fog on December 1 to 6; Asteroids on September 16. Thermometer at 3° below freezing, (Fahrenheit) on May 15.

Geographical Discovery.

NEW LAND IN THE SOUTHERN OCEAN.

In the *Year-Book of Facts*, 1839, (p. 271,) will be found a notice of the Expedition to the Antarctic Ocean, chiefly fitted out under the direction of Mr. Charles Enderby; and which sailed in July, 1838, and returned in September last, with a most successful issue.

Those who take an interest in Antarctic discovery will remember that in the years 1831-2, Mr. John Biscoe, R.N., in command of the *Tula*, a brig belonging to the Messrs. Enderby, of London, discovered two portions of land, about 110 deg. of longitude apart, in the parallel of the Antarctic Circle, which were respectively named Graham Land, and Enderby Land. In the following year, Mr. Biscoe was again despatched by these spirited owners, but the vessel was wrecked. Nothing discouraged by this failure, and by the heavy loss already incurred, Messrs. Enderby, in conjunction with seven other merchants, (Messrs. G. F. Young, W. Borradaile, J. W. Buckle, T. Sturge, W Brown, J. Row, and W. Beale,) determined on another South Sea sealing voyage, giving special instructions to the commander of the expedition to push as far as he could to the south, in hopes of discovering land in a high southern latitude.

The schooner, *Eliza Scott*, of 154 tons, commanded by Mr. John Balleny, and the dandy-rigged cutter, *Sabrina*, of 54 tons, Mr. H. Freeman, master — the vessels selected for this purpose, having three chronometers on board, and being well equipped, sailed from the port of London on July 16, 1838.

Sighting the island of Madeira, the two vessels crossed the equator in 22 deg. 40 min. W. longitude, touched at the island of Amsterdam, and, on December 3, anchored in Chalky Bay, near the south-western angle of the southern island of New Zealand: here they refitted, watered, and prepared for their sealing voyage to the Frozen Ocean.

There is not the vestige of a hut in Port Chalky. Preservation Bay, to the southward, is a picturesque spot, full of islands covered with wood.

On Jan. 7, 1839, the vessels sailed for the southward; and on the 11th anchored at Campbell Island, where, by a curious coincidence, they met Mr. John Biscoe, R.N., (already named,) in command of the *Emma*, on a sealing voyage. On the 17th, they again made sail to the south-eastward: on the 19th, in lat. 54 deg., with weather calm and fine, the Aurora Australis was very brilliant. On the 23rd, in lat. 59 deg. 16 min., long. 173 deg. 20 min. E. of Greenwich, the indications of the vicinity of land, as large quantities of sea-weed, mutton birds, &c., were so strong, that the weather being very thick, the vessels were hove to. On the following day they passed the branch of a tree; but, as it cleared, neither land nor ice were in sight, and they continued standing to the S.S.E. till the 27th, when in lat. 63 deg. 37 min., long. 176 deg. 50 min. E., they crossed Capt. Bellingshausen's route of the Russian corvette, the *Vostok*, in December, 1820, and here they saw the first iceberg. Continuing to the southward, over the very spot where compact ice had forced the Russian navigator to alter his course to the eastward, the vessels, on the 28th, reached their extreme eastern longitude, namely, 178 deg. 13 min. E.; and on the following evening, in the parallel of 66 deg. 40 min, the variation observed by azimuth was 28 deg. E. At this time field-ice bounded their southern horizon, and numerous large icebergs were in sight. At sunset, on the 30th, in lat. 67 deg., and long. 170 deg., the variation observed by amplitude was found to be 33 deg. 25 min. E. They were now surrounded by icebergs, and small drift-ice; the wind during the last week had been constantly from the westward, varying from N.W. to S.W.

At noon, on the 1st February, the sun broke out, and the weather cleared — lat., by observation, 68 deg. 45 min. At this time no ice was in sight from the mast-head; and they stood to the southward, with a fresh breeze, till three o'clock, P.M., when they found themselves near the edge of a large body of packed ice, and were obliged to tack to the northward to avoid it. This, then, was their extreme south point, as they had now reached the parallel of 69 deg. in long., 172 deg. 11 min. E., full 220 miles to the southward of the point which Bellingshausen had been able to attain about this meridian; thus adding one proof more, that ice in these regions, even in the immediate neighbourhood of land, is very far from stationary.

On February 2, they were still

THE BALLENY ISLANDS. Discovered 9th February, 1839.

embayed in field-ice: on the 5th, the water was much discoloured, and many feathers were seen floating; and several whales, sea-leopards, and penguins were descried. They gradually worked to the N.W., to clear the ice, against a strong westerly wind, which, contrary to the received opinion, was found to prevail in these high latitudes.

They sailed onward till Feb. 9, when, at 8h. clear, steering west by compass, Capt. Balleny got sights for his chronometers, which gave the ship, by the Port Chalky rate, in long. 164 deg. 29 min. E. At 11, A.M., a darkish appearance was noticed to the S.W.; lat. 66 deg. 37 min. S. by mer. alt.: wind, north. At noon, the sun shone brightly; and the appearance of land was seen to the S.W., extending from west to about south—ran for it: at 4h. it was distinctly made out to be land. At 8 h., P.M. (having run S.W. 22 m.), they got within five miles of it, when was seen another piece of land of great height, bearing W. by S. At sunset, three separate islands, of good size, were made out, the western one being longest. On Feb. 10, after running through much drift ice, within half a mile, the middle island was found completely ice-bound, with high perpendicular cliffs: from this island to the eastern one, S.W., the sea was in one firm and solid mass, without a passage. On Feb. 11th, the land was seen bearing about W.S.W. to be of a tremendous height; Capt. Balleny supposes about 12,000 feet, and covered with snow: at noon, lat. 66 deg. 30 min.; wind, N.W.; temp. 42 deg.

Next day, they got abreast of the eastern island: lat. by acc. 66 deg. 22 min.; long. 163 deg. 49 min. E. The cutter's boat went ashore, though there was no landing or beach; but for the bare rocks whence the icebergs had broken, it would not have been known for land at first; still, as they stood in for it, smoke was plainly seen rising from its peaks. Its stone, or rather cinders, also prove this island to be volcanic. They returned on board, and got the vessels safely through the drift-ice before dark, and ran along the land.

On the 13th, were seen numerous whales, penguins, a few Cape pigeons, and a small white bird; but no albatrosses nor mollymawks. P.M., came on a thick fog; but many whales and seals were seen, with icebergs and drift-ice. At midnight, light variable winds, and cloudy dark weather.

This was the last time that the land, now appropriately named the *Balleny Isles*, was seen. The group consists of five islands, three large and two small; the highest of which, named Young Island, is estimated at 12,000 feet above the sea. It rises in a beautiful peak, which may be called Peak Freeman, as being on the island whereon the commander of the cutter, Sabrina, landed.*

When at the distance of from eight to ten miles from the centre island, with the extremes of the land bearing from W. round southerly to E. by S., the accompanying sketch was made by Mr. John M'Nab, second mate of the schooner; the outline of the islands is evidently

* These islands and peaks are named respectively after Messrs. G. F. Young, W. Borradaile, J. W. Buckle, T. Sturge, W. Brown, J. Row, and W. Beale, the spirited merchants who united with Mr. Enderby in sending out this Expedition.

volcanic, and the smoke which arose from the second island to the E., or Buckle Island, and the stones brought away from Young Island, by Mr. Freeman, which prove to be scoriæ and basalt, with crystals of olivine, leave no doubt on the subject. These, then, are, with the exception of that discovered by Bellingshausen, in 69 deg. S., the most southerly volcanoes known. The easternmost, or Sturge Island, rises also to a peak, named Brown's Peak, but it is not half the height the former. Immediately off the eastern end of the centre, or Borradaile Island, is a remarkable pinnacle of rock, called Beale Pinnacle, which is described as rising like a tall lighthouse from the waters. The westernmost, or Row Island, is low, and offers no remarkable feature.

Obituary

OF PERSONS EMINENT IN SCIENCE OR ART, 1839.

Count Montlosier, one of the most striking writers in that great controversy respecting the origin of Basaltic Rock, which occupied the attention of mineralogists during the latter half of the last century; and to which, in so large a degree, the progress and present state of Geology are to be ascribed.

Anselme Gaetans Desmarest, Professor of Zoology at the Royal Veterinary College of Alfort; author of several works on Fossil Zoology and Botany.

Professor Rigaud, F.R.S., to whom was confided the care of the Radcliffe Observatory. He was the author of many valuable communications to the Transactions of the Royal Astronomical Society, and to other scientific journals, on subjects connected with physical and astronomical science. There was, probably, no other person of his age who was equally learned on all subjects connected with the history and literature of astronomy.—*Anniv. Address, Royal Soc.*

Mr. W. Wilkins, F.R.S., Professor of Architecture to the Royal Academy.

Mr. George Saunders, F.R.S., architect; and a diligent and learned antiquary.

The Baron de Prony, one of the most distinguished engineers of France; and one of the most voluminous writers of his age, generally upon mathematical and other subjects connected with his professional pursuits.

The venerable Pierre Prevost, formerly Professor of Natural Philosophy in the University of Geneva. The range of his philosophical researches was unusually extensive and various; and his discoveries on Heat constitute a most important epoch in a branch of science which has recently received so extraordinary a development in the hands of Fourier, Forbes, Melloni, and other philosophers.—*Anniv. Address, Royal Soc.*

Mr. Benjamin Bevan, F.G.S., civil engineer.

Mr. W. Salmond, F.G.S., of York; one of the persons who was most actively engaged in the examination of the celebrated Kirkdale Cavern.

Baron von Jacquin, whose house at Vienna was, for a long series of years, the rendezvous of all the most eminent characters in literature and science.

His Grace the Duke of Bedford, F.S.A., F.L.S., and a skilful botanist; the author of *Pinetum Woburnense, &c.*

John Lander, the brother of Richard Lander, and the attendant of Clapperton, in his African Expedition.

William Hilton, R.A., historical painter.

Allan Cunningham, the colonial botanist; than whom few men of his time have done more for botany and geography.

Prof. Nibi, the distinguished antiquary, at Rome.

Dr. Blumenbach, Professor of Natural History in the University of Göttingen. The most popular of his numerous works is his *Manual of the Elements of Natural History*, a translation from the tenth German edition of which was published in London in 1825.

Mr. Charles Tennant, of Glasgow; the eminent practical chemist, and patentee of chloride of lime for bleaching.

Mr. Hunneman, the botanical bookseller.

Samuel Brookes, Esq., F.L.S., author of an *Introduction to the Study of Conchology*.

John Hall, M.D., author of a *British Flora*, and *Elements of Botany*.

William Younge, M.D., the companion of Sir J. E. Smith in his tour on the continent in 1786 and 1787.

Dr. Arthur Kochen, archaiologist.

Landriani, architect and painter, Milan.

James Lonsdale, portrait-painter.

Sir W. Beechey, R.A., portrait-painter.

Joseph Valadier, architect, Rome.

John Vendramini, engraver.

Charles Rossi, R.A., sculptor.

Otto Sigism. Runge, (St. Petersburgh), sculptor.

John Bromley, engraver.

Dr. B. F. Fries, Swedish naturalist.

Dan. Ohlmuller, architect.

Sam. Holberg, (Russia,) sculptor.

Davies Gilbert, D.C.L., President of the Royal Society, from the resignation of the office, by Sir Humphry Davy, in 1827, to the election of the Duke of Sussex to the chair, in 1830. Mr. Gilbert was likewise a Fellow of the Society of Antiquaries, and of the Royal Society of Edinburgh; and a member of the Royal Irish Academy.

Michael Jerome Lalande, a distinguished astronomer, author of several articles in the *Connaisance des Tems*, and nephew of the illustrious astronomer of the same name, who died in 1807.

GENERAL INDEX.

Clarke, Printers, Silver Street, Falcon Square, London.